CIRCADIAN CLOCKS AND THEIR ADJUSTMENT

The Ciba Foundation is an international scientific and educational charity (Registered Charity No. 313574). It was established in 1947 by the Swiss chemical and pharmaceutical company of CIBA Limited —now Ciba-Geigy Limited. The Foundation operates independently in London under English trust law.

The Ciba Foundation exists to promote international cooperation in biological, medical and chemical research. It organizes about eight international multidisciplinary symposia each year on topics that seem ready for discussion by a small group of research workers. The papers and discussions are published in the Ciba Foundation symposium series. The Foundation also holds many shorter meetings (not published), organized by the Foundation itself or by outside scientific organizations. The staff always welcome suggestions for future meetings.

The Foundation's house at 41 Portland Place, London W1N 4BN, provides facilities for meetings of all kinds. Its Media Resource Service supplies information to journalists on all scientific and technological topics. The library, open five days a week to any graduate in science or medicine, also provides information on scientific meetings throughout the world and answers general enquiries on biomedical and chemical subjects. Scientists from any part of the world may stay in the house during working visits to London.

Ciba Foundation Symposium 183

CIRCADIAN CLOCKS AND THEIR ADJUSTMENT

1995

JOHN WILEY & SONS

Chichester · New York · Brisbane · Toronto · Singapore

©Ciba Foundation 1995

Published in 1995 by John Wiley & Sons Ltd
Baffins Lane, Chichester
West Sussex PO19 1UD, England
Telephone (+44) (243) 779777

All rights reserved.

No part of this book may be reproduced by any means,
or transmitted, or translated into a machine language
without the written permission of the publisher.

Suggested series entry for library catalogues:
Ciba Foundation Symposia

Ciba Foundation Symposium 183
ix + 337 pages, 74 figures, 5 tables

Library of Congress Cataloging-in-Publication Data
Circadian clocks and their adjustment / [editors, Derek J. Chadwick
 (organizer), and Kate Ackrill].
  p. cm.—(Ciba Foundation symposium; 183)
 "Symposium on Circadian Clocks and Their Adjustment, held at the
Ciba Foundation, London, 7–9 September 1993."
 Includes bibliographical references and index.
 ISBN 0 471 94305 3
 1. Circadian rhythms—Congresses. I. Chadwick, Derek.
II. Ackrill, Kate. III. Symposium on Circadian Clocks and Their
Adjustment (1993: Ciba Foundation) IV. Series.
QP84.6.C55 1995
612′.022—dc20   94-37788
                                                              CIP

British Library Cataloguing in Publication Data
A catalogue record for this book is
available from the British Library

ISBN 0 471 94305 3

Phototypeset by Dobbie Typesetting Limited, Tavistock, Devon.
Printed and bound in Great Britain by Biddles Ltd, Guildford.

Contents

Symposium on Circadian clocks and their adjustment, held at the Ciba Foundation, London, 7–9 September 1993

The topic of this symposium was proposed by Dr Peter H. Redfern and Dr James Waterhouse

Editors: Derek J. Chadwick (Organizer) and Kate Ackrill

J. M. Waterhouse Introduction 1

J. C. Dunlap, J. J. Loros, B. D. Aronson, M. Merrow, S. Crosthwaite, D. Bell-Pedersen, K. Johnson, K. Lindgren and **N. Y. Garceau** The genetic basis of the circadian clock: identification of *frq* and FRQ as clock components in *Neurospora* 3
Discussion 17

L. Rensing, A. Kallies, G. Gebauer and **S. Mohsenzadeh** The effects of temperature change on the circadian clock of *Neurospora* 26
Discussion 41

G. Block, M. Geusz, S. Khalsa, S. Michel and **D. Whitmore** Cellular analysis of a molluscan retinal biological clock 51
Discussion 60

M. R. Ralph and **M. W. Hurd**   Circadian pacemakers in vertebrates 67
Discussion 81

R. Y. Moore   Organization of the mammalian circadian system 88
Discussion 100

General discussion I 107

T. Roenneberg   The effects of light on the *Gonyaulax* circadian system 117
Discussion 128

v

M. U. Gillette, M. Medanic, A. J. McArthur, C. Liu, J. M. Ding, L. E. Faiman, E. T. Weber, T. K. Tcheng and E. A. Gallman Intrinsic neuronal rhythms in the suprachiasmatic nuclei and their adjustment 134
Discussion 144

N. Mrosovsky A non-photic gateway to the circadian clock of hamsters 154
Discussion 167

M. H. Hastings, F. J. P. Ebling, J. Grosse, J. Herbert, E. S. Maywood, J. D. Mikkelsen and A. Sumova Immediate-early genes and the neuronal bases of photic and non-photic entrainment 175
Discussion 190

S. M. Reppert Interaction between the circadian clocks of mother and fetus 198
Discussion 207

F. W. Turek, P. Penev, Y. Zhang, O. Van Reeth, J. S. Takahashi and P. Zee Alterations in the circadian system in advanced age 212
Discussion 226

B. Lemmer Clinical chronopharmacology: the importance of time in drug treatment 235
Discussion 247

C. A. Czeisler The effect of light on the human circadian pacemaker 254
Discussion 290

A. J. Lewy, R. L. Sack, M. L. Blood, V. K. Bauer, N. L. Cutler and K. H. Thomas Melatonin marks circadian phase position and resets the endogenous circadian pacemaker in humans 303
Discussion 317

General discussion II 322

Index of contributors 327

Subject index 329

Participants

J. Arendt School of Biological Sciences, University of Surrey, Guildford, Surrey GU2 5XH, UK

S. M. Armstrong Department of Psychology, La Trobe University, Bundoora, Victoria 3083, Australia

G. D. Block NSF Science and Technology Center for Biological Timing and Department of Biology, University of Virginia, Gilmer Hall, Charlottesville, VA 22901, USA

V. M. Cassone Department of Biology, Texas A & M University, College Station, TX 77843, USA

C. A. Czeisler Section on Sleep Disorders/Circadian Medicine, Brigham & Women's Hospital, 221 Longwood Avenue, Boston, MA 02115, USA

S. Daan Zoology Laboratory, University of Groningen, PO Box 14, NL-9750 AA Haren, The Netherlands

J. C. Dunlap Department of Biochemistry, Dartmouth Medical School, Hanover, NH 03755-3844, USA

L. N. Edmunds Jr Division of Biological Sciences, State University of New York, Stony Brook, NY 11794-5200, USA

M. U. Gillette Department of Cell & Structural Biology, University of Illinois, 506 Morrill Hall, 505 S Goodwin Avenue, Urbana, IL 61801, USA

M. H. Hastings Department of Anatomy, University of Cambridge, Downing Street, Cambridge CB2 3DY, UK

R. E. Kronauer Division of Applied Sciences, 324 Pierce Hall, Harvard University, Cambridge, MA 02138, USA

B. Lemmer Zentrum der Pharmakologie, J. W. Goethe-Universität, Theodor-Stern-Kai 7, D-60590 Frankfurt/M, Germany

A. J. Lewy Sleep & Mood Disorders Laboratory, L-469, Departments of Psychiatry, Ophthalmology and Pharmacology, Oregon Health Sciences University, 3181 S W Sam Jackson Park Road, Portland, OR 97201-3098, USA

J. J. Loros Department of Biochemistry, Dartmouth Medical School, Hanover, NH 03755-3844, USA

J. Meijer Department of Physiology, Division of Medical Chronobiology, University of Leiden, PO Box 9604, NL-2300 RC Leiden, The Netherlands

M. Menaker NSF Center for Biological Timing, Department of Biology, University of Virginia, Gilmer Hall, Charlottesville, VA 22901, USA

J. D. Mikkelsen (*Bursar*) Institute of Medical Anatomy, Department B, University of Copenhagen, The Panum Institute, 3 Blegdamsvej, DK-2200 Copenhagen N, Denmark

J. D. Miller Department of Biological Sciences, Stanford University, Palo Alto, CA 94305, USA

R. Y. Moore Center for Neuroscience, BST W1656, University of Pittsburgh, Pittsburgh, PA 15261, USA

N. Mrosovsky Department of Zoology, University of Toronto, Toronto, Ontario, Canada M5S 1A1

M. R. Ralph Department of Psychology, University of Toronto, 100 St George Street, Toronto, Ontario, Canada M5S 1A1

P. H. Redfern School of Pharmacy and Pharmacology, University of Bath, Claverton Down, Bath BA2 7AY, UK

L. Rensing Department of Biology, University of Bremen, PO Box 33 04 40, D-28334 Bremen, Germany

S. M. Reppert Laboratory of Developmental Chronobiology, Children's Service, Massachusetts General Hospital, Boston, MA 02114, USA

T. Roenneberg Institut für Medizinische Psychologie, Ludwig-Maximilians-Universität, Goethestrasse 31, D-80336 München, Germany

Participants

D. S. Saunders Division of Biological Sciences, ICAPB, University of Edinburgh, Ashworth Laboratories, King's Buildings, West Mains Road, Edinburgh EH9 3JT, UK

J. S. Takahashi NSF Science and Technology Center for Biological Timing, Department of Neurobiology & Physiology, Northwestern University, 2153 North Campus Drive, Evanston, IL 60208-3520, USA

F. W. Turek Department of Neurobiology & Physiology, Northwestern University, 2153 North Campus Drive, Evanston, IL 60208-3520, USA

J. M. Waterhouse (*Chairman*) School of Biological Sciences, University of Manchester, Stopford Building, Oxford Road, Manchester M13 9PT, UK

Introduction

Jim M. Waterhouse

Department of Physiological Sciences, University of Manchester, Stopford Building, Oxford Road, Manchester M13 9PT, UK

The amount of research into circadian rhythms has increased markedly in recent years. Important developments in our understanding of the mechanisms involved have taken place through the use of many different animal models and technological advances. As examples, consider the developments in the genetics of *Drosophila* and *Acetabularia*, the neurophysiology of the eye of *Bulla*, the neurophysiology, function and physiology of the rodent suprachiasmatic nuclei, and the adjustment of body clocks in humans by light and melatonin. We have learned about the pathways by which environmental stimuli can adjust the clock and have increasing knowledge of what such zeitgebers are. The ontogeny of the clock and the changing nature of the zeitgebers affecting it have become better defined. There has been a parallel increase in clinical and applied interests, as illustrated by work on shift-work, blindness and jet lag.

Such advances have been covered in scientific meetings and in the literature, of course, but in spite of an ever-increasing number of meetings dealing with chronobiological topics, two limitations remain to the process of integrating recent findings into the general body of knowledge. First, there is rarely enough time devoted to the discussion of new work. Second, the specialist nature of many meetings prevents the cross-fertilization of ideas and methodologies between the many disciplines with an interest in the subject. It is rare to have a meeting in which there is the necessary combination of authority and extended constructive and organized debate.

This symposium gives us an opportunity to discuss recent advances in the field unhampered by these limitations. I hope that the ensuing discussions will be lively and constructive, with continual interaction between experts from many different disciplines, so as to suggest possible approaches to current problems. We have chosen to start at one end of the scale, at the molecular level, and work up to the clinical end, but this does not mean that the fungal geneticists should be involved only at the beginning and the human chronobiologists only at the end; I hope all of you will ask questions over and over again. Even if we are working at markedly different levels of complexity, we are nevertheless dealing ultimately with the same thing, working towards some presumably common mechanisms. I also hope that some of the intellectual stimulation that

I know will be evident will be readily inferred by readers of the papers and recorded discussions.

Finally, I would like to pay tribute to my colleague Dr Peter Redfern, who played an integral part in making this symposium the success I am sure it will be, for his continual support and wisdom.

The genetic basis of the circadian clock: identification of *frq* and FRQ as clock components in *Neurospora*

Jay C. Dunlap, Jennifer J. Loros, Benjamin D. Aronson, Martha Merrow, Susan Crosthwaite, Deborah Bell-Pedersen, Keith Johnson, Kristin Lindgren and Norman Y. Garceau

Department of Biochemistry, Dartmouth Medical School, Hanover, NH 03755-3844, USA

Abstract. Genetic approaches to the identification of clock components have succeeded in two model systems, *Neurospora* and *Drosophila*. In each organism, genes identified through screens for clock-affecting mutations (*frq* in *Neurospora*, *per* in *Drosophila*) have subsequently been shown to have characteristics of central clock components: (1) mutations in each gene can affect period length and temperature compensation, two canonical characteristics of circadian systems; (2) each gene regulates the timing of its own transcription in a circadian manner; and (3) in the case of *frq*, constitutively elevated expression will set the phase of the clock on release into normal conditions. Despite clear genetic and molecular similarities, however, the two genes are neither molecular nor temporal homologues. The timing of peak expression is distinct in the two genes, *frq* expression peaking after dawn and *per* expression peaking near midnight. Also, although expression of *per* from a constitutive promoter can rescue rhythmicity in a fly lacking the gene, constitutive expression of *frq* will not rescue rhythmicity in *Neurospora* *frq*-null strains, and in fact causes arrhythmicity when expressed in a wild-type strain. These data suggest that *frq* is and/or encodes a state variable of the circadian oscillator. Recent molecular genetic analyses of *frq* have shed light on the origin of temperature compensation and strongly suggest that this property is built into the oscillatory feedback loop rather than appended to it. It seems plausible that clocks are adjusted and reset through adjustments in central clock components such as *frq*, and, by extension, *per*.

1995 Circadian clocks and their adjustment. Wiley, Chichester (Ciba Foundation Symposium 183) p 3-25

Living things generally have the capacity for endogenous temporal organization of cellular processes over the course of an approximately 24 h period. The cellular machinery that generates this ability is collectively known as the biological clock, and its output as a circadian rhythm. Investigators have gone to great lengths to prove that these circadian rhythms are in fact endogenous. In the case of *Neurospora*, for example, such controls have included placing the organism on a rotating platform on the South Pole so that the position of the organism in cosmic space could be controlled (Hamner et al 1962) and even a trip into outer space (Sulzman et al 1984). Despite such machinations, the take-home message from work to date remains the same: that clocks consist of a biochemical–genetic feedback loop, the components of which are active at the intracellular level.

In order to understand, at the level of genetics and biochemistry, how cells keep time and regulate their metabolism, we have been concentrating our efforts on the analysis of the simplest rhythmic system in which both genetics and molecular biology can be used, *Neurospora crassa* (see Dunlap 1993 for review). In this system, 15 mutations associated with seven genetic loci have been identified as affecting the biological clock (Table 1). The genetic characteristics of the *Neurospora* clock suggest several things. There are a number of genes involved in or capable of affecting the operation of the clock, and it is likely

TABLE 1 *Neurospora crassa* **mutations identified in screens for circadian clock genes**

Allele	Linkage	Period length (h)	Dominance/ recessivity	Other clock properties affected
frq^1	VII R	16	Semi-dominant	
frq^2	VII R	19.3	Semi-dominant	
frq^3	VII R	24	Semi-dominant	Temperature compensation
frq^4	VII R	19.3	Semi-dominant	
frq^6	VII R	19.3	Semi-dominant	
frq^7	VII R	29	Semi-dominant	Temperature compensation, cycloheximide resetting
frq^8	VII R	29	Semi-dominant	Temperature compensation
frq^9	VII R	Uncompensated, conditionally arrhythmic	Recessive	Temperature compensation, nutritional compensation[a], entrainment
frq^{10}	VII R	Uncompensated, conditionally arrhythmic	Recessive	Temperature compensation, nutritional compensation[a], entrainment
chr	VI L	23.5	Semi-dominant	Temperature compensation
prd-1	III C	25.8	Recessive	Temperature compensation
prd-2	V R	25.5	Recessive	
prd-3	I C	25.1	Recessive	Temperature compensation
prd-4	I R	18	Dominant	Temperature compensation
cla-1	I R/VII R	27	Semi-dominant	Temperature compensation

[a]Period length of the rhythm, when expressed, is known to be dependent on the growth medium in the null mutant frq^9, and is inferred to be so dependent in the other null strain, frq^{10}.

that many more clock-affecting loci have yet to be identified, because if one were approaching the point at which most of the clock genes had been found, one would expect most genes to have been identified more than once, yet few of the clock genes found so far have been identified more than once. Also, clearly, the *frequency* (*frq*) locus stands out as being important, because mutations at this single locus can result in both long and short period lengths and loss of temperature compensation.

Our current view of the *Neurospora* circadian programme is shown in Fig. 1, which makes several points worth noting. First, everything of interest at the most basic level of the oscillator is happening *within* the cell; cell–cell communication is not a factor in timekeeping at its simplest level. The problem of organismal rhythmicity is then broken up, at least metaphorically, into two separate issues—that of how the oscillator itself works, and that of how

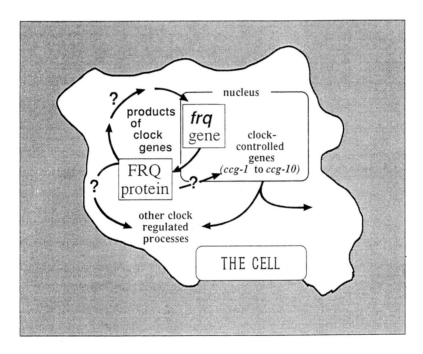

FIG. 1. The components of the circadian system in a cell. This figure implies more than we really know (for example, the location of the nuclear boundary), while leaving out some recognized effects such as the effects of light on the clock and on *frq* and the independent effects of light on the morning-specific clock-controlled genes, *ccgs* (which are known to be direct rather than being simply transmitted to these genes through the clock). Question marks refer to pathways that are possible but not proven (e.g., FRQ directly affecting the *ccgs*) or to pathways which are known but which contain an unknown number of steps (e.g., the feedback loop connecting *frq* to itself via FRQ).

information about the time leaves the oscillator and effects modulation of the metabolism and behaviour of the cell and, eventually, the organism. Our view of this grows more and more complex as we fill in the arrows in the figure. Now, on the basis of evidence discussed here, we can firmly place *frq* and its protein product FRQ in the oscillatory loop. It appears likely that the capacity for temperature compensation also lies within the loop, not appended to it. It is still somewhat difficult to decide exactly where in this figure to draw the nuclear boundary, and whether to include FRQ in it sometimes or never. At the level of regulation of metabolism, some time ago we undertook the targeted isolation of a bank of genes specifically regulated on a daily basis by the clock, the *ccg* or *c*lock-*c*ontrolled *g*enes (Loros et al 1989); it is possible that at least some of these clock-controlled genes are regulated directly by FRQ. I shall have little to say about these genes beyond what is in Fig. 1: they were all identified by either subtractive or differential hybridization and, somewhat surprisingly, are generally morning-specific genes.

Because on the basis of its genetics alone, *frq* might have been predicted to be a central component of the circadian clock, we have undertaken the analysis of its structure, function, and regulation. *frq* was cloned several years ago (McClung et al 1989) and is now known to encode several transcripts including at least the following: a short one going leftward, a long one going rightward, and another lying at least partially on top of the latter (Fig. 2). Through the isolation and analysis of cDNA clones corresponding to these products of *frq*,

The *frq* Locus

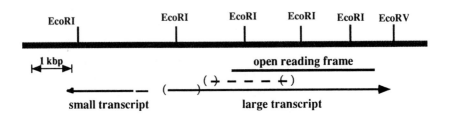

FIG. 2. A physical map of the region of DNA required for the rescue of recessive loss-of-function mutations in the *frq* gene; marks above the line represent restriction sites (a sizing bar gives an estimate of distance in kbp) (adapted from McClung et al 1989). Two *frq* transcripts (arrows) are drawn underneath a fragment capable of complementing the recessive *frq*^{9} allele, and the dotted line above the large rightward transcript represents a third transcript whose presence was inferred from the cDNAs. Parentheses indicate uncertainty as to the end points of transcripts. See text for details (Aronson et al 1994a).

we learned that the small transcript has no open reading frames (ORFs) running through it, suggesting that it does not encode a protein. It does, however, contain an intron that is spliced out. The large transcript region yields at least two transcripts, a long rightward-directed one being intron free and an overlapping one, which goes leftward, having an intron. Thus, of these transcripts, only one, the approximately 4.5 kb transcript transcribed to the right in the figure, has the potential to encode a protein.

Which of these transcripts are required for clock functioning? To answer this question we have used three separate approaches: physiological analysis of gene knock-outs, molecular analysis of the existing *frq* mutants and analysis of phylogenetic conservation. Knock-outs in either the large or the small transcript region, brought about by replacement or by methylation and disruption, can disrupt rhythmicity (Aronson et al 1994a). This result is consistent with the results of resection from either end (K. Johnson, unpublished work; McClung et al 1989), which has shown that all of the region encoding the large transcript, and most of that encoding the small transcript, is required for normal clock functioning. We (K. Johnson, unpublished work) have shown that knocking out the small transcript region alone or making deletions in the small transcript region disrupts rhythmicity. However, prevention of normal splicing of the intron (by destroying the 5′ splice donor site) has no effect on rhythmicity. Finally, knocking out the large transcript region disrupts rhythmicity, as does a frame shift point mutation in the long ORF that results in the production of a truncated protein. These findings focus attention on the large transcript region, but leave open the question of whether one or both of the transcripts are essential.

To get at this question we have studied the existing *frq* mutants, because mutational analysis offers an unbiased way of finding out which parts of a region are important. When Ben Aronson looked to see where in the entire *frq* region the existing mutations mapped, and what aspects of *frq* they disrupted, he found that they all lie within the single long FRQ ORF; none of the mutations lies within *frq* regulatory regions, the small transcript, or the promoter region driving expression of the transcripts. Surprisingly, there is considerable redundancy within the allelic series. Alleles with identical phenotypes are, in fact, genetically identical; a meaningful distinction can no longer be made between frq^2, frq^4 and frq^6, or between frq^7 and frq^8. Finally, there is distinct clustering of the alleles around one region in the centre of FRQ. The three alleles with the most extreme differences in period length as compared with wild-type (frq^1, frq^3 and frq^7) map to an interval of just 12% (124 codons) of the protein, and the two most extreme alleles, frq^1 and frq^7, lie within just 24 amino acids of each other.

The third approach we undertook in an effort to understand the function of *frq* was the study of its phylogenetic conservation. We (Merrow & Dunlap 1994) have now cloned and sequenced all or part of *frq* from five different

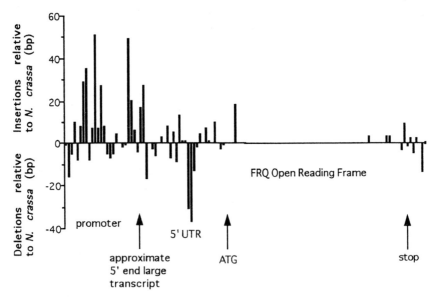

FIG. 3. Comparison of DNA sequences corresponding to the *frq* region from two different fungi, *Neurospora crassa* and *Sordaria fimicola*, reveals that the protein-coding region of *frq* is highly conserved. The *frq* gene, as defined by the region of DNA required to complement recessive loss-of-function mutations in *N. crassa*, is shown schematically as a horizontal line. Hallmarks of the different regions of *frq* are noted below the line. Insertions in the *Sordaria fimicola* genome as compared with the *N. crassa* genome are noted as bars above the horizontal line, and deletions as bars below the line. The open reading frame encoding the *frq* gene product (FRQ) is clearly the most highly conserved region (Merrow & Dunlap 1994). 5′ UTR, untranslated region.

fungi representing three genera (*Neurospora crassa*, *N. discreta*, *N. intermedia*, *Gelasinospora cerealis* and *Sordaria fimicola*) and have found that there is considerable conservation within the long *frq* ORF, but not so much outside it (Fig. 3). Additionally, within the ORF there are preferentially conserved regions. Specifically, all of the sites identified in *N. crassa* as important to *frq* functioning are conserved, as are many of the hallmark features of FRQ, including the PEST regions (sequences highly enriched in the amino acids proline, aspartic acid, glutamic acid, serine and threonine, often fond in proteins showing rapid turnover), the putative nuclear localization signal and the hyperacidic region. Among the parts of *frq* that are most highly conserved is the region surrounding the *frq*^{1} and *frq*^{7} mutations in which even between the most highly diverged species (*Sordaria fimicola* and *Neurospora crassa*) there is just one conservative change over 75 codons. Despite about 15% overall divergence of sequence between the *frq* genes of these two fungi, there is also conservation of function. To examine this, we cloned the *Sordaria fimicola*

FRQ ORF into the context of the *N. crassa* gene so that the *N. crassa* regulatory sequences would control the expression of the *Sordaria* protein. This construct was then targeted back into *N. crassa* bearing loss-of-function mutations in *frq*. Under these conditions, the *S. fimicola* gene will rescue, i.e., restore, rhythmicity in the *N. crassa* mutant. This is particularly interesting in light of the fact that the rhythm being monitored in *Neurospora* was the typical rhythm in conidial production, but *Sordaria* does not produce conidia. Thus, what is conserved is not a 'developmental gene' in the classical sense, but rather a central control gene that is required for the operation of a circadian clock which can be used to initiate diverse developmental pathways.

Several conclusions can be drawn from our studies of functional and mutational analysis and phylogenetic conservation. FRQ is a large protein, of about 1000 amino acids, whose general structure and role in the operation of the clock are conserved. Single amino acid substitutions within it are sufficient to affect both period length and temperature compensation, and loss-of-function mutations result in conditional arrhythmicity and complete loss of clock compensation.

One can also speculate about *frq*. It seems likely that it is tightly regulated; a very long 5' untranslated region preceding the ORF may play an important role in regulating *frq* functioning post-transcriptionally. The small *frq* transcript may have no explicit role in running the clock, but the region encoding it apparently must be present, perhaps to regulate the transcription of the large ORF-containing region. The role of the third transcript remains a mystery also; it is of very low abundance and may be irrelevant, or it may play a role in regulating the level of the large ORF-containing transcript. However, it seems clear that there is a single ORF encoding a protein, FRQ, that is the *trans*-acting product of the locus. Although we can only guess at the real biochemical role FRQ plays in the oscillator, it is noteworthy that the FRQ ORF contains a strong hyperacidic region in its C-terminal part that is reminiscent of known transcriptional activators from several systems (Leuther et al 1993), and that we (M. Merrow) have recently shown that FRQ, when tethered 5' to a yeast promoter by fusion to the DNA-binding domain of GAL4, is capable of activating transcription in yeast.

The regulation of *frq* suggests that FRQ is a component of the clock

Although the results presented thus far are consistent with a role of FRQ in the oscillator, they can also be explained in other ways. An understanding of the regulation of this gene is therefore essential. It has long been appreciated that if FRQ is a component of the clock, and if the clock is a feedback loop, then there is a real possibility that the synthesis of FRQ is regulated by the clock; i.e., if there is a loop defining the clock, in it FRQ could control the clock and the clock control *frq*. We tried for several years to find evidence for control of *frq* expression by the clock under the conditions we commonly use for biochemical analysis of clock-related processes, namely, starvation conditions in the dark. We chose these conditions because under them the clock runs normally but differentiation is drastically curtailed. Thus, we had reasoned, we should be able to focus on a normal clock without being distracted by developmental regulation. Although under these conditions, *frq* transcript(s) are extremely inabundant, we (J. Loros) recently succeeded in adapting conditions using very highly labelled riboprobes and showing that the synthesis of *frq* is regulated by the clock (Fig. 4). In this regard then, *frq* is a morning-specific *ccg*. In separate experiments, we (K. Johnson) have shown that the level of *frq* transcript in the loss-of-function strain *frq*^{9} remains high at all times of day, consistent with *frq* and FRQ regulating their own expression and synthesis through negative feedback (Aronson et al 1994).

These findings clearly put *frq* in the feedback loop, because mutations in the gene affect the timing of its own expression. Interestingly, *frq* expression peaks in the morning and bottoms out in the early evening, generally around dusk. In this regard, *frq* is clearly different from *per*, the transcription of which peaks in the early night. This is an important observation: the first two clock genes to be cloned and analysed at the molecular level both show regulation of the levels of their transcripts (and can therefore be placed with some, if not complete, certainty within the feedback loop), but they are out of phase with one another. Finally, as an aside, it should be noted that this observation provides a ready explanation for the phase-resetting effects of inhibitors of translation (Dunlap & Feldman 1988, Khalsa & Block 1992) and transcription (Dunlap et al 1994, Raju et al 1991): *frq* must be transcribed and the transcript translated on a daily basis for the clock to run, and the time of maximum sensitivity to inhibitors of both transcription and translation is in the subjective morning, about the time of the maximum *frq* transcript level.

The other transcription that is going on in the subjective morning, of course, is the transcription of the various *ccgs*. More than 10 of these genes have now been identified and nearly all are morning specific; this is a surprise, because we had expected the times of maximum *ccg* transcript concentration to be scattered throughout the day rather than being heavily clustered within just one part of the day. This finding provides the basis for several speculations. First, because we have shown that FRQ is capable of activating transcription from a heterologous promoter, it is possible that the protein might be synthesized rapidly and then act immediately to turn on the morning-specific genes. In this manner, it could be viewed as an activator of conidiation, a role that fits several of the phenotypes associated with loss-of-function mutations at *frq*. Alternatively, because *frq* expression is being turned on in the morning, it is possible that whatever is turning on the *ccgs* might also be turning on *frq*.

Several testable predictions can be made to verify the hypothesis that *frq* and FRQ are central components of the feedback-based oscillator: (1) just

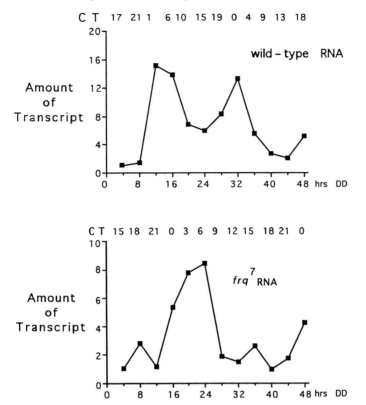

FIG. 4. *frq* is a morning-specific gene which regulates the timing of its own synthesis. Relative amounts of the large (approximately 4500 nucleotide) *frq* transcript found in *N. crassa* mycelia isolated at different times are plotted. Data were collected in two genetic backgrounds, *frq*⁺ (*top*) and the long period mutant *frq*⁷ (*bottom*). In each case, the *frq* message is seen to oscillate with the period length of the clock appropriate to the strain. The RNAs peak at different times in darkness but at the same subjective clock times, in the early subjective morning (Aronson et al 1994b). CT, circadian time.

as loss-of-function mutations eliminate rhythmicity, constitutive elevated expression of FRQ should result in arrhythmicity; and (2) reducing FRQ concentrations in a step-wise fashion from these elevated levels to normal concentrations should reset the clock to a unique phase.

To test these predictions, we (Aronson et al 1994b) developed an inducible promoter system for *Neurospora*, using as a basis the *qa-2* gene and regulatory system (Giles et al 1985). The *qa-2* gene is regulated in a positive manner by the *qa-1*^{F} gene, whose activation is normally blocked by complexation with the product of the *qa-1*^{S} gene. In the presence of inducer (a non-essential metabolite, quinate), the repressor dissociates from the activator, which then turns on the structural genes, including *qa-2*. With this system, a 300-fold induction can be achieved within about six hours (Chaleff 1974). We generated a construct (pBA40) in which the FRQ ORF was cloned immediately following the initiating ATG of the *qa-2* gene, so that in the presence of inducer *frq* message (but without the long 5′ untranslated leader) would be synthesized and FRQ made. RNA analysis confirmed that FRQ-encoding RNA was made in the presence but not in the absence of inducer.

Prediction 1: constitutively elevated expression of frq will result in arrhythmicity

Strains of *N. crassa* wild-type with respect to the clock were transformed with the pBA40 construct and grown in the presence of increasing concentrations of inducer. In this experiment, then, the experimental strain (*frq*$^+$ transformed with pBA40) is its own control. At zero or at low concentrations of inducer (up to about 10^{-6} M), conidial banding and the clock appear normal. [The operation of the *Neurospora* circadian clock is expressed as a daily clock-regulated change in developmental potential and hyphal morphology. Each subjective morning, the mycelia at the growing front of a culture are endowed with the capacity to differentiate from vegetative surface mycelia into aerial hyphae that can eventually give rise to asexual spores (conidia). Mycelia that are laid down several hours later, near noon and throughout the rest of the circadian day, however, do not have this strong potential to differentiate. Thus, following a week of growth across an agar surface, the periods of subjective morning are represented as yellow/orange zones or 'bands' of conidia and differentiated or differentiating mycelia, interspersed with interbands of predominantly vegetative surface mycelia which were laid down at other times of day.] At higher concentrations of quinate, banding appears to broaden and become obscure after several days, and at the highest concentrations (greater than 10^{-5} M) the banding rhythm is completely lost after the first day. The first prediction, that increased constitutive expression of *frq* should cause arrhythmicity, is thus confirmed.

There are several points worth making about this experiment. This is a conditional phenotype resulting from the manipulation of a known clock component. Rhythmicity is not lost instantaneously, but rather gradually, and more rapidly at higher concentrations of inducer, and therefore FRQ. Also, unexpectedly, there are no major effects on period length, although there could have been small effects, of up to a few hours, that were obscured by the broadening and coarsening of the rhythm. This may reflect tight compensation of the rhythm, which is held steady at higher and higher concentrations of FRQ until compensation fails and arrhythmicity ensues. Finally, it should be noted that we have been unable to rescue normal rhythmicity in an *frq*-null strain through constitutive expression of FRQ at any level. This suggests that FRQ must not only be present, but also that it must be regulated in the correct manner, i.e., because we have exquisite control of FRQ synthesis, we predict that the missing aspect of expression is not quantitative but rather temporal control.

These findings also suggested to us that *frq* and FRQ might have two roles in the cell, one as a component of the clock and the other simply as a transcriptional activator of the developmental process resulting eventually in conidiation. Such a role for FRQ in regulating conidiation would be consistent with the known phenotype of *frq*-null strains which produce fewer conidia than *frq*$^+$ strains. It is possible that FRQ's developmental effect might be exerted at lower concentrations than that on the clock, with the result that at concentrations of FRQ at which we would see effects on period, the observation of the rhythm is already obscured.

Prediction 2: in a wild-type strain constitutively expressing frq under the influence of an inducer, step-wise decreases in the concentration of the inducer should completely reset the phase of the clock

The transformed strains described above were grown in liquid medium in constant light in the presence of 10^{-4} M inducer, the minimal concentration necessary for the full phenotypic effect. Controls, either untransformed strains with and without inducer, or transformed strains without inducer, were grown under the same light regimen. Every five hours over a period of 20 h, a full circadian cycle, groups were transferred from light to dark to set the clock to phases covering the cycle. Three hours after the final transfer into the dark, all of the cultures, experimental and control, were transferred out of liquid medium, washed, and transferred to race tubes. The first conidial bands (signifying expression of a functional clock) appeared one or more days later, and the phase of the clock at the time of transfer onto the race tubes was inferred from the phase of these bands (Fig. 5).

There are several points to note about this experiment. First, the transition from light to dark clearly sets the phase of all cultures except those bearing the inducible construct in the presence of inducer. Second, in the experimental strain, a stepped decrease in *frq* expression clearly sets the phase of the rhythm absolutely; prediction 2 is thus confirmed. Third, all experimental cultures (i.e. all cultures of *frq*$^+$/*qa-2*FRQ ORF + 10^{-4} M quinic acid, irrespective of their light-to-dark transfer time) are set (by the shift out of inducer) to the same unique phase point corresponding to around circadian time (CT) 9–11 (approximately the same phase reached following strong resetting by light). This third point represents an unexpected bonus in this experiment, because only the resettability of the clock by the high FRQ to low FRQ shift was required for the model to pass the test, and not the ability to predict the phase (in this case to CT 9–11) following the release from quinic acid induction. This predictability of phase thus deserves further discussion.

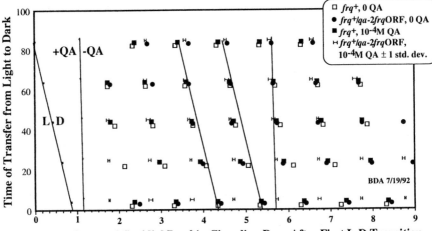

FIG. 5. Steps from high to low expression levels of an FRQ-encoding transcript set the clock to a unique phase. Cultures (*bd; frq*⁺ and *bd; frq*⁺ [pBA40 transformant]) were grown in liquid medium in constant light in the presence or absence of 10^{-4} M inducer (quinic acid, QA). Starting at time 0, at staggered five-hour intervals, groups of cultures (six individual for each treatment and transfer time) were transferred from constant light to darkness to set the clocks out-of-sync to phases scanning the circadian cycle. Five hours after the final light-to-dark (L–D) transfer, all cultures were washed free of old medium and inducer, and placed onto race tubes where they could express their rhythm in conidiation. The first bands of conidiation appeared one or two days later, and, given the knowledge that the centre of a conidial band corresponds to approximately CT 0, the phase of the clock at the time of transfer onto the race tubes (and therefore out of inducing conditions) was inferred from the phase of these bands. A more detailed explanation of the experiment follows. The figure can be viewed as a schematic representation of a series of five sets of race tubes running from left to right, where the position of each symbol (□, ●, ■ and H) corresponds to the position of the conidial band (the point representing CT 0) for each tube on that day. The *x*-axis has units of time, but biological time rather than sidereal time is used; one circadian day (approximately 23 sidereal hours for the strains in this experiment) contains 24 circadian hours, so each division represents four circadian hours. Each symbol is plotted at the average time of day of the centre of a conidial band on a set of six replicate race tubes on one circadian day, such that the same symbol recurs at intervals of one circadian day. The positioning of the symbols, however, differs among the sets because the clocks in each tube were set to different times of day by the different pretreatments of light/dark and ±QA. Standard deviation is shown only for the experimental samples; in all other cases, the error bars were contained within the width of the symbol used. In the three control strains—*frq*⁺ without QA (uninducible strain without inducer), *frq*⁺ with QA (uninducible strain with inducer), and *frq*⁺/*qa-2frq*ORF without QA (inducible strain without inducer)—the clocks should be set solely by the time of the L–D transition. The experimental strain was *frq*⁺/*qa-2frq*ORF with QA in the liquid culture medium (inducible strain with inducer); the third control above verifies that the clock in this strain can respond to a light-to-dark transfer, but the clock in this strain also should be

If the feedback loop constituting the clock is made up of a number of state variables (i.e., more than just *frq* and FRQ) that oscillate with respect to one another, and whose phases with respect to one another yield the final 'overt' phase, then it might have been supposed that the phase to which the clock would be reset following the high to low FRQ shift would, *a priori*, not be predictable, since this final phase would depend not only on the level of FRQ and *frq* but also on the values of other state variables at the time the oscillation was re-initiated. None the less, the data show that the clock is reset to the low point of the *frq* transcript cycle, the same point as would be predicted if the oscillator were made up of only *frq* and FRQ. Does this mean there are no additional components awaiting discovery? This is possible, but seems highly unlikely given the complex properties attributable to the oscillator, including the limits of entrainment and temperature and nutritional compensation. Thus, we favour the alternative, that there are several state variables yet to be identified, and suggest the following interpretation for our data: the extremely strong clock-resetting stimulus represented by the high concentration of FRQ may artificially drive not only *frq* and FRQ, but also the other state variables, to the unique phase points they assume when *frq* is at its low point in the normal cycle. The analogy could easily be made to resetting by light. A short pulse of light, of minutes in duration, will reset (advance or delay) the clock to the daytime part of the cycle no matter where in the cycle the clock is when it sees the light. However, the clock can end up anywhere in the day—a span of some 10 h—depending on where it was before the pulse. A long

---

influenced by the presence of a high level of FRQ (brought about by the presence of the inducer QA) and *may* be reset by the step decrease in QA when the culture is transferred out of QA-containing medium (this is the experiment). The *y*-axis plots the time of transfer from light to dark, and a line drawn through the L–D transfer times (marked L–D) for all five sets of cultures runs, approximately 30° from vertical, obliquely across the first day. The line connnecting the times at which all of the cultures were washed and plated onto race tubes (five hours after the final L–D transfer) is nearly vertical and is marked +QA−QA. After several days of growth, a line connecting all of the band centres for all the controls would run obliquely across the figure (at an angle of about 30°), nearly parallel to the line marking the time of the L–D transfer; this is because the light-to-dark transfer has started the clocks of those cultures at five-hour intervals. To illustrate this, lines actually paralleling the L–D transfer line are plotted four and five days into the experiment; the data points fit closely to these lines. However, if the +QA to −QA shift has reset the clocks in the experimental strains, then a line connecting the centres of these experimental cultures should be close to vertical, paralleling the +QA to −QA line; clearly this is the case, and to illustrate this, a line actually parallel to the +QA−QA line has been drawn on Day 5. This line runs through all of the experimental points (H), verifying that the +QA to −QA shift, not the L–D shift, was the predominant determinant of the phase in these samples. Thus, the shift from high levels of FRQ-encoding transcript to the normal feedback-regulated level has completely reset the clock. See text for more details (Aronson et al 1994b).

pulse of light (over 12 h), however, will drive the clock to the unique phase point corresponding to subjective dusk—a span of about one hour—regardless of where the clock was when the long pulse of light began. In our case, the 'pulse' of FRQ was both quite strong (in comparison with the normal level of *frq* transcript and FRQ in the cell) and more than 12 h long, so it may be that this high level of FRQ is driving all of the state variables, rather than just *frq*. If so, then our results are the results expected and are easily interpreted: elevated FRQ represses its own synthesis and eliminates feedback. Release from the effects of high FRQ concentrations resets the clock to the low point in the oscillation, a point corresponding very generally to the region surrounding dusk, the point noted above to be the nadir in the daily *frq* cycle.

Conclusion: *frq* and FRQ have characteristics of components, very probably state variables, in the circadian oscillator

We have reported data showing that: (1) point mutations in the *frq* gene affect both period length and temperature compensation; (2) loss-of-function mutations result in loss of stable rhythmicity and loss of compensation; (3) the protein-coding portion of the *frq* gene is highly conserved and the function of the gene in the operation of the clock is also conserved: *frq* is a clock gene, not just a conidiation gene; (4) *frq* is within the feedback loop: it regulates the timing of its own expression by negatively feeding back, at least indirectly, to repress its own expression; (5) constitutive expression of *frq* and FRQ appears to stop the clock (resulting in arrhythmicity); (6) step reductions in the concentration of FRQ set the clock to a unique phase, CT 9–11, around late day to dusk.

By now concentrating on the identification of the factors governing the expression of *frq*, we hope to complete the roster of clock genes in this circadian system.

Acknowledgements

This work was supported by NIH grants to J.C.D. (GMS34985, MH44651), an NSF grant to J.J.L. (MCB-9307299) and the Norris Cotton Cancer Center Core grant. B.D.A. was supported by NIH Postdoctoral Fellowship GM14465.

References

Aronson BD, Johnson KA, Dunlap JC 1994a Circadian clock locus *frequency*: protein encoded by a single open reading frame defines period length and temperature compensation. Proc Natl Acad Sci USA 91:7683–7687

Aronson BD, Johnson KA, Loros JJ, Dunlap JC 1994b Negative feedback defining a circadian clock: autoregulation of the clock gene *frequency*. Science 263:1578–1584

Chaleff J 1974 Induction of quinate metabolizing enzymes in *Neurospora crassa*. J Gen Microbiol 81:357–372

Dunlap JC 1993 Genetic analysis of circadian clocks. Annu Rev Physiol 55:683–728

Dunlap JC, Feldman JF 1988 On the role of protein synthesis in the circadian clock of *Neurospora crassa*. Proc Natl Acad Sci USA 85:1096–1100

Dunlap JCL, Crosthwaite S, Feldman JF, Loros JJ 1994 An inhibitor of transcription resets the phase and lengthens the period of the *Neurospora* clock. J Biol Rhythms, under revision

Giles NH, Case ME, Baum J et al 1985 Gene organization and regulation in the *qa* (quinic acid) gene cluster of *Neurospora crassa*. Microbiol Rev 49:338–358

Hamner K, Finn J, Sirohi G, Hoshizaki T, Carpenter B 1962 A biological clock on the South Pole. Nature 195:476–480

Khalsa SBS, Block GD 1992 Stopping the biological clock with inhibitors of protein synthesis. Proc Natl Acad Sci USA 89:10862–10866

Leuther KK, Salmeron JM, Johnston SA 1993 Genetic evidence that an activation domain of GAL4 does not require acidity and may form a β-sheet. Cell 72:575–585

Loros JJ, Denome S, Dunlap JC 1989 Molecular cloning of genes under the control of the circadian clock in *Neurospora*. Science 243:385–388

McClung CR, Fox BA, Dunlap JC 1989 The *Neurospora* clock gene *frequency* shares a sequence element with the *Drosophila* clock gene *period*. Nature 339:558–562

Merrow MW, Dunlap JC 1994 Intergeneric complementation of a circadian rhythmicity defect: phylogenetic conservation of structure and function of the clock gene *frequency*. EMBO (Eur Mol Biol Organ) J 13:2257–2266

Raju U, Koumenis C, Nunez-Reguiero M, Eskin A 1991 Alteration of the phase and period of a circadian oscillator by a reversible transcription inhibitor. Science 253:673–675

Sulzman F, Ellman D, Fuller C, Moore-Ede M, Wassmer G 1984 *Neurospora* rhythms in space: a reexamination of the endogenous–exogenous question. Science 225:232–234

DISCUSSION

Menaker: Would you accept a correction to Fig. 1, since it's recently been clearly demonstrated that prokaryotes have circadian rhythms (Kondo et al 1993)? Figure 1 seems to me to be too complex to represent the minimal unit; a much simpler organization can sustain circadian rhythmicity. That's absolutely fascinating and has some interesting possible consequences. It means that we have pushed back the origin of circadian rhythmicity a whole kingdom's worth, and perhaps to one and a half billion years ago when we have the earliest fossil record of blue-green algae. That has two potential further consequences. One is that it is possible that the clocks of eukaryotic cells were introduced by prokaryotic symbiosis—that is, that the clock mechanism itself comes from prokaryotes. The second potential consequence, which is more relevant to this discussion, is that if in fact clocks evolved as long ago as this finding suggests, and if there has been a common line of descent—which is of course a big if—they have had to adjust over that one and a half billion years to a remarkable slowing down of the earth's rotation. The mechanism which evolved under those early conditions has been modified evolutionarily over the last one and a half billion years into what we think of as a circadian rhythm. Perhaps the easiest way to have done that would have been by adjusting the delay region of a system comprising feedback with delay. If that's what happened, we would expect to see mutations causing quantal changes in period, which might act through insertion of successive delay elements.

Dunlap: I would agree that Fig. 1 could be modified; let's call it the clock in a eukaryotic cell rather than just the cell. The results of Kondo et al (1993) are right and support Beatrice Sweeney's last major contribution before her death, which was to show that there were circadian period lengths and temperature compensation in the cell division rhythms of *Synechococcus*, the same prokaryotic system (Sweeney & Borgese 1989). These findings really do push the origin of clocks further back.

The issue of quantal elements is more dicey. Many of the arguments about such quantal elements in the clock were originally made on the basis of Feldman's mutants in *Neurospora* (Feldman 1982). He had reported independently isolated strains with mutations altering the normal 21.5 h period length of the *Neurospora* clock. One had a period of 16 h, and three strains with ostensibly different mutations had a period of 19 h, suggesting that a quantal element of about three hours could be added or subtracted to change the period length. Similarly, going from wild-type with a period of 21 h to a long period mutant of 24 h and two nominally different mutations altering the period to 29 h suggested there were quantal elements adding, here, 2.5 and three times 2.5 (7.5) h to the period length. However, one of the primary assumptions was wrong: Feldman's mutations were almost certainly not independent after all. All three supposedly independently isolated 19 h period alleles are genotypically identical; they have the same single base pair alteration giving rise to the same amino acid change in FRQ, a change from alanine to threonine at position 895. Similarly, the two ostensibly different 29 h period alleles have the same base change resulting in a glycine to aspartic acid substitution at position 459 of FRQ. The odds of this happening by chance are exceedingly low, suggesting that the phenotypically identical isolates were simply reisolates of the same, single, original mutant strains.

Given this, the whole 'quantal element' picture has fallen apart. There is one mutant at 16 h and one at 19 h. There could also be short period *frq* mutants with period lengths of 17 and 18 h that we haven't found yet, but there is now no reason to assume that they cannot be found. Similarly, with just one mutant with a period of 24 h and one with a period of 19 h, we can't be sure that there are no potential alleles giving period lengths between these two. That doesn't mean that your argument about adding delays is wrong, but rather that there is no genetic support for the idea that period lengths can be changed by adding or subtracting quantal elements.

Loros: It's important to point out that the mutants with long periods, 24 and 29 h, are also altered in temperature compensation. You have to remember the period length is not 24 and 29 h invariably, but only at 25 °C. *Neurospora* has a physiological range from 18–20 °C up to 30 °C.

Edmunds: Some years ago Klevecz (1976) proposed that there might be a quantal element in the length of the cell division cycle. He based this idea on his analysis of the variance of the published generation times of mammalian cell lines in culture, as determined by time-lapse cinematography of individual cells. He discovered what appeared to be a quantal element (G_q) of about 3.75 h. Although in retrospect the evidence may have been a bit shaky, the idea of building up longer period lengths (that is, generation times) up to about 26 h by adding on successive integer multiples of this basic quantal period (which he took to reflect a fundamental limit cycle oscillator underlying the cell division cycle) remains interesting and has its analogue in the possibility that the circadian oscillator derives its period from a multiplication (frequency demultiplication) of some basic ultradian period (see Edmunds 1988, p 111–118). This notion came long before we knew anything much about the supposed quantized array of *Neurospora* circadian τs. Now that an entirely new area in cell cycle regulation has opened up, replete with restriction and execution points that mediate the transit of a cell across the division cycle and help to explain the variability in the length of the G1 (and other) phase(s)—with a little help from the cyclin oscillator and the *cdc2* or *cdc28* gene products—perhaps the concept of quantal elements should be revisited.

Takahashi: Because Jay Dunlap was too modest to say this, I would like to point out that the experiment shown in Fig. 5 is the most definitive to show that any gene is actually part of a feedback loop. Dr Dunlap may have said that, but I'm not sure that everyone appreciated the significance of this experiment. For none of the clock genes that have been studied, *per* in particular, has such a definitive experiment yet been done. The inducible regulation of *frq* and its resetting of the phase of the conidiation rhythm is remarkable. A similar approach has been taken in *Drosophila* using a heat-shock promoter construct to drive *per* expression (Edery et al 1994). The complication with that experiment, however, is that the heat pulse resets the rhythm in control animals, which is not the case with the inducer, quinic acid, in Dr Dunlap's *Neurospora* system.

In *Neurospora*, constitutive expression of *frq* stops the clock in a wild-type background and also does not rescue the *frq*-null mutant. That is not the case in *Drosophila*. What are your thoughts on that?

Dunlap: The results in *Drosophila* to which you are referring, which supposedly show that constitutive expression of *per* from a heat-shock promoter is capable of rescuing rhythmicity in a *per* loss-of-function fly (Ewer et al 1988)—if taken as correct and at face value—are not consistent with feedback-regulated transcription from *per* being a part of the clock. Either oscillatory *per* transcription isn't necessary for the clock, or the supposedly constitutive expression isn't constitutive after all. In contrast, our experiments suggest just what we would have predicted, that the clock gene (*frq*) transcript and clock protein (FRQ) must not only be present, but must also be present at the right time of day—having them there all the time isn't sufficient for a functional clock. There is a caveat to our experiments, though. As we turned on *frq* expression constitutively the rhythm got worse and worse over about two days until it finally degraded into arrhythmicity. *frq*, acting through FRQ, appears to regulate its own transcription, but we (M. Merrow & J. Dunlap, unpublished work) also have some results suggesting that FRQ can activate transcription; as well as being a component of the feedback loop, FRQ may actually turn on conidiation genes. If it turns on conidiation genes at concentrations lower than those required for its operation in the clock, we may be masking the effect on the clock; that is, the clock could, conceivably, have been running at intermediate levels of induction with an altered period length. By the time we increase the concentration of *frq* mRNA enough to see arrhythmicity in conidiation, we haven't yet got long period lengths. However, if we could look at the operation of the clock at even higher concentrations of FRQ, we might see long periods. It was to disprove this possibility that we carried out the FRQ-induced clock resetting experiments shown in Fig. 5.

The results in *Drosophila* are somewhat difficult to understand. On the one hand, elegant experiments by Hardin et al (1990) have shown that *per* is transcriptionally regulated and so is similar to *frq* except that the time of day of the *per* RNA peak is different—*per* is an evening gene whereas *frq* is a morning gene. On the other hand, supposed constitutive expression of *per* under the control of a heat-shock promoter is sufficient to rescue the rhythm. That suggests that the transcriptional rhythm, which we now know to be a fact in two out of two regulated systems (*Neurospora* and *Drosophila*), isn't critical for the clock in *Drosophila*. It is essential to determine whether transcription from the heat-shock-driven *per* promoter is actually constitutive or whether it's rhythmic. If the transcription is rhythmic, then there must be a second loop, one loop built around the clock gene—*frq* or *per*—feeding back on itself, and the other built around the clock protein feeding back on itself, and either of those two loops should be sufficient for rhythmicity.

Takahashi: A third possibility is that there is an intronic regulatory element in the *per* gene.

Dunlap: If that is the case, heat-shock-driven *per* expression is not really constitutive. I have a hard time understanding why there should be rhythmic transcription if it is not necessary; I believe it is, and that the heat-shock-driven *per* expression is not, after all, completely constitutive.

Block: I'm not convinced that Dr Dunlap's elegant experiment shows definitively that *frq* expression is an intrinsic part of the timing loop. Consider, for example, an oscillator that we know something about, such as membrane potential oscillations in cardiac cells. In this case we know that calcium influx controls the potassium channels which are responsible for membrane potential oscillations. We can now do a thought experiment: if you dramatically increase a membrane conductance such as chloride conductance and, for the sake of argument, let's say that chloride conductance does not normally vary rhythmically (and therefore cannot be a state variable in the oscillator), the increase in chloride conductance will clamp the membrane potential and thus stop rhythmic changes in membrane potential. If you allow chloride conductance to return to its normal state, the oscillation will almost certainly start from some fixed point in the cycle. In other words, you would get many of the effects that Dr Dunlap observed, but the difference is that we know the chloride conductance is not a rhythmic element in the feedback loop. However, the issue is more complicated: chloride conductance might well be rhythmic, because of voltage sensitivity of the chloride channels, and yet still not be responsible for generating the rhythmic membrane potential changes.

My feeling is that the experiment Dr Dunlap described is not a definitive experiment, but part of an argument, one of a set of observations that can be taken together along with the fact that *frq* expression is rhythmic to argue that rhythmic *frq* expression is part of the timing loop. However, for this argument to be convincing, it will have to be shown that the oscillation stops if *frq* expression is held constant at any level, not just at a high level of expression.

Takahashi: This experiment by itself is not sufficient, but the aggregate of all the experiments that Jay Dunlap has done together with the constitutive expression of *frq* are quite convincing.

Dunlap: The *frq* gene encodes a product which regulates its own transcription and therefore is involved in feedback. The loss-of-function mutations at *frq* result in high-level expression of *frq* mRNA, which implies negative feedback. Finally, artificially driving constitutive high levels of FRQ-encoding transcript, via the *qa-2* promoter in an *frq*$^+$ background, results in arrythmicity, and a stepped reduction from these high levels both allows rhythmicity to resume and sets the phase of the subsequent rhythm, starting it from the unique phase point defined by the low point in the *frq* transcript oscillation. Those four things suggest that *frq* must be a clock component. Additionally, we have done just the experiment that you suggested, holding the level of *frq* transcript constant, in an *frq* loss-of-function background, at many intermediate levels; in no case could we rescue rhythmicity in this *frq*-null strain. Thus, *frq* cannot simply be supplying a support function as modelled in Dr Block's thought experiment. *per* could be doing this, but not *frq*. There may be ways of interpreting all these results that don't put *frq* in the feedback loop but I haven't come across them yet.

Mrosovsky: I wonder whether it is really true that temperature compensation is intrinsic to the rhythmic mechanism itself. There are some tropical plants, discussed in Bünning's book (1973, p 76), which apparently don't have temperature compensation, but have circadian rhythms.

Dunlap: It would be interesting to find out if they really have a circadian rhythm, which depends somewhat on what we define as a circadian rhythm. I have a slide, which I often use, outlining the characteristics of clocks—a period length close to 24 h which is reset by stepped changes in light and temperature, and temperature compensation. Among the fungi, there are some 'rhythmic' strains which are completely uncompensated. In the physiological temperature range they have period lengths of around a day, but when you push them up to 30 °C or to below 15 °C you can get period lengths that run from six to 40 h. These are typically viewed as non-circadian oscillatory systems.

Loros: If these plants anticipated a change from light to dark, or dark to light, this would help you to define their rhythms as real circadian rhythms that may have lost the capacity for temperature compensation because of the tropical conditions.

Edmunds: This discussion could be extended to ultradian rhythms and their relation to circadian rhythmicity. Jay Dunlap asked whether these plants' rhythms are truly circadian rhythms if they lack a temperature-compensated free-running period. Could they be 'half-way houses' in the evolutionary development of a fully formed circadian clock? This has been suggested for the circadian rhythm of change between resting and actively moving phases in the marine plasmodial rhizopod *Thalassomyxa australis*, which shows a pronounced temperature dependence (Q_{10} of 2.7 between 12 and 24 °C) (Silyn-Roberts et al 1986). On the other hand, some ultradian rhythms have been reported to have temperature-compensated periods (see Lloyd & Rossi 1992, p 1–117). One's hypothesis might be, then, that there has been a gradual evolution of temperature-compensated and entrainable circadian rhythms from temperature-compensated ultradian periodicities (see Edmunds 1988).

Waterhouse: Dr Dunlap, you suggested that in your *Neurospora* mutants there is a correlation between the loss of rhythmicity and the loss of temperature compensation. How strong is this correlation? Does one part of the genome have multiple functions, or can they be split?

Dunlap: Out of the series of seven single gene mutations affecting the *Neurospora* clock, well over half result in alterations in both period length and temperature compensation of the oscillator. There are no cases in which temperature compensation alone is affected, although there are cases where period length has been altered without an effect on temperature compensation.

Waterhouse: Can you say anything more quantitative about this, about the degree or percentage to which they have been changed?

Dunlap: The number of different alleles affecting both periodicity and compensation is too small. frq^3 has a period length of 24 h, frq wild-type one of 21 h, and frq^7 and frq^8 29 h. 24 divided by 21 is about 1.15, which is the change in Q_{10} in frq^3. So, the data fit quite well with the prediction that period length and temperature compensations are altered to the same extent, but the sample size is so small that this calculation is almost meaningless.

Turek: You say that no one has found a *Neurospora* mutant with an unchanged period that does not show temperature compensation. Has there been intensive screening for mutants with altered temperature compensation as there has been for period mutants?

Dunlap: No.

Turek: So the fact that you haven't found any may not be significant.

Dunlap: That's correct. Absence of evidence is not evidence of absence.

Kronauer: If the period is shortened, does the Q_{10} become less than one?

Dunlap: No. In fact, long period length is not required for a change in Q_{10}, and there can be a long period length without a change in Q_{10}. One of the striking things about the mutant strains with short periods is that they do not have altered Q_{10} values.

Menaker: Could one look at this question more directly by asking whether *frq* or FRQ are involved in the oscillations in other fungi that do not show temperature compensation? In an old paper, Jerebzoff (1965) describes a whole series of uncompensated oscillations in fungi. Why not investigate whether *frq* is involved in any of these?

Dunlap: That is certainly possible and would be interesting.

Rensing: You raised the question of whether temperature compensation is a built-in component of the loop or is appended to the loop. What do you consider would demonstrate that a component was built in, for example, that the rates of synthesis and degradation of the mRNA are not dependent on temperature? Could you measure the amount of *frq* mRNA and determine the half-life at different temperatures?

Dunlap: We could do that. It's too soon, at least in this system and probably also in *Drosophila*, to try to produce a quantitative model. We know what some of the components are and we are beginning to describe how they act on one another, but we can't do that quantitatively yet. All I can say at the moment is that there's a lot more mRNA for *frq* in a mutant stain with a long period than there is in one with a short period. We looked for four and a half years for an oscillation of *frq* transcript under starvation conditions, and couldn't find one because there was very little mRNA for *frq* under those growth conditions. We then found that if you add glucose you suddenly see a large amount of *frq* mRNA, yet the clock runs along perfectly normally in either high or low glucose, with lots of *frq* mRNA or with very little. It stunned me that you can add glucose to a starving culture and not reset the clock although in the presence of glucose there's a lot more *frq* mRNA. That says that the level of *frq* transcript itself contains no information about timing. It is apparently the level of *frq* mRNA as compared with something else that's important for keeping time.

Hastings: Can I ask a question about the phase response curves and the nature of the altered periods? Is it possible to talk about the short period as being the result of deletion of part of the temporal programme and the long period the result of an insertion, or do you view the short period as an equal acceleration across all phases and the long period as an equal slowing down of the oscillator across all phases? Does information from phase response curves allow you to differentiate between those two possibilities?

Dunlap: We don't have any good data speaking to that one way or the other. Studies in *Drosophila* (Konopka 1979) suggest that part of the cycle, the dead zone, is shortened in *per* mutants with shortened periods and lengthened in those with lengthened periods. Also, Nakashima (1985) has data in *Neurospora* that he has interpreted similarly to say that there is a zone of the *Neurospora* clock that is specifically lengthened or shortened. The data are good in *Drosophila*, but the error bars in the *Neurospora* phase response curves are larger and the distinction between parametric lengthening and specific lengthening of a particular zone is difficult to make.

Ralph: You said that there are no identified mutations in the promoter regions. The feedback control of the *frq* transcript implicates the promoter or regulatory regions of the long transcript as a logical place to look for both new mutations and entrainment pathways into the underlying biochemical loop generating the rhythm. You could go so far as to say that the promoter region is also a possible site of connection with the temperature compensation mechanism. What is known about this promoter region?

Dunlap: Almost nothing is known about it, and all your points are well taken. It's interesting that none of the mutations has turned up there.

Takahashi: Is it fair to say that the short transcript covers the promoter of the long open reading frame and that your deletion of the short transcript region is a promoter deletion?

Dunlap: That's how we are looking at it now, but to say that is tantamount to saying that the short transcript, which contains an intron and is polyadenylated, is superfluous to clock functioning. To establish this as the most likely alternative took a great deal of work, but I think we have done this now. Our interpretation is that FRQ is encoded by the long open reading frame within the large transcript and that the necessary *frq* regulatory regions extend across the region encoding the small transcript.

References

Bünning E 1973 The physiological clock, 3rd edn. Springer-Verlag, New York
Edmunds LN Jr (1988) Cellular and molecular bases of biological clocks. Springer-Verlag, New York
Edery I, Rutila JE, Rosbash M 1994 Phase shifting of the circadian clock by induction of the *Drosophila period* protein. Science 263:237–240
Ewer J, Rosbash M, Hall J 1988 An inducible promoter fused to the *period* gene of *Drosophila* conditionally rescues adult *per*-mutant rhythmicity. Nature 333:82–84
Feldman J 1982 Genetic approaches to circadian clocks. Annu Rev Plant Physiol 33:563–608

Hardin P, Hall J, Rosbash M 1990 Feedback of the *Drosophila period* gene on cycling of its mRNA. Nature 343:536-540

Jerebzoff S 1965 Manipulation of some oscillating systems in fungi by chemicals. In: Aschoff J (ed) Circadian clocks. North-Holland, Amsterdam (Proc Feldafing Summer Sch) p 183-194

Klevecz RR 1976 Quantized generation time in mammalian cells as an expression of the cellular clock. Proc Natl Acad Sci USA 73:4012-4016

Kondo T, Strayer CA, Kulkarni RD et al 1993 Circadian rhythms in prokaryotes: luciferase as a reporter of circadian gene expression in cyanobacteria. Proc Natl Acad Sci USA 90:5672-5676

Konopka R 1979 Genetic dissection of the *Drosophila* circadian system. Fed Proc 38:2602-2605

Lloyd D, Ross EL 1992 Ultradian rhythms in life processes. Springer-Verlag, London

Nakoshima H 1985 Biochemical and genetic aspects of the conidiation rhythm in *Neurospora crassa*: phase shifting by metabolic inhibitors. In: Hiroshige T, Honma K (eds) Circadian clocks and zeitgebers. Hokkaido University Press, Sapporo, p 35-44

Sweeney BM, Borgese MB 1989 A circadian rhythm in cell division in a prokaryote, the cyanobacterium *Synechucoccus* WH7803. J Phycol 25:183-186

Silyn-Roberts H, Engelmann W, Grell KG 1986 *Thalassomyxa australis* rhythmicity. I. Temperature dependence. J Interdiscip Cycle Res 17:181-187

The effects of temperature change on the circadian clock of *Neurospora*

Ludger Rensing, Andreas Kallies, Gerd Gebauer and Saadat Mohsenzadeh

Department of Biology, University of Bremen, PO Box 33 04 40, D-28334 Bremen, Germany

Abstract. The phase resetting of the circadian oscillatory system by pulses of increased temperature (zeitgebers) and the temperature compensation of its period length during longer exposures are major features of the system, but are not well understood in molecular terms. In *Neurospora crassa*, the effects of pulses of increased temperature on the circadian rhythm of conidiation were determined and possible inputs to the oscillatory system tested, including changes in cyclic 3′,5′-adenosine monophosphate (cAMP), inositol 1,4,5-trisphosphate and H^+ concentrations, as well as changes of phosphorylation, synthesis and degradation of proteins. Following the kinetics of these parameters during exposure to increased temperature showed transient changes. Experimental manipulation of cAMP, Ca^{2+} and H^+ levels, and of the synthesis and, possibly, degradation of proteins, resulted in phase shifts of the oscillatory system. It is assumed that the temperature signal affects the oscillator(s) by multiple pathways and shifts the whole state of the oscillatory system. Second messenger levels, protein synthesis and protein degradation show adaptation to longer exposures to elevated temperature which may be involved in the temperature compensation of the period length. The temperature compensation is also proposed to involve a shift in the state of all or most oscillator variables.

1995 Circadian clocks and their adjustment. Wiley, Chichester (Ciba Foundation Symposium 183) p 26–50

Changes of temperature (except in homeothermic organisms), and changes of light intensity, are the strongest natural time cues (zeitgebers) for the circadian oscillator. Pulses of raised temperature can cause strong type 0 phase response curves (Winfree 1970, Cornelius & Rensing 1982). However, the cellular pathways conveying the temperature signal to the oscillator are not known, mainly because of the enormous range of different temperature effects and because the oscillatory mechanism is only just beginning to be unravelled (Dunlap et al 1994, this volume). The cellular oscillatory system, in addition, seems to consist of an ensemble of oscillators (von der Heyde et al 1992), whose coupling may also be affected by temperature changes.

An approach often applied in the analysis of the oscillatory mechanism and of the input pathways to the clock is exposure to pulses of agents which might affect the mechanism or input elements of the oscillator; a subsequent phase shift is then taken as evidence for the involvement of the particular processes affected (Drescher et al 1982, Rensing & Schill 1987). This approach generated several models of the organization of the circadian oscillator(s) (for reviews see Edmunds 1988, Rensing & Hardeland 1990, Takahashi et al 1993). However, the identification of oscillatory and input elements by this approach is seriously hampered by the many different effects most of these agents have, and by the on–off nature and often unknown kinetics of pulsatile treatments. Many signals, temperature changes in particular, do not affect only a specific element of the oscillator, but shift a whole set of state variables and thus determine a new state of the system (Gooch 1985).

In spite of these difficulties, we consider this approach to reveal relevant and interesting information about the properties of the oscillatory system. In the first section we shall discuss cellular responses to temperature signals, such as changes in second messenger concentrations and protein turnover and translocation. In the second section we shall discuss another characteristic feature of circadian oscillators, their frequency adaptation (compensation) during temperature shifts of longer duration (Sweeney & Hastings 1960). For the same reasons as mentioned above, these features are not well understood in molecular terms. Because the turnover rate of oscillatory proteins (Per in *Drosophila*, FRQ in *Neurospora*) as well as their nuclear translocation (Liu et al 1992) may be involved in determining the frequency of the oscillator, we analysed protein synthesis and degradation and protein translocation to the nucleus during exposures to higher temperatures. For both aspects of the temperature effects—phase shifts and frequency compensation—we shall focus on the fungus *Neurospora crassa*, which is a well-established model system in circadian rhythm research (Feldman & Dunlap 1983, Lakin-Thomas et al 1990, Dunlap et al 1995, this volume).

Effects of pulses of increased temperature on the circadian oscillators

Pulses of increased temperature shift the phase of the circadian oscillator in a phase-dependent way. The amplitude and duration of the pulses determine the character of the resulting phase response curve (type 1 or 0; Winfree 1970). In most non-homeothermic organisms, including *Neurospora crassa* (Rensing et al 1987), maximal shifts occur when pulses of increased temperature are applied at the end of the subjective day or during the subjective night (Cornelius & Rensing 1982). In *Neurospora* the extent of the phase shift (5–10 h) correlates with the extent of the temperature increase of the pulses (5–15 °C); the phase response curves of frequency mutants (frq^1, frq^7; Gardner & Feldman 1981) show some differences in the relationship between the advance and delay portions and in the position of the break-point (Rensing et al 1987).

Putative pathways for the temperature signal within cells

The chain of events by which a pulse of increased temperature affects the circadian oscillatory system is unknown. A conventional investigative approach is screening of temperature-induced changes in various variables, subsequent simulation of these changes by other means and a test of whether or not a particular altered variable causes phase shifts similar to those induced by heat pulses. Such an approach has also been applied in the analysis of the pathways for light signals.

Temperature changes influence almost all cellular variables, including membrane functioning, ion and second messenger concentrations, cytoskeletal structures, enzyme activities, gene expression, protein turnover and the cell cycle, to name but a few (Alexandrov 1977). A temperature change thus shifts the cell from one state to another state. Therefore, it is likely that temperature changes also shift the whole state of the oscillatory system, that is, they affect various oscillatory components. Consequently, the input pathway for a temperature signal consists of multiple transmission chains rather than a single one.

Putative inputs are second messengers, such as cyclic $3',5'$-adenosine or cyclic guanosine monophosphate (cAMP, cGMP), inositol 1,4,5-trisphosphate (InsP$_3$), Ca^{2+} ions and diacylglycerol. They play important roles in animals and have been shown to be involved in the transmission of signals to the circadian clock (Gillette & Prosser 1988, Takahashi et al 1993). Second messenger systems also exist in fungi such as *Neurospora* (Hanson 1991). Manipulations of cAMP and Ca^{2+}-calmodulin in this organism resulted in phase shifts of the circadian oscillator (Nakashima 1986, Techel et al 1990). In the light of these results, we analysed changes in cAMP and InsP$_3$ concentrations in *Neurospora* caused by exposure to increased temperatures for different lengths of time. We then compared the observed changes in second messenger concentrations with the phase-shifting effects of experimental manipulations of second messenger concentrations. If second messengers are involved in the transmission of temperature signals to the oscillatory system, this would imply that the transmission also involves protein kinases and changes in the phosphorylation state of proteins; they play key roles in cellular regulation including the control of gene expression, protein synthesis and activity.

Because experimental inhibition of protein synthesis produces strong phase-shifting responses, and because genetic and molecular analyses have provided evidence for oscillatory changes of a period-determining protein (Per) in *Drosophila* (Rosbash & Hall 1989), we focused on protein synthesis as an important target of raised temperature's effects. We also investigated temperature-dependent changes in protein degradation rates, because they have often been neglected in the analysis of the circadian clock, although the degradation of cyclins, for example, has been shown to be a critical factor in the control of the cell cycle (see Lewin 1990).

Second messengers and protein phosphorylation

Increasing temperature from 25 to 44 °C for various times raises the cAMP concentration in *Neurospora* hyphae by a factor of 1.7–3 within 90–120 s of the beginning of the treatment. When the temperature remains at 44 °C, the cAMP concentration returns to control levels after 30–60 min (Fig. 1a). Similar responses are measured after short (10 s, 10 min) raised-temperature pulses and after treatments of corresponding length with ethanol (2.5 M). This concentration of ethanol shifted the phase of the circadian rhythm by about four to five hours. These results were reproduced with another strain of *Neurospora*, white collar (wc-2), a 'blind' mutant which was used mainly to test the normal functioning of second messengers in this strain (G. Gebauer, A. Kallies & L. Rensing, unpublished work).

The same temperature treatments caused even greater changes in the $InsP_3$ concentration. The rise in $InsP_3$ concentration occurred about 30–60 s after the start of the treatment. With continued exposure to the higher temperature, the $InsP_3$ concentration returned to control levels (Fig. 1b). Again, ethanol induced similar responses, with the $InsP_3$ concentration decreasing below control values after the first rise. All responses were again reproduced in the wc-2 mutant. $InsP_3$ in turn was shown to induce a release of Ca^{2+} from isolated vacuoles and thus should lead to an increased cytoplasmic Ca^{2+} concentration, which is further enhanced by the effects of increased temperature on membrane permeability (Cornelius et al 1989). Thus, in *Neurospora*, temperature signals induce considerable changes in the concentrations of cAMP and $InsP_3$, an early increase and a later decrease in both messengers, with the changes in the $InsP_3$ concentration generally occurring simultaneously with the cAMP changes. A concomitant increase and decrease in cytoplasmic Ca^{2+} may be inferred from the changes in the concentration of $InsP_3$ and their effects on the vacuolar membrane, and have been directly shown in other cells (Stoklosinski et al 1992). These results demonstrate the possible involvement of several second messenger systems in the intracellular transmission of the temperature signal. They also suggest that the transient change in the messenger levels, a rise and a fall, may transmit the information.

We then analysed the phase-shifting effects of cAMP pulses (50 mM) of corresponding lengths given at circadian time (CT) 12. In earlier experiments, 8-parachlorophenyl-cAMP (4×10^{-4} M for 1.5 h) given at this phase produced phase advances of up to six hours (Techel et al 1990). We found small but statistically insignificant phase delays after short pulses and increasing phase advances with longer exposures to cAMP (Fig. 1c). One possible explanation for a duration–response relationship of this kind might be found in the kinetics of the intracellular concentration changes of the second messenger, which may be responsible for the different responses to short and longer exposures. We therefore measured the intracellular concentration of cAMP at different times of exposure to exogenous 50 mM cAMP. The concentration of cAMP increased after short

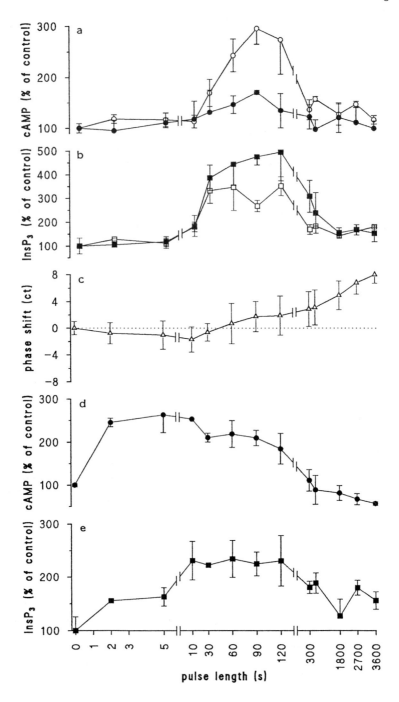

exposures and decreased after long exposures (Fig. 1d). The concentration of InsP$_3$ changed similarly after exposure to cAMP (Fig. 1e). Similar results have been reported recently in human epidermoid A-431 cells (Kiang & McClain 1993).

One interpretation of these results is that the adaptation kinetics leading to a decrease in the concentrations of the two messenger molecules may actually be the relevant advancing signal to the circadian oscillator at this phase. Furthermore, ethanol caused similar phase advances at this circadian time and a decrease in the concentrations of the two second messengers on longer exposure. This would be in line with results of Techel et al (1990) showing phase-shifting effects of a cAMP antagonist and of calmodulin inhibitors (Nakashima 1986).

Temperature-induced and cAMP- and Ca^{2+}-calmodulin-dependent as well as endogenous circadian changes in the phosphorylation state of proteins have been documented (Techel et al 1990, Zwartjes & Eskin 1990; for review see Rensing et al 1993). In *Neurospora*, for example, temperature-induced and circadian changes occur in proteins of 47–51 kDa containing the regulatory subunit of protein kinase A (Techel et al 1990, Trevillyan & Pall 1982) and in a Ca^{2+}-calmodulin-dependent protein (Van Tuinen et al 1984). Circadian changes in the phosphorylation of a 95 kDa protein have also been observed in *Gonyaulax polyedra* (Schroeder-Lorenz & Rensing 1987). That protein phosphorylation changes may, in fact, transmit signals to the basic clock mechanism was already indicated by sequence analysis of *frq* revealing phosphorylation sites for protein kinases (McClung et al 1989). Circadian phosphorylation changes in Per may actually be a component of the basic mechanism itself (J. Edery, personal communication).

FIG.1. Temperature-dependent changes in the concentrations of intracellular second messengers and phase shifts of the circadian conidiation rhythm in *Neurospora*. (a) Changes in cAMP concentration in response to pulses of increased temperature (from 25 to 44 °C) of different length in the wild-type (●) and in a mutant, wc-2 (○). The cAMP concentrations were determined by means of a binding assay (Amersham) immediately after the exposure to increased temperature. (b) Changes in InsP$_3$ concentration at 44 °C in the wild-type (■) and in the 'blind' mutant (□) strain as in (a). The InsP$_3$ concentrations were determined by a binding assay (NEN) at the same time as in (a). High performance liquid chromatography measurements revealed that InsP$_3$ is 13–17% of the total bound inositol phosphates. (c) Phase shifts induced by cAMP (50 mM) pulses of different durations at circadian time (CT) 12. Mycelial discs were exposed to cAMP one circadian period (21.5 h) after a transfer from constant light to constant darkness (CT 12). They were then transferred from liquid culture onto agar. The phase shifts (△) of the first two conidial bands on agar were determined with respect to untreated controls. (d) Changes in cAMP concentration (●) and (e) changes in InsP$_3$ concentration (■) in response to incubation for various lengths of time in medium containing 50 mM cAMP. Intracellular concentrations of cAMP and InsP$_3$ were determined immediately after the incubation. Error bars, standard deviation; $n=4$.

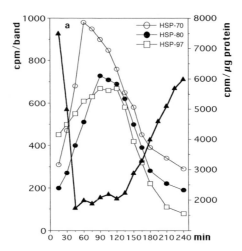
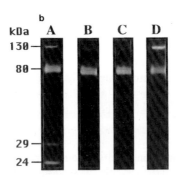

FIG. 2. Temperature-dependent synthesis of heat-shock proteins (Hsps), inhibition of total protein synthesis and inhibition of protease activities in *Neurospora*. (a) [^{35}S]methionine labelling of Hsp70 (○), Hsp80 (●), Hsp97 (□) and of total protein (▲; right ordinate) at different times after a shift in temperature from 25 to 42 °C. Every 15 min, samples were removed and incubated for 20 min with 0.18 MBq/ml [^{35}S]-methionine before washing and freezing. HSP bands were cut out of polyacrylamide gels after one-dimensional electrophoresis and their radioactivity as well as that of total cytoplasmic protein was determined. (b) Activity of proteases measured on renaturing gels. Mycelia were kept at 25 °C (A, control), for two hours at 42 °C (B) or for two hours at 4 °C (C). They were then homogenized and the cytoplasmic proteins separated on polyacrylamide–gelatine gels (Heussen & Dowdle 1980). During the subsequent incubation in buffer (pH 8.2) the renatured proteases digest the gelatine, producing white bands. Sample D was treated as in A. The renatured gel was incubated without Ca^{2+}, the result showing the dependency of the low molecular mass proteases on this ion.

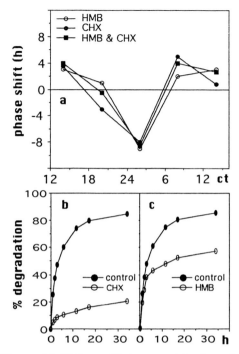

FIG. 3. Phase shifting and inhibitory effects on protein degradation resulting from incubation with inhibitors of proteases and protein synthesis in *Neurospora*. (a) Phase response curves of the circadian conidiation rhythm produced by two-hour pulses of 100 μM hydroxymercuribenzoate (HMB, ○), 1.6 μg/ml cycloheximide (CHX, ●) and 100 μM HMB plus 1.6 μg/ml CHX (■). Phase advances (+) or delays (−) are shown in hours against the circadian time (ct) of pulse application. (b) Effect of 1.6 μg/ml CHX on protein degradation rates (control, ●; CHX-treated, ○). (c) Effect of 100 μM HMB on protein degradation rates (control, ●; HMB-treated, ○). In (b) and (c) mycelia were labelled for two hours with [^{35}S]methionine then washed and incubated in medium (control) or medium plus HMB or CHX. At defined intervals samples were taken and the radioactivity of cytoplasmic proteins was determined. Protein degradation is shown as the percentage decrease in protein radioactivity against time in hours.

Synthesis, degradation and localization of proteins

In general, temperature rises ('heat shock') and many other agents, such as ethanol, produce considerable changes in the transcription of genes. A temperature rise activates a number of heat-shock genes ubiquitously present in prokaryotes and eukaryotes (for review see Nover 1991) including *Neurospora* (Kapoor 1983, Fracella et al 1993). At the level of translation, a reduction in total protein synthesis is observed, together with increased rates of synthesis of heat-shock proteins (Hsps, Fig. 2a). The heat-shock response in *Neurospora* shows strong adapation to longer exposures to increased temperature (40 °C):

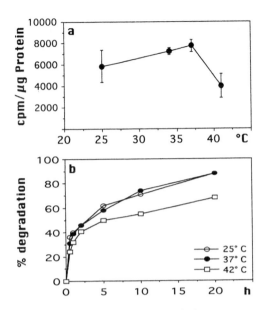

FIG. 4. Adaptation of protein synthesis and degradation rates to temperature change. (a) Radioactive labelling of total protein after two hours incubation with [^{35}S]methionine 24 h after a temperature shift from growth temperature (25 °C) to a higher temperature. Error bars represent standard deviations of the means from three independent series of experiments (U. Schroeder & L. Rensing, unpublished results). (b) Degradation rates of radioactively labelled cytoplasmic proteins (labelled as in Fig. 3c).

after about 45 min total protein synthesis decreases to about 12% of the control level, thereafter increasing to reach about 80% of the initial rate after four hours. Hsp synthesis, on the other hand, increases during the first 60–90 min after the temperature rise and then decreases to constitutive levels or below. The transcriptional and translational control mechanisms responsible may be influenced considerably by temperature-dependent changes in intracellular pH (Weitzel et al 1987) and concomitant changes of protein conformation (see Fig. 5); they also seem to involve changes in the phosphorylation state of transcription and translation factors such as eIF-2 (Murtha-Riel et al 1993). The adaptation of Hsp synthesis is controlled by an autoregulatory system, possibly based on binding of Hsp70 to the heat-shock transcription factor (HSF; see Nover 1991).

A rise in temperature from 25 to 42 °C also affects protein degradation: the activity of three proteases, of 130, 29 and 24 kDa, was considerably inhibited or blocked (Fig. 2b, C). The activity of these proteases was minimal about 45–90 min after the temperature rise.

When the inhibitory effects of a temperature rise on protein synthesis are compared with the temperature pulse's phase-shifting effects in *Neurospora* a good correlation of the effects at different temperature steps is evident (Rensing

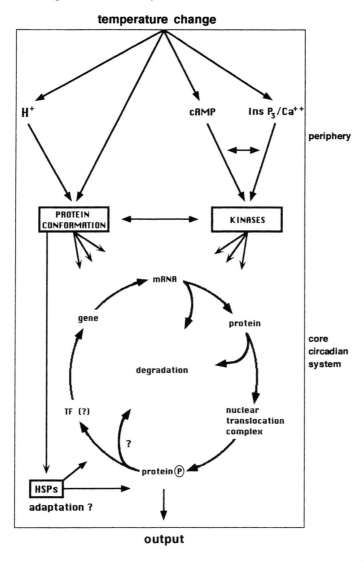

FIG. 5. A hypothetical scheme showing the intracellular pathways by which a temperature signal reaches the circadian oscillator. Of the many possible pathways, we should emphasize that via second messenger and kinases and that via pH and protein conformation changes. The core circadian oscillatory system is still speculative (see Dunlap et al 1994, this volume). Heat-shock proteins (HSPs) may play a role in the adaptation of protein synthesis and degradation and hence the circadian period. TF, transcription factor.

et al 1987). However, the phase response curves produced by treatment with the protein synthesis inhibitor cycloheximide (CHX) and increased temperature are quite different, for example in the position of maximal positive or negative phase shifts (Schulz et al 1985, Rensing et al 1987). This discrepancy led us to consider the effects of temperature on protein degradation as an additional signal for phase shifting. Theoretically, a pulsed inhibition of protein degradation should result in a phase response curve phased differently from that produced by a protein synthesis inhibitor (Drescher et al 1982, Rensing & Schill 1987).

We analysed the phase-shifting effects of artificial serine protease inhibitors, such as phenylmethylsulphonylfluoride (PMSF), N-tosyl-L-lysinechloromethylketone (TLCK) N-tosyl-L-phenylalaninechloromethylketone (TPCK) and a cysteine protease inhibitor, hydroxymercuribenzoate (HMB). Pulses of PMSF, TLCK or TPCK (100 or 150 μM) produced no significant phase shifts, but two-hour pulses of 100 μM HMB produced strong (type 0) phase shifts (Fig. 3). This phase response curve, however, is very similar to that generated by CHX (1.6 μg/ml) or a combination of HMB and CHX, indicating that these two agents have multiple and perhaps similar effects.

The 24 and 27 kDa temperature-sensitive proteases (see Fig. 2) were also inhibited by 100 μM HMB. In addition, the overall rate of protein degradation decreased when *Neurospora* was incubated with 100 μM HMB (Fig. 3b). Cycloheximide, however, inhibits not only protein synthesis, but also protein degradation, as demonstrated in Fig. 3c and in experiments measuring the activity of proteases, in which the 24 and 29 kDa proteases were inhibited completely. Since we also have evidence for an inhibitory influence of HMB on protein synthesis, each of the three phase response curves may arise as a result of a complex interference with protein turnover affecting the oscillatory system.

High-temperature pulses affect not only synthesis and degradation of proteins but also their post-translational modifications and their nuclear translocation. Nuclear translocation often requires modification of the protein's structure (e.g., oligomerization, phosphorylation state; for review see Silver 1991). In *Neurospora* we found an immediate translocation of HSPs into the nucleus after a temperature rise together with an increase in nuclear actin and other proteins (K. Helmbrecht, F. Fracella & L. Rensing, unpublished work). About one to four hours after the onset of the temperature rise (to 40 °C), a portion of nuclear Hsp70 returned to the cytoplasm. It is not known whether this temperature-induced protein translocation also affects the nuclear location of the product of the *frq* (or *per*) gene (FRQ) and thus the circadian oscillatory process.

Conclusions

The effects of a temperature rise on concentrations of second messengers in *Neurospora* suggest that these molecules are involved in the input pathway(s) for the zeitgeber signal, because changes in the cAMP concentration in particular cause phase shifts when it is manipulated by other means (Fig. 5). Manipulation of Ca^{2+} levels has so far produced ambiguous results. Temperature-dependent changes in the phosphorylation state of proteins have been observed, and in the case of Per in *Drosophila* these changes seem to have relevance to the input or oscillatory mechanisms. The effects of a temperature rise on intracellular pH, gene expression, RNA processing, protein synthesis and protein degradation are apparently additional contributors to phase shifting. Whether these changes are mediated by the phosphorylation state of proteins or independently affected by way of, for example, conformational changes ('misfolding') remains an open question. We propose that there are multiple pathways by which the temperature change reaches the oscillatory system which is then shifted to a different state (Gooch 1985).

Compensation of the oscillatory frequency after longer exposures to higher temperatures

In a *Neurospora* wild-type (frq^+) strain, the oscillatory frequency shows almost complete compensation (adaptation) between 20 and 30 °C, whereas the low-frequency mutant frq^7 increases its frequency with increasing temperatures over a larger temperature range (Gardner & Feldman 1981). A defect in temperature compensation has also been described in a fatty acid-deficient mutant (*cel*, Mattern et al 1982). Several theoretical attempts have been made to explain the basis of temperature compensation, which is regarded as a characteristic property of circadian oscillators (Zimmermann et al 1968, Rensing & Schill 1987, Lakin-Thomas et al 1991).

Temperature-adaptable cellular processes

Many processes in cells adapt to different temperatures (Alexandrov 1977), for example, the fluidity of membranes (homeoviscous adaptation), feedback-controlled metabolic pathways such as energy metabolism (energy charge, Atkinson 1977), ion and second messenger concentrations, and gene expression and protein turnover, as discussed above. Cells are more resistant to damage by high temperatures when they have been pre-exposed to raised temperatures (for review see Nover 1991). Thus, adaptation of the circadian frequency to temperature change may not be an extraordinary feature among other cellular processes. Below, we focus on adaptation processes in protein turnover, because the involvement of at least one protein (Per) in the oscillatory mechanism appears to be well established (for review see Rosbash & Hall 1989).

Synthesis and degradation of proteins

In *Neurospora*, total protein synthesis rates were determined at different temperatures. Cultures were moved from the growth temperature (25 °C) to various higher temperatures and allowed to adapt for 24 or 48 h. During this adaptation time we kept the cultures in reduced constant concentrations of sucrose (0.01%) to simulate the conditions under which the conidiation rhythm is observed. Incubation with a higher concentration of sucrose (2%), however, did not alter the results significantly. The rates of protein synthesis adapted to a more or less constant level over a temperature range between 25 and 37 °C, and to a decreased level at higher temperatures (Fig. 4a). Similarly, total protein degradation rates adapted to relatively constant values between 25 and 37 °C and to decreased values at 42 °C (Fig. 4b). Synthesis and degradation rates thus change concomitantly, a result also obtained in earlier investigations with *Gonyaulax* (Cornelius et al 1985).

Adaptation of protein synthesis during continued exposure to high temperature is considered to be partly dependent on an increase in the synthesis and translocation of heat-shock proteins (Hayahshi et al 1991). It is tempting to speculate that such a process also plays a role in the period compensation of the circadian oscillator (Fig. 5).

The main question here is whether the observed processes of adaptation of total protein synthesis and degradation have any significance for the responses of the oscillatory protein(s) and the temperature compensation of the circadian frequency, or, more specifically, whether one can explain the features of frq^7—long period length, low phase-shifting sensitivity to cycloheximide (Dunlap & Feldman 1988) and lower temperature compensation—on the basis of altered synthesis and degradation rates of FRQ. This alteration may be due to a change in the phosphorylation and the turnover of the oscillatory protein.

Let us assume that oscillatory changes in the activity of the oscillatory protein(s) are based on concentration changes. A net increase in the concentration of an oscillatory protein will be achieved when the rate of synthesis is higher than the degradation rate, whereas a net concentration decrease will result when degradation dominates over synthesis. It is not the absolute synthesis and degradation rates which are important but rather their relationship. Changing both rates in one direction need not result in a frequency change; a change in one of them will alter frequency and amplitude. Thus, the adaptation of both synthesis and degradation rates as well as their concomitant change may contribute to frequency compensation. The frequency changes of frq^+ at higher temperatures and of frq^7 over the whole temperature range may result from a change in the relationship between synthesis and degradation.

Lowering the rate of degradation of the oscillatory protein in a model oscillator (Rensing & Schill 1987) considerably lengthens the period and increases the amplitude of the oscillation. The lower sensitivity of frq^7 to the phase-shifting effects of cycloheximide may also be explained by a lower degradation rate because the degradation of the oscillatory protein during the time of inhibition of its synthesis would be less. This explanation, however, is complicated by the fact that cycloheximide also inhibits degradation.

Conclusions

Total protein synthesis and degradation rates show adaptation to temperature change similar to the temperature compensation of the circadian frequency. Second messenger concentrations also adapt to different temperatures. Protein phosphorylation and dephosphorylation thus may be involved in the temperature compensation of the period as well. We assume that temperature compensation consists of a shift in most or all oscillator variables from a particular altered state shortly after the temperature rise to the adapted state. The adapted state of the circadian oscillator is based on general cellular mechanisms of temperature adaptation as well as on the particular properties of the oscillatory elements which are subject to these mechanisms. A mutational change in one of these properties may bring about an altered adapted state.

References

Alexandrov VY 1977 Cells, molecules and temperature. Springer-Verlag, Heidelberg

Atkinson DE 1977 Cellular energy metabolism and its regulation. Academic Press, New York

Cornelius G, Rensing L 1982 Can phase response curves of various treatments of circadian rhythms be explained by effects on protein synthesis and degradation? Biosystems 15:34–47

Cornelius G, Schroeder-Lorenz A, Rensing L 1985 Circadian-clock control of protein synthesis and degradation in *Gonyaulax polyedra*. Planta 166:365–370

Cornelius G, Gebauer G, Techel D 1989 Inositol trisphosphate induces calcium release from *Neurospora crassa* vacuoles. Biochem Biophys Res Commun 162:852–856

Drescher K, Cornelius G, Rensing L 1982 Phase response curves obtained by perturbing different variables of a 24 h model oscillator based on translational control. J Theor Biol 94:345–353

Dunlap JC, Feldman J 1988 On the role of protein synthesis in the circadian clock of *Neurospora crassa*. Proc Natl Acad Sci USA 85:1096–1100

Dunlap JC, Loros JL, Aronson B et al 1995 The genetic basis of the circadian clock: identification of *frq* and FRQ as clock components in *Neurospora*. In: Circadian clocks and their adjustment. Wiley, Chichester (Ciba Found Symp 183) p 3–25

Edmunds LN Jr 1988 Cellular and molecular bases of biological clocks. Springer-Verlag, Heidelberg

Feldman J, Dunlap JC 1983 *Neurospora crassa*: a unique system for studying circadian rhythms. Photochem Photobiol Rev 7:319–368

Fracella F, Mohsenzadeh S, Rensing L 1993 Purification and partial amino acid sequence of the major 70 000 Dalton heat shock protein in *Neurospora crassa*. Exp Mycol 17:362–367

Gardner GF, Feldman J 1981 Temperature compensation of circadian period length in clock mutants of *Neurospora crassa*. Plant Physiol 68:1244–1248

Gillette MU, Prosser RA 1988 Circadian rhythm of rat suprachiasmatic brain slice is rapidly reset by daytime application of cAMP analogs. Brain Res 474:348–353

Gooch V 1985 Effects of light and temperature steps on circadian rhythms of *Neurospora* and *Gonyaulax*. In: Rensing L, Jaeger NI (eds) Temporal order. Springer-Verlag, Heidelberg, p 232–237

Hanson BA 1991 The effects of lithium on phosphoinositides and inositol phosphates of *Neurospora crassa*. Exp Mycol 15:76–90

Hayashi Y, Tohnai J, Kaneda T, Kobayashi T, Ohtsuka K 1991 Translocation of HSP70 and protein synthesis during continuous heating at mild tempertures in HeLa cells. Radiat Res 125:80–88

Heussen C, Dowdle EB 1980 Electrophoretic analysis of plasminogen activator in polyacrylamide gels containing sodium dodecyl sulfate and copolymerized substrates. Anal Biochem 102:196–202

Kapoor M 1983 A study of heat shock response in *Neurospora crassa*. Int J Biochem 15:636–649

Kiang JG, McClain DE 1993 Effect of heat shock, $[Ca^{2+}]_i$, and cAMP on inositol trisphosphate in human epidermoid A-431 cells. Am J Physiol 264:C1561–C1569

Lakin-Thomas PL, Coté GG, Brody S 1990 Circadian rhythms in *Neurospora crassa*: biochemistry and genetics. Crit Rev Microbiol 17:365–416

Lakin-Thomas PL, Brody S, Coté GG 1991 Amplitude model for the effects of mutations and temperature on period and phase resetting of the *Neurospora* circadian oscillator. J Biol Rhythms 6:281–297

Lewin B 1990 Driving the cell cycle: M phase kinase, its partners and substrates. Cell 61:743–752

Liu X, Zwiebel LJ, Hinton D, Benzer S, Hall JC, Rosbash M 1992 The period gene encodes a predominantly nuclear protein in adult *Drosophila*. J Neurosci 12:2735–2747

Mattern DL, Forman LR, Brody S 1982 Circadian rhythms in *Neurospora crassa*: a mutation affecting temperature compensation. Proc Natl Acad Sci USA 79:825–829

McClung CR, Fox BA, Dunlap JC 1989 The *Neurospora* clock gene *frequency* shares a sequence element with the *Drosophila* clock gene *period*. Nature 339:558–562

Murtha-Riel P, Davies MV, Scherer BJ, Choi SY, Hershey JWB, Kaufman RJ 1993 Expression of a phosphorylation-resistant eukaryotic initiation factor 2α-subunit mitigates heat shock inhibition of protein synthesis. J Biol Chem 268:12946–12951

Nakashima H 1986 Phase shifting of the circadian conidiation rhythm in *Neurospora crassa* by calmodulin antagonists. J Biol Rhythms 1:163–169

Nover L (ed) 1991 Heat shock response. CRC Press, Boca Raton, FL

Rensing L, Hardeland R 1990 The cellular mechanism of circadian rhythms—a view on evidence, hypotheses and problems. Chronobiol Int 7:353–370

Rensing L, Schill W 1987 Perturbation of cellular circadian rhythms by light and temperature. In: Rensing L, An der Heiden U, Mackey M (eds) Temporal disorder in human oscillatory systems. Springer-Verlag, Heidelberg, p 233–234

Rensing L, Bos A, Kroeger J, Cornelius G 1987 Possible link between circadian rhythm and heat shock response in *Neurospora crassa*. Chronobiol Int 4:543–549

Rensing L, Kohler W, Gebauer G, Kallies A 1993 Protein phosphorylation and circadian rhythms. In: Battey NJ, Dickinson HG, Hetherington AM (eds) Posttranslational modifications in plants. Cambridge University Press, Cambridge, p 171–185

Rosbash M, Hall J 1989 The molecular biology of circadian rhythms. Neuron 3:387–398

Schroeder-Lorenz A, Rensing L 1987 Circadian changes in protein synthesis rate and protein phosphorylation in cell-free extracts of *Gonyaulax polyedra*. Planta 170:7–13

Schulz R, Pilatus U, Rensing L 1985 On the role of energy metabolism in *Neurospora* circadian clock function. Chronobiol Int 2:223–233

Silver PA 1991 How proteins enter the nucleus. Cell 64:489–497

Stoklosinski A, Kruse H, Richter-Landsberg C, Rensing L 1992 Effects of heat shock on neuroblastoma (N1E115) cell proliferation and differentiation. Exp Cell Res 200:89–96

Sweeney B, Hastings WJ 1960 Effects of temperature upon diurnal rhythms. Cold Spring Harbor Symp Quant Biol 25:87–104

Takahashi JS, Kornhauser JM, Koumenis C, Eskin A 1993 Molecular approaches to understanding circadian oscillations. Annu Rev Physiol 55:729–753

Techel D, Gebauer G, Kohler W, Braumann T, Jastorff B, Rensing L 1990 On the role of Ca^{2+}-calmodulin-dependent and cAMP-dependent protein phosphorylation in the circadian rhythm of *Neurospora crassa*. J Comp Physiol B Biochem Syst Environ Physiol 159:695–706

Trevillyan JM, Pall ML 1982 Isolation and properties of a cyclic AMP-binding protein from *Neurospora*. J Biol Chem 257:3978–3986

Van Tuinen D, Perez RO, Marme D, Turian G 1984 Calcium calmodulin-dependent protein phosphorylation in *Neurospora crassa*. FEBS (Fed Eur Biochem Soc) Lett 176:317–320

von der Heyde V, Wilkens A, Rensing L 1992 The effects of temperature on the circadian rhythms of flashing and glow in *Gonyaulax polyedra*: are the two rhythms controlled by two oscillators? J Biol Rhythms 7:115–123

Weitzel G, Pilatus U, Rensing L 1987 The cytoplasmic pH, ATP content and total protein synthesis rate during heat shock protein inducing treatments in yeast. Exp Cell Res 170:64–79

Winfree AT 1970 Integrated view of resetting a circadian clock. J Theor Biol 28:327–374

Zimmerman WF, Pittendrigh CS, Pavlidis T 1968 Temperature compensation of the circadian oscillation in *Drosophila pseudoobscura* and its entrainment by temperature cycles. J Insect Physiol 14:669–684

Zwartjes RE, Eskin A 1990 Changes in protein phosphorylation in the eye of *Aplysia* associated with circadian rhythm regulation by serotonin. J Neurobiol 21:376–383

DISCUSSION

Block: In the *Bulla* eye preparation two different classes of phase shift are produced by temperature change (Bogart 1992). At reduced temperatures there are both phase advances and delays, whereas at higher temperature steps, where heat-shock proteins begin to appear, we get only delays.

Rensing: We find similar responses to different temperatures in *Neurospora*. Phase shifts in response to temperatures above 45 °C are all delays.

Block: Do you mean that below 45 °C you don't get any shift?

Rensing: Below 45 °C we get the typical phase response curve similar to that produced in response to light.

Edmunds: Is there a circadian rhythm in the overall level of cAMP under constant conditions in *Neurospora*, and, if there is, is it bimodal?

Rensing: There is a bimodal rhythm in constant darkness. The difference between the minimum and maximum concentrations is about 40%.

Edmunds: I asked the question because we have shown a similar rhythm in *Euglena*, a 12 h, or bimodal, rhythm, with peaks corresponding to key transition points in the cell division cycle (Carré et al 1989). Are the peaks in your system at around dawn and dusk, as they are in *Euglena*, and do these relate in any way to cell cycle transition points in *Neurospora*?

Rensing: *Neurospora* has a short cell cycle varying between 100 min and three or four hours depending on the nutritional state. We find an oscillation of inositol phosphates with a period of about four hours, which may be related to the cell cycle.

Gillette: In the rat suprachiasmatic nucleus *in vitro* in minimal conditions which should not stimulate cell division (although we haven't directly addressed that) we also see bimodal changes in [cAMP] corresponding to the times that you and Dr Edmunds spoke of in *Euglena* and *Neurospora*, around dusk and dawn. Additionally, we find a single rise in [cGMP] around dawn. These messengers are oscillating with the clock oscillations in the apparent absence of cell division. Post-translational modification of proteins could have two roles in clock regulation, one of which is in the running of the clock. In *Drosophila*, Per lasts for about 12 h and is hyperphosphorylated. The two mutations resulting in shortened period in *Neurospora* are changes from Gly and Ala to Ser (frq^1, period 16 h) and Thr (frq^2, period 19 h), respectively, potential phosphorylation sites. Is that significant?

Dunlap: It's too soon to tell. They could well be phosphorylated and this might be a causal factor. However, amino acid phosphorylation site consensus sequences are neither very long nor particularly stringent, so by sequence gazing alone it's impossible to tell whether or not those sites are phosphorylated. For Per, the number of potential phosphorylation sites predicted by the amino acid sequence is less than the apparent number of phosphorylation sites.

Rensing: One interesting question is whether the phosphorylation affects nuclear translocation. This seems to be the case with some transcription factors.

Lemmer: Is there any information on the formation or degradation of cAMP and $InsP_3$, which might be temperature dependent? Also, to what extent are the different *Neurospora* mutants affected by, for example, changes in the formation or degradation of cAMP or $InsP_3$?

Rensing: This question could be addressed using the *crisp* mutant, which is defective in the adenylate cyclase.

Dunlap: This mutation reduces intracellular cAMP concentrations to about 1% of normal. The clock has a normal period length, but I don't think temperature compensation has been examined.

Edmunds: This was the original evidence that Feldman et al (1979) used to argue against the participation of cAMP in the clock.

Rensing: It has not been checked whether or not the inositol phosphate concentrations are different from normal in this mutant.

Lemmer: The endogenous concentration tells you little about the turnover. Even if the concentration is reduced by 99%, the turnover might still be high. The interesting question is whether the cyclase is affected by stimulatory or inhibitory receptors coupled to the adenylate cyclase and whether there are different types of phosphodiesterase to degrade cAMP in the different mutants.

Rensing: There is no information on that.

Edmunds: In *Euglena* there are antiphase bimodal circadian rhythms in adenylate cyclase and phosphodiesterase, although we learnt to our horror that there are a whole family of phosphodiesterases which have to be looked at individually (Tong et al 1991).

Rensing: There is bimodal activity of phosphodiesterase, but we haven't checked this with the adenylate cyclase.

Edmunds: What about cGMP? We find a bimodal rhythm in *Euglena*, which leads that of cAMP by about two hours (Tong & Edmunds 1993).

Rensing: We tested several analogues of cGMP but found no striking effects on the rhythm.

Edmunds: Dr Gillette, do you have any idea why the rhythm of cGMP concentration should be unimodal in the suprachiasmatic nucleus, whereas cAMP concentration is bimodal?

Gillette: No, I wish I did. We have also found a bimodal oscillation of total phosphodiesterase activity, but adenylate cyclase activity does not oscillate.

Rensing: The difficulty in your system and in *Neurospora* is that there seems to be both an output and an input of cAMP, which creates another loop in addition to the basic oscillatory loop.

Gillette: Our system is complicated, because there are many cell types in the suprachiasmatic nucleus and we don't have any way of of knowing where the actual critical sites of action are.

Hastings: Is there any evidence about whether the changes in the nucleotide concentrations are reflected by changes in nucleotide-dependent kinase activities? They would be the important end-points, not the presence of nucleotide itself.

Rensing: We are doing experiments with antibodies against the catalytic subunit of protein kinase A. We want to find out whether the distribution of this catalytic subunit in the nucleus differs according to the level of cAMP, which is the case in mammalian systems (Trinczek et al 1993).

Edmunds: In *Euglena* there are at least two cAMP-dependent kinases, with different affinities for cAMP and for several cAMP analogues, which correspond nicely to the times of the two peaks in the bimodal rhythm of cAMP concentration (Carré & Edmunds 1992).

Hastings: So you would predict different substrates for those two enzymes.

Edmunds: Yes. Activation of each of the kinases would cause the phosphorylation of a different set of targets and perturb different cell cycle control pathways.

Waterhouse: I am convinced that a change in temperature influences cyclic nucleotide concentrations and induces heat-shock proteins, but what is known in detail about the way they are affecting the clock?

Rensing: The change in cAMP concentration brings about a phase shift. Whether or not FRQ is phosphorylated by protein kinase A is not yet known.

We have no direct evidence for the involvement of the heat-shock proteins in temperature compensation. Heat-shock proteins guide the translocation and also the folding of proteins to the right conformation, and may even help to restore the correct conformation to transcriptional factors after a temperature change.

Dunlap: The *frq* mutants *frq*^{3} and *frq*^{7} have altered temperature compensation. Is the phase response curve produced by a temperature change in these mutants different from that of the wild-type?

Rensing: There are differences in the position of the break-point and in the relative proportion of the delay and advance parts of the phase response curve between *frq*^{1}, *frq*^{7} and wild-type.

Redfern: You discussed the involvement of Ca^{2+}, and I understand that you have carried out experiments with dantrolene and nifedipine, which gave different results (Techel et al 1990). How much is known about the pharmacology of Ca^{2+} channels in *Neurospora*? How far can one extrapolate from mammalian pharmacology?

Rensing: We couldn't measure changes in cytoplasmic Ca^{2+} concentration because fura-2 does not enter the cell. We used nickel, cobalt and other known blockers of plasma membrane Ca^{2+} channels to see whether they affect the phase of the conidiation rhythm. With dantrolene we tested phase shifts and the inhibitory effects on Ca^{2+} release from isolated vacuoles (Cornelius et al 1989).

Lemmer: Have you set up dose response curves? At what concentration did you test the effects of nifedipine and dantrolene?

Rensing: We tested the phase response at several concentrations and observed shifts in the micromolar range. In muscle cells dantrolene interferes with $InsP_3$-induced Ca^{2+} release from intracellular stores and it does something similar with the Ca^{2+} release from vacuolar stores in *Neurospora* (Techel et al 1990).

Block: If you are concerned about non-specific effects of the drug, have you tried simply lowering the extracellular [Ca^{2+}]? Does that have the same effect?

Rensing: Ca^{2+} chelators had no effect on the phase. The channel blockers nickel and cobalt known to act on the plasma membrane also had no effects on the phase, whereas dantrolene affects intracellular Ca^{2+} release and phase.

Waterhouse: You have talked about heat-shock proteins and temperature compensation in the wild-type and in mutants. One would predict that these mechanisms would be of varying importance in different species. The obvious examples would be homeotherms and creatures whose natural environments vary little in temperature. Is there any evidence on this point?

Takahashi: We have done some work like that, and perhaps I can comment on the general area of heat-shock proteins, temperature and cyclic nucleotides in other species. I've worked on both *Aplysia* and chick pineal. In *Aplysia*, cGMP and cAMP have profound effects on the clock. cAMP produces a phase response curve like that produced by serotonin, and cGMP produces a light-type phase response curve. There is a circadian rhythm of cAMP in cultured chicken pineal cells, but cAMP analogues and cyclase-stimulating agents have no phase-resetting effects. We concluded that these cyclic nucleotides were probably not working in either system at the level of the clock's core mechanism, but that in *Aplysia* they were mediating inputs to the clock—serotonin being the major input transmitter using cAMP—and that in the pineal, cAMP was probably mediating an output from the clock, being controlled by the clock but not part of the clock mechanism itself. I don't know whether that's a general conclusion that applies to all the systems that we work on.

In cultured chicken pineal cells, using the melatonin rhythm, we obtain a beautiful phase response curve to six-hour temperature pulses, raising the temperature from 37 °C to 42 °C that is very similar to that produced by light. However, if we increase the temperature to 43 °C for six hours, the circadian melatonin rhythm is abolished. A six-hour pulse to 42 °C stimulates a transient induction of heat-shock proteins. At 43 °C, there is a different kind of response, with prolonged expression of heat-shock proteins and a shut-down of protein synthesis. The clock doesn't run under those conditions in pineal cells. So, temperature clearly affects the circadian clock in chicken pineal cells.

The other issue we addressed was whether the clock is temperature compensated in a homeotherm, i.e., in a situation in which it might not *have* to be compensated. In chicken pineal cells the clock is actually well compensated; the Q_{10} is less than 1, around 0.8.

Block: You said that at 43 °C the clock doesn't run in pineal cells. Do you really mean that there is no rhythm expressed?

Takahashi: When we increased the temperature to 43 °C at any phase the circadian melatonin rhythm was abolished; the abolition could have been at the level of the output.

Hastings: Did the rhythm restart from a unique phase?

Takahashi: No, it didn't restart.

Miller: We have done some work on temperature compensation in the suprachiasmatic nucleus (Miller et al 1994). In slice preparations we found that the Q_{10} values for temperature compensation of action potentials in rats and hibernating species such as squirrels all exceed 2.0. That action potentials are not well temperature compensated in the suprachiasmatic nucleus is one more nail in the coffin for the idea that synaptic transmission is an important part of the clock in the mammalian suprachiasmatic nucleus.

In monolayer cultures we virtually always see rhythmicity in vasopressin concentrations (V. Cao & J. D. Miller, unpublished work). Other peptides such as VIP (vasoactive intestinal peptide) and somatostatin, however, are not rhythmic unless we use a synchronizing pulse. The synchronizing pulse that we have found most effective is a 24 h decrease in temperature down to about 25 °C. David Earnest also has some results along these lines with cAMP as a synchronizing pulse (D. Earnest, personal communication). We feel that a temperature pulse can have a major synchronizing influence on mammalian clock cells.

Rensing: A temperature increase can also increase c-*fos* expression in HeLa cells (Andrews et al 1987) analogous to the light-induced expression.

Gillette: David Earnest's results (D. Earnest, personal communication) are preliminary, and he is using treatments that increase the concentration of cAMP not as synchronizers for the oscillator, but to induce differentiation in his cell lines by inhibiting their mitotic activity.

Edmunds: Do we, as a whole, agree that cAMP is not an element, a gear of the oscillator itself? We have direct evidence that it is not part of the oscillator in *Euglena*, but is working downstream from the clock, modulating the cell division cycle (Carré & Edmunds 1993). In the chick pineal, cAMP acts on the melatonin-synthesizing system downstream from the pacemaker, working in a parallel pathway of some sort (Zatz 1992). Does cAMP produce lasting, permanent phase shifts?

Gillette: cAMP can certainly produce permanent phase shifts in the molluscan eye system at least, and can stimulate shifts in the rat suprachiasmatic nucleus *in vitro* that are stable, as far as we can tell.

Edmunds: I asked the question because I wondered whether there is a difference between nervous excitable tissue (such as the eye and the suprachiasmatic nucleus) and non-excitable tissue (such as *Euglena*, where permanent, cAMP-induced phase shifts of the cell division rhythm are not observed).

My other worry concerns compartmentalization of cAMP (or Ca^{2+}, or whatever); even though we think of a peak in the level of cAMP as one signal, there could be a cellular sequestration, or localization, meaning that different pools of cAMP would exist. It is possible that only a small portion of the total cAMP would be involved in a given cascade scheme. Thus, it is possible that only a small portion of the total cAMP (which you are measuring) is what actually matters.

Gillette: That's absolutely true, but there's no way to address that unless you can visualize local changes, as was done in the recent study by Bacskai et al (1993), in which cAMP-dependent protein kinase A subunits labelled with both fluorescein and rhodamine were used to find out where in the cell the changes took place.

Edmunds: Even that might not be enough, because there might be a rapid exchange between compartments, which you would never be able to detect by those techniques.

Miller: If you think cAMP is part of the endogenous clock mechanism, the experiment that you must do is to see whether antagonizing cAMP over the 24 h period disrupts your marker rhythm.

Edmunds: That would be consistent but would not necessarily be conclusive.

Gillette: That's a hard experiment to do.

Edmunds: We have done some work with pulses of cAMP antagonists that appears to rule out cAMP as an element of the clock in *Euglena*, but again, these experiments are open to the same criticisms (Carré & Edmunds 1993).

Rensing: I think that cAMP is one of several input-transducing molecules for the temperature signal in *Neurospora*. There is no evidence for the involvement of inositol phosphates in the transmission of light signals in *Neurospora*. We are completely in the dark about what light does!

Dunlap: I think everyone is completely in the dark. There are two 'blind' mutants of *Neurospora*, wc-1 and wc-2 (wc, white collar). The *Neurospora* blue light photoreceptor appears to be flavin-mediated, as are plant blue light photoreceptors. The gene for a putative blue light receptor has just been cloned from *Arabidopsis* by Tony Cashmore (Ahmed & Cashmore 1993). We have sent him cDNA libraries so that he can look for it in *Neurospora*, so there's some hope that the gene at least will be available soon. There's some pharmacological evidence that this blue light photoreceptor works via an inhibitory G protein; sexual development is inhibited in some fungi by pertussis toxin. As far as other photoreceptors go, the red light photoreceptor seen in *Gonyaulax*, for example, might be chlorophyll, but little is known about that. The third known cirdadian photoreceptor is phytochrome. I don't know how it works.

Roenneberg: Kippert has recently reported a red-sensitive cytochrome in yeast (Kippert et al 1991).

Takahashi: So there's no evidence for an opsin-like pigment in a microorganism other than *Chlamydomonas*?

Dunlap: That's correct.

Loros: Pat Lakin-Thomas has spent some years examining inositol phosphate metabolism, as it had been proposed to be part of the blue light signal-transduction pathway responsible for resetting the phase of the circadian oscillator in *Neurospora*. She has also investigated the effects of inositol depletion and supplementation on the *Neurospora* clock, as it had been suggested that the effects of lithium on the period of the clock were the result of its inhibition of inositol monophosphate phosphatase and subsequent inositol lipid depletion. Her conclusion is that inositol phosphate metabolism is unlikely to be involved in either the blue-light signal transduction pathway itself or light input to the clock (Lakin-Thomas 1993).

Block: In the *Bulla* eye there is good evidence that Ca^{2+} is the second messenger in the light-entrainment pathway. That's probably true in *Aplysia* too, although it's not as well documented.

Edmunds: All we can say in *Euglena* is that pulses of Ca^{2+} at either high or low concentration can permanently shift the circadian oscillator (Tamponnet & Edmunds 1990). There has been a suggestion that our high Ca^{2+} pulses actually produce a type 2 (as opposed to type 0) phase response curve (Coté 1991). I don't think anyone has observed type 2 phase resetting in a biological system. A type 2 curve would thread the hole of the torus twice when plotted in Winfree's (1975) new-phase versus old-phase format and would require at least three state variables to describe it mathematically.

Rensing: Did you follow changes in the overall Ca^{2+} concentration and in the rather intricate concentrations along the plasma membrane?

Edmunds: We, as others do, have trouble using fura-2 because either it doesn't get into the cell, or, when it gets in, it's difficult to localize. The process of localization probably modifies its sequestration, not to mention circadian time. We have real technical difficulties there, so the answer to your question is no. Even if we could follow changes in $[Ca^{2+}]$, only a small portion of the total Ca^{2+} flux, in a particular compartment, is likely to be involved (see Edmunds & Tamponnet 1990). Some of the new tracer techniques for visualizing waves of Ca^{2+} flux hold some promise but they are used primarily for studying intercellular communication, not for monitoring fluxes within sub-compartments of a cell.

Dunlap: The involvement of inositol trisphosphates and calcium metabolism in signal transduction is highly conserved phylogenetically. Heterotrimeric G proteins are doing the same sorts of things in yeast as they are in human liver cells. Changes need to be acute, and fairly massive, in order for signal transduction to work. Given that cells are relatively small places, one might not expect these Ca^{2+} fluxes or G proteins, for example, to be involved in a stable, well-compensated clock mechanism, because while the clock is running in a stable manner, the cell has to respond to all these acute changes by undergoing large changes in, for instance, the intracellular Ca^{2+} concentration. Philosophically, if you were designing a clock, you wouldn't pick something as a component that undergoes massive changes spontaneously all the time.

Miller: There's another way to look at that. You might say that what you really want is a refractory intracellular calcium store.

Dunlap: It would have to be well compartmentalized.

Miller: Yes, but such a store might be able to function in clock mechanisms in isolation from the 1000-fold changes in intracellular $[Ca^{2+}]$ that occur when cell membrane Ca^{2+} channels open.

Edmunds: The concepts are much like those in the cell cycle where two or three sharp changes in cAMP or other factors are needed before something happens. To avoid saturating the entire system the trick is to have a compartmentalized, short, sharp signal.

Gillette: In cells or tissues that are directly phototransductive, the signalling mechanisms may be different from those in cells which don't have direct communication with the environment but have to rely on relayed chemical signals transduced by second messengers when they impinge on clock cells. Nevertheless, both mechanisms have to intercept with the clock at some point, and that may be a common point between the systems.

Lemmer: Anticancer drugs can interfere with DNA synthesis and its transcription into RNA and this may lead to inhibition of protein synthesis. Is there any evidence that anticancer drugs can induce mutations that affect the clock, temporarily or permanently? I ask because cancer cell lines and cancer cells *in vivo* are arrhythmic, indicating a loss of clock control.

Dunlap: There is a connection but it's probably not causal. Anticancer drugs do cause mutations, and some mutations can affect the clock, but there's no reason to believe that anticancer drugs work through the clock.

Rensing: In any case, there is no convincing documentation of a rhythm in transformed or normal mammalian cells or tissue in culture, except for pacemaker cells.

Lemmer: Why not?

Rensing: Possibly because most mammlian cells have lost the ability to produce autonomous oscillations.

References

Ahmed M, Cashmore AR 1993 *HY4* gene of *A. thaliana* encodes a protein with characteristics of a blue-light photoreceptor. Nature 366:162–166

Andrews GK, Harding MA, Calvet JP, Adamson EG 1987 The heat shock response in HeLa cells is accompanied by elevated expression of the c-*fos* proto-oncogene. Mol Cell Biol 7:3452–3458

Bacskai BJ, Hochner B, Mahaut-Smith M et al 1993 Spatially resolved dynamics of cAMP activity and protein kinase A subunits in *Aplysia* sensory neurons. Science 260:222–226

Bogart B 1992 Cellular mechanisms of circadian pacemaker entrainment in the mollusc *Bulla*. PhD thesis, University of Virginia, Charlottesville, VA, USA

Carré IA, Edmunds LN Jr 1992 cAMP-dependent kinases in the algal flagellate *Euglena gracilis*. J Biol Chem 267:2135–2137

Carré IA, Edmunds LN Jr 1993 Oscillator control of cell division in *Euglena*: cyclic AMP oscillations mediate the phasing of the cell division cycle by the circadian clock. J Cell Sci 104:1163–1173

Carré IA, Laval-Martin DL, Edmunds LN Jr 1989 Circadian changes in cyclic AMP levels in synchronously dividing and stationary-phase cultures of the achlorophyllous ZC mutant of *Euglena gracilis*. J Cell Sci 94:267–272

Cornelius G, Gebauer G, Techel O 1989 Inositol trisphosphate induces calcium release from *Neurospora crassa* vacuoles. Biochem Biophys Res Commun 162:852–856

Coté GG 1991 Type 2 resetting of the *Euglena gracilis* circadian rhythm? J Biol Rhythms 6:367–369

Edmunds LN Jr, Tamponnet C 1990 Oscillator control of cell division cycles in *Euglena*: role of calcium in circadian timekeeping. In: O'Day DH (ed) Calcium as an intracellular messenger in eukaryotic microbes. American Society for Microbiology, Washington, DC, p 97–123

Feldman JF, Gardner GF, Denison RA 1979 Genetic analysis of the circadian clock of *Neurospora*. In: Suda M, Hayaishi O, Nakagawa H (eds) Biological rhythms and their central mechanism. Elsevier/North-Holland Biomedical Press, Amsterdam, p 57–66

Kippert F, Ninnermann H, Engelmann W 1991 Photosynchronization of the circadian clock of *Schizosaccharomyces pombe*: mitochondrial cytochrome b is an essential component. Curr Genet 19:103–107

Lakin-Thomas PL 1993 Evidence against a direct role for inositol phosphate metabolism in the circadian oscillator and the blue-light signal transduction pathway in *Neurospora crassa*. Biochem J 292:813–818

Miller JD, Cao VH, Heller HC 1994 Thermal effects on neuronal activity in suprachiasmatic nuclei of hibernators and non-hibernators. Am J Physiol 266:R1259–R1266

Tamponnet C, Edmunds LN Jr 1990 Entrainment and phase-shifting of the circadian rhythm of cell division by calcium in synchronous cultures of the wild-type Z strain and of the ZC achlorophyllous mutant of *Euglena gracilis*. Plant Physiol 93:425–431

Techel D, Gebauer G, Kohler W, Braumann T, Jastorff B, Rensing L 1990 On the role of Ca^{2+}-calmodulin-dependent and cAMP-dependent protein phosphorylation in the circadian rhythm of *Neurospora crassa*. J Comp Physiol B Biochem Syst Environ Physiol 159:695–706

Tong J, Edmunds LN Jr 1993 Role of cyclic GMP in the mediation of circadian rhythmicity of the adenylate cyclase cyclic AMP phosphodiesterase system in *Euglena*. Biochem Pharmacol 45:2087–2091

Tong J, Carré IA, Edmunds LN Jr 1991 Circadian rhythmicity in the activities of adenylate cyclase and phosphodiesterase in synchronously dividing and stationary-phase cultures of the achlorophyllous ZC mutant of *Euglena*. J Cell Sci 100:365–369

Trinczek B, Robert-Nicoud M, Schwoch G 1993 *In situ* localization of cAMP-dependent protein kinases in nuclear and chromosomal substructures: relation to transcriptional activity. Eur J Cell Biol 6:196–202

Winfree AT 1975 Unclocklike behaviour of biological clocks. Nature 253:315–319

Zatz M 1992 Does the circadian pacemaker act *through* cyclic AMP to drive the melatonin rhythm in chick pineal cells? J Biol Rhythms 7:301–311

Cellular analysis of a molluscan retinal biological clock

G. Block, M. Geusz, S. Khalsa, S. Michel and D. Whitmore

NSF Science and Technology Center for Biological Timing and Department of Biology, Gilmer Hall, University of Virginia, Charlottesville, VA 22901, USA

Abstract. The eye of the opisthobranch mollusc *Bulla gouldiana* expresses a circadian rhythm in optic nerve impulse frequency. The circadian rhythm is generated among approximately 100 neurons at the base of the retina referred to as basal retinal neurons. These cells are electrically coupled to one another and fire spontaneous action potentials in synchrony. Basal retinal neurons recorded intracellularly exhibit a circadian rhythm in membrane potential that appears to be driven by a circadian modulation of membrane conductance. Membrane conductance is relatively high during the subjective night and decreases after subjective dawn. Recent experiments in our laboratory indicate that individual basal retinal neurons in culture can express circadian rhythms in membrane conductance. When completely isolated, these cells continue to show circadian conductance changes. These studies provide the first direct demonstration that individual neurons can act as circadian pacemakers. Although the precise details of the mechanism generating the circadian periodicity remain obscure, our research indicates that several transmembrane ionic fluxes are not involved in rhythm generation, but that a transmembrane Ca^{2+} flux is critical for entrainment. Both transcription and translation appear to play critical roles in generating the circadian cycle.

1995 Circadian clocks and their adjustment. Wiley, Chichester (Ciba Foundation Symposium 183) p 51-66

The retinas of several opisthobranch molluscs contain biological oscillators. These oscillators or pacemakers play a critical role in timing rhythmic locomotor behaviour patterns (Lickey et al 1977, Block & Davenport 1982). The traditional virtues of the gastropod nervous system for electrophysiological and biochemical study have provided a special opportunity for detailed cellular study of the circadian pacemaker. In the brief review that follows we summarize what is known about the molluscan eye preparation, with primary emphasis on *Bulla gouldiana*, the cloudy bubble snail. The eyes of these marine snails have become an important *in vitro* model for the study of the generation and expression of circadian rhythms and their adjustment to local time (entrainment) (for review see Block et al 1993). Other chapters in this volume address the issue of circadian

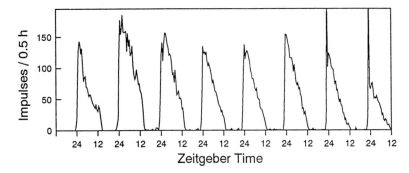

FIG. 1. The circadian rhythm in spontaneous optic nerve impulses recorded from the isolated eye of *Bulla gouldiana* in continual darkness at 15 °C. Zeitgeber time 24 is the time of the previous dawn.

clocks and their adjustment at many levels from molecular analysis, through mathematical treatments to human behavioural studies. The research on the molluscan eye provides important insights at the cellular level regarding the generation of rhythmicity by circadian pacemakers and the mechanisms of entrainment.

The ocular rhythm

When the eye is removed from *Bulla* and optic nerve activity recorded with a suction electrode a robust circadian rhythm is expressed in the frequency of spontaneously generated impulses (Fig. 1). The rhythm persists for more than seven cycles *in vitro* with a period length of about 23.7 h (Block et al 1993). The circadian rhythm is temperature compensated, with a Q_{10} of 0.98 determined over a range of 12–25 °C (Bogart 1992). When the eye is exposed to 24 h light cycles *in vivo*, effective phase control is established, with half-maximum impulse frequency (the conventional phase reference point for the *Bulla* ocular rhythm) occurring about 40 min before projected dawn (Block et al 1993). Two eyes from the same animal show a similar circadian wave-form of impulse frequency in isolation, which usually enables accurate comparisons of phase differences to be made between experimental and control eye rhythms.

Localization of the retinal pacemaker

The retina of *Bulla* consists of about 1000 large and perhaps as many small photoreceptors surrounding a spheroidal lens (Jacklet & Colquhoun 1983). Spatially distinct from the photoreceptor layer is a group of about 100 neurons that surround a neuropil region at the base of the retina. We refer to these neurons as basal retinal neurons (Block et al 1984). Several electrophysiological studies attempting to characterize the functional relationships among retinal elements (Block et al 1984, Block & McMahon 1983, Geusz & Block 1992a) have revealed that the large impulses recorded from the optic nerve that express a circadian rhythm are compound action potentials produced by the basal retinal neurons. Pair-wise intracellular recording and stimulation showed that these cells are interconnected with strong electrical coupling (Geusz & Block 1992b). Basal retinal neurons appear to inhibit a class of small neurons—the H-type cells (Geusz & Block 1992a)—and also appear to be inhibited by other cells within the retina (Block et al 1984). Several observations indicate that the basal retinal neurons are responsible for generating the circadian rhythm in optic nerve impulse activity. Firstly, surgical reduction of the retina to the extreme base, leaving primarily basal retinal neurons (and the processes of other ablated retinal cells), does not prevent expression of the circadian rhythm in impulse activity (Block & Wallace 1982, Block & McMahon 1984). Secondly, basal retinal neurons continue to express circadian rhythms in membrane conductance when dispersed in cell culture and when individually isolated in a microwell dish (Michel et al 1993). The persistence of circadian rhythms in isolated basal retinal neurons confirms, as had been suspected (Block & McMahon 1984), that individual neurons act as circadian pacemakers. Thus, the *Bulla* retina contains about 100 circadian pacemaker neurons that generate impulses in synchrony as compound action potentials and act in unity as a single 'clock' by virtue of their strong electrical coupling. It is presently unclear whether all of the 'circadian properties' associated with biological clocks are contained within individual basal retinal neurons or whether some of these properties, such as temperature compensation and entrainment by environmental cycles, are conferred at the tissue level. Our observation of circadian rhythms in isolated *Bulla* basal retinal neurons is the first demonstration in any organism that individual neurons can generate circadian periodicities.

Entrainment of the circadian rhythm

Over the past few years we have been attempting to identify processes involved in the entrainment of the pacemaker by periodic environmental signals. Processes in the entrainment pathway are interesting in themselves, but this effort is also a strategy for identifying components of the circadian pacemaker mechanism, because the entrainment pathway must ultimately converge on elements of the 'causal loop' generating the circadian oscillation. Identified elements of this entrainment pathway are shown in Fig. 2. Briefly, light pulses lead to membrane depolarization of pacemaker neurons. This depolarization persists in retinal tissue surgically reduced to the basal retinal neuron population, suggesting that the photoreception critical for entrainment occurs within these cells rather than in the photoreceptor layer that surrounds the lens (Block & Wallace 1982, Block

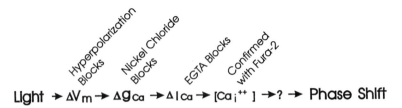

FIG. 2. Proposed entrainment pathway for the *Bulla* ocular pacemaker. Light induces depolarization of pacemaker (basal retinal) neurons, which alters the conductance (g) of voltage-activated Ca^{2+} channels. Intracellular $[Ca^{2+}]$ increases for the duration of the depolarization. The intermediate steps leading to phase shifting are not yet known. V_m, membrane potential.

et al 1984). Depolarization is a necessary and sufficient step in the entrainment pathway: depolarization caused either by direct injection of current (McMahon & Block 1987) or by ionic substitution (McMahon & Block 1987, Khalsa & Block 1988a) leads to phase shifts of the pacemaker similar to those induced by light pulses; blocking depolarization by simultaneous injection of a hyperpolarizing current blocks light-induced phase shifts (McMahon & Block 1987).

Depolarization appears to act through a change in the conductance of voltage-activated Ca^{2+} channels. Blocking Ca^{2+} channels with nickel chloride, a potent Ca^{2+} channel blocker in invertebrate neurons, prevents phase shifts induced by depolarization (Khalsa & Block 1988a). A similar result is obtained if Ca^{2+} is omitted from the bathing medium and the Ca^{2+} chelator EGTA added (McMahon & Block 1987, Khalsa & Block 1988a). If a transmembrane Ca^{2+} flux couples environmental timing information to an intracellular oscillator, it should be possible to measure increases in intracellular Ca^{2+} during depolarization. Furthermore, if a transmembrane Ca^{2+} flux is a series element in a simple entrainment cascade, we would expect the intracellular Ca^{2+} concentration to reflect the duration of the entrainment stimulus.

Using the Ca^{2+} indicator dye fura-2 we have been able to measure changes in intracellular $[Ca^{2+}]$ in response to membrane depolarization in cultured neurons. We find that Ca^{2+} concentrations are increased following depolarization and this elevation remains for the duration of the depolarization (at least with measurements of up to one hour) (Geusz et al 1994). Digital imaging of Ca^{2+} using a silicon-intensified target camera and image analysis system reveals a relatively uniform increase in $[Ca^{2+}]$ in the cytoplasm and in the nucleus. We are uncertain about the next steps in the entrainment cascade (Khalsa & Block 1988b).

The mechanisms underlying expression of the rhythm

What little we know about the mechanisms underlying expression of the activity of the ocular pacemaker is shown in Fig. 3, and is best dealt with by working backwards from the output. The circadian rhythm is expressed in the spontaneous impulse activity of basal retinal neurons. The optic nerve impulses

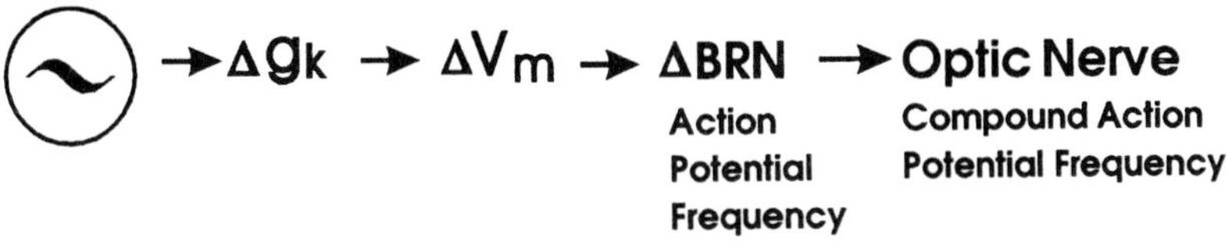

FIG. 3. A proposed pathway for expression of the activity of the ocular pacemaker. The pacemaker modulates K^+ conductance (g) through an action at as yet unidentified K^+ channels. This K^+ conductance is responsible for driving the rhythm in membrane potential (V_m) in basal retinal neurons (BRN). The resultant synchronous firing of these neurons generates compound action potentials.

are compound action potentials generated by the synchronous firing of the electrically interconnected basal retinal neurons. Intracellular recordings from basal retinal neurons reveal an approximately 15 mV circadian rhythm in membrane potential, with the membrane being relatively hyperpolarized during the subjective night and then becoming depolarized spontaneously near subjective dawn (McMahon et al 1984). Inspection of the phase relationship between impulse activity and membrane potential indicates that the membrane potential depolarizes at or just before impulse production, suggesting that the spontaneous rhythm in impulse activity is being driven by the circadian rhythm in membrane potential (McMahon et al 1984, Ralph & Block 1990).

The circadian cycle in membrane potential appears to be controlled by a rhythm in membrane conductance. Measurements of membrane conductance before and after subjective dawn (Ralph & Block 1990, Michel et al 1993) and before and after subjective dusk (Michel et al 1993) show that membrane conductance decreases near subjective dawn and increases again near subjective dusk. A tetraethylammonium (TEA)-sensitive K^+ conductance appears to be responsible for driving the rhythm in membrane potential. Application of 100 mM TEA before dawn, a treatment which would be expected to close a large proportion of the K^+ channels, significantly reduces membrane conductance. A similar treatment after projected dawn has no measurable effect on membrane conductance. We do not know the exact identity of these K^+ channels or how they are modulated by the circadian pacemaker.

Mechanisms underlying rhythm generation

We still know relatively little about the mechanisms underlying the generation of circadian rhythms. Evidence is emerging, however, that transcription and translation are most probably involved in circadian timing, whereas membrane conductances act primarily to couple an intracellular 'clock' to the environment.

Membrane conductances

As mentioned, changes in Ca^{2+} conductance appear to be involved in pacemaker entrainment in *Bulla*, whereas a K^+ conductance is probably responsible for the membrane potential rhythm. We have systematically evaluated the role of several ionic conductances in circadian rhythm generation. Chelation of extracellular Ca^{2+} blocks expression of the rhythm in optic nerve impulse frequency but does not affect the underlying oscillation. Severe reduction of extracellular Ca^{2+} concentration for up to 76 h has little effect on the projected phase of the rhythm during the treatment as assessed by the phase following the return to normal bathing medium (Khalsa et al 1993a). Similarly, removal of extracellular Na^+ prevents expression but not generation of the circadian cycle (Khalsa et al, unpublished observations). Removal of extracellular Cl^-, however, does have an impact on the free-running period. Replacement of Cl^- with SO_4^{2-} or Hepes and treatments that block Cl^- channels (1 mM anthracene-9-carboxylic acid) shorten the period by up to 2.5 h (Khalsa et al 1990). Pulses of low-Cl^- seawater generate an all-advance phase response curve (PRC), the phase shifts occurring in the late subjective night–early subjective day (Michel et al 1992). Although it is not yet clear how low Cl^- shortens the period, it is evident that Cl^- conductance is not essential for the circadian rhythm to be generated. Together, these experiments lead us to believe that transmembrane conductances, although important for coupling the pacemaker to environmental timing information (the input pathway) and to downstream effector processes (the output pathway), are not directly involved in generation of the rhythm. This 'transmembrane loop', however, may play a role in determining the shape of the phase response curve (see below).

Transcription and translation

Although we have failed to find compelling evidence for a role of membrane conductances in rhythm generation, there is good evidence that translation and transcription are involved in rhythm generation. Continuous inhibition of transcription or translation lengthens the pacemaker period to up to 50 h (Khalsa et al 1992, Khalsa et al 1993b). At higher concentrations of translation or transcription inhibitors there is no impulse activity and therefore it is not possible to determine whether or not a circadian cycle is being generated. However, if *pulses* of high concentrations of transcription and translation inhibitors of increasing durations are used, it is possible to determine whether a circadian cycle is being generated (although not expressed) by examining the phase of the circadian rhythm on its resumption after termination of the inhibitor pulse. The analyses of data so produced show that the pacemaker is stopped by these higher concentrations of inhibitors and that the pacemaker's operation stops in a phase-dependent manner. The phases of sensitivity to inhibitors of translation are between the late subjective night and early subjective day (Khalsa et al 1992) and those to inhibitors of transcription fall from the late subjective night to the late subjective day (Khalsa et al 1993b). These findings suggest a critical

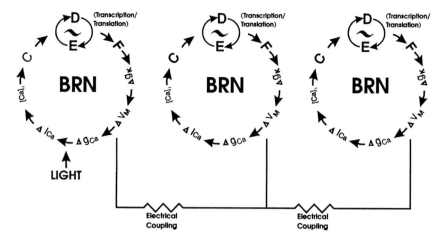

FIG. 4. A two-loop model for the basal retinal neuron (BRN) pacemaker. Each BRN generates a circadian periodicity through an intracellular feedback loop involving both transcription and translation. Membrane potential (V_m) is on both the entrainment and output pathways of the pacemaker, such that the intracellular pacemaker is coupled to a transmembrance input–output loop. The pacemaker is entrained by a light-induced membrane depolarization and the resultant Ca^{2+} flux. C, D, E and F represent unidentified components of the loops. g, conductance.

role for transcription and translation in the timing of the circadian cycle within basal retinal neurons.

Two loops: a synthesis

Presented below is our working hypothesis for the organization of the circadian pacemaker system in the *Bulla* retina (see Block et al 1993 for additional details). Each basal retinal neuron generates a circadian periodicity via an intracellular feedback loop that involves both transcription and translation (Fig. 4). The intracellular pacemaker is coupled to a 'transmembrane' input–output loop formed by virtue of the fact that membrane potential is on both the entrainment and output pathways of the pacemaker. The pacemaker is entrained via a light-induced membrane depolarization and a resultant Ca^{2+} influx.

Although the transmembrane input–output loop is not required for generation of the circadian periodicity (Khalsa et al 1993a), this loop may influence the pacemaker's response to entrainment signals and, specifically, may be responsible for the portion of the phase response curve commonly referred to as the dead zone, where no phase shifts of the pacemaker occur. A possible explanation for the dead zone is as follows: during the subjective day the membranes of pacemaker neurons are depolarized and generating action potentials. This depolarization and the resultant impulses may lead to a persistent Ca^{2+} flux during this phase of the cycle. Light pulses, which further depolarize the membrane, may be ineffectual against this background Ca^{2+} flux. In contrast, during the subjective night, when the pacemaker membrane is relatively hyperpolarized and no spontaneous impulses are generated, membrane depolarization would be expected to cause a larger change in Ca^{2+} entry.

Some support for this hypothesis comes from experiments in which the membrane is hyperpolarized by intracellular injection of current or by substitution of ions in the bathing medium (McMahon & Block 1987, Khalsa & Block 1990). Unlike depolarization, which leads to phase shifts primarily during the subjective night, hyperpolarizing pulses produce phase shifts largely during the subjective day and the phase response curve is displaced by about 12 h. One interpretation of these results is that membrane hyperpolarization reduces a Ca^{2+} flux that is caused by the circadian rhythm in membrane depolarization during the subjective day, thereby leading to a phase shift in the pacemaker. In this scheme, the phase of the pacemaker is perturbed if the normal diurnal transmembrane Ca^{2+} flux is disrupted.

Additional support for this Ca^{2+} flux hypothesis comes from experiments in which the presumptive diurnal Ca^{2+} flux is disrupted by reduction of bath $[Ca^{2+}]$. Pulses of low $[Ca^{2+}]$ bathing solutions containing EGTA generate a phase response curve nearly identical to that produced by membrane hyperpolarization and displaced by about 180° on the time axis from depolarization-induced phase shifts (Khalsa & Block 1990). The existence of these two classes of PRC, one to depolarizing treatments and one to hyperpolarizing treatments, raises the possibility that modulation of a single process—a transmembrane Ca^{2+} flux—mediates phase shifts during *both* the subjective day and the subjective night. This is particularly interesting in the light of the fact that a number of approximately antiphasic PRCs have also been generated in mammalian circadian systems; one example is the phase response curves to light pulses and to cyclic AMP analogues (see Klein et al 1991). Indeed, it is quite reasonable that opposite changes in pacemaker cell membrane potential may underlie the two families of response curves. This is an experimentally addressable issue that should permit further evaluation of the utility of the *Bulla* ocular pacemaker model for informing us about the mammalian circadian timing system.

Conclusions

The molluscan eye is proving to be an exceptional preparation for the cellular study of the generation, entrainment and expression of circadian rhythms. The accessibility of the retina for intracellular recording and the relative simplicity of long duration extracellular recording has allowed us to build up an increasingly detailed picture of the processes underlying circadian timing. The recent demonstration of circadian rhythms in fully isolated neurons should greatly facilitate the elucidation of the fundamental clock mechanism.

Acknowledgements

Support for this research is provided from a number of sources including the NSF Center for Biological Timing, NIH grant NS15264 to G.D.B.; NS08806 to M.E.G.; NS09621 to S.B.K.

References

Block GD, Davenport PA 1982 Switch from nocturnal to diurnal behavior in the cloudy bubble snail *Bulla gouldiana*. J Exp Zool 224:57–63

Block GD, McMahon DG 1983 Localized illumination of the *Aplysia* and *Bulla* eye reveals new relationships between retinal layers. Brain Res 265:134–137

Block GD, McMahon DG 1984 Cellular analysis of the *Bulla* ocular circadian pacemaker system. III. Localization of the circadian pacemaker. J Comp Physiol A Sens Neural Behav Physiol 155:387–395

Block GD, Wallace SF 1982 Localization of a circadian pacemaker in the eye of a mollusc, *Bulla*. Science 217:155–157

Block GD, McMahon DG, Wallace SF, Friesen WO 1984 Cellular analysis of the *Bulla* ocular circadian pacemaker system. I. A model for retinal organization. J Comp Physiol A Sens Neural Behav Physiol 155:365–378

Block GD, Khalsa SBS, McMahon DG, Michel S, Geusz ME 1993 Biological clocks in the retina: cellular mechanisms of biological timekeeping. Int Rev Cytol 146: 83–144

Bogart B 1992 Cellular mechanisms of circadian pacemaker entrainment in the mollusc *Bulla*. PhD thesis, University of Virginia, Charlottesville, VA, USA

Geusz ME, Block GD 1992a The retinal cells generating the circadian small spikes in the *Bulla* optic nerve. J Biol Rhythms 7:255–268

Geusz ME, Block GD 1992b Measurements of electrical coupling between circadian pacemaker cells of the *Bulla* eye. Soc Neurosci Abstr 18:1

Guesz ME, Michel S, Block GD 1994 Intracellular calcium responses of circadian pacemaker neurons measured with fura-2. Brain Res 638:109–116

Jacklet JW, Colquhoun W 1983 Ultrastructure of photoreceptors and circadian pacemaker neurons in the eye of a gastropod, *Bulla*. J Neurocytol 12:673–696

Khalsa SBS, Block GD 1988a Calcium channels mediate phase shifts of the *Bulla* circadian pacemaker. J Comp Physiol A Sens Neural Behav Physiol 164:195–206

Khalsa SBS, Block GD 1988b Phase-shifts of the *Bulla* ocular circadian pacemaker in the presence of calmodulin antagonists. Life Sci 43:1551–1556

Khalsa SB, Block GD 1990 Calcium in phase control of the *Bulla* circadian pacemaker. Brain Res 506:40–45

Khalsa SBS, Ralph MR, Block GD 1990 Chloride conductance contributes to period determination of a neuronal circadian pacemaker. Brain Res 520:166–169

Khalsa SBS, Whitmore D, Block GD 1992 Stopping the circadian pacemaker with inhibitors of protein synthesis. Proc Natl Acad Sci USA 89:10862–10866

Khalsa SBS, Ralph MR, Block GD 1993a The role of extracellular calcium in generating and in phase shifting the *Bulla* ocular circadian rhythm. J Biol Rhythms 8: 125–139

Khalsa SBS, Whitmore D, Bogart B, Block GD 1993b Evidence for the direct involvement of transcription in the timing mechanism of the circadian pacemaker. Soc Neurosci Abstr 19:1703

Klein DC, Moore RY, Reppert SM 1991 The suprachiasmatic nucleus: the mind's clock. Oxford University Press, New York

Lickey ME, Wozniak JA, Block GD, Hudson DJ, Augter GK 1977 The consequences of eye removal for the circadian rhythm of behavioral activity in *Aplysia*. J Comp Physiol 118:121–143

McMahon DG, Block GD 1987 The *Bulla* ocular circadian pacemaker. I. Pacemaker neuron membrane potential controls phase through a calcium dependent mechanism. J Comp Physiol A Sens Neural Behav Physiol 161:335–346

McMahon DG, Wallace SF, Block GD 1984 Cellular analysis of the *Bulla* ocular circadian pacemaker system. II. Neurophysiological basis of circadian rhythmicity. J Comp Physiol 155:379–385

Michel S, Khalsa SBS, Block GD 1992 Phase shifting the circadian rhythm in the eye of *Bulla* by inhibition of chloride conductance. Neurosci Lett 146:219–222

Michel S, Geusz ME, Zaritsky JJ, Block GD 1993 Circadian rhythm in membrane conductance expressed in isolated neurons. Science 259:239–241

Ralph MR, Block GD 1990 Circadian and light-induced conductance changes in putative pacemaker cells of *Bulla gouldiana*. J Comp Physiol A Sens Neural Behav Physiol 166:589–595

DISCUSSION

Mrosovsky: Your work with this preparation is elegant and convincing, but there is some danger that this idea of there being two families of PRCs will be overgeneralized. It's already achieving almost dogma status. If you look closely at the PRCs produced by dark pulses and light pulses in hamsters (Boulos & Rusak 1982), you will see that they are not mirror images of each other, although they have been described as such. Also, in the *tau* mutant hamster we have *three* phase response curves that seem different: the photic phase response curve with a peak advance at about circadian time (CT) 18, the nonphotic with a peak advance at about CT 24 and the anisomycin phase response curve with a peak advance somewhere about CT 9. We shouldn't think that this simplifying model of two families will apply to all systems. People are applying this notion too readily without scrutinizing the actual data. If you look at the PRCs in the paper by Smith et al (1992), you will see that there can be differences of several hours within a family in the times of a peak or a trough in the curves. Some of these differences are probably due to methodological differences—how long the drug lasts, how the shifts are measured, how many days are allowed for transients, etc. I would be interested to know whether everybody believes that there are two families of PRC and that this is a valid simplifying notion.

Block: I agree that we have to be careful about grouping all PRCs into two families as I have done for agents that change membrane potential in *Bulla* pacemaker cells. Certainly, the PRCs produced by light pulses and dark pulses in mammals may not be identical in shape or displaced exactly 180° from each other on the time axis. How much of the difference in shape or the exact phase relationship of these two treatments is due to methodological differences needs to be determined. I thought it useful, however, to offer the hypothesis for mammals, based on our work in *Bulla*, that dark pulse PRCs are generated through pacemaker cell membrane depolarization. This is a testable hypothesis.

Czeisler: Going from *Bulla*, to humans may be quite a leap, but it's remarkable that the PRC produced by melatonin that Dr Lewy has found does appear to be inverse to the one we have described for light. The notion that one could begin to understand that in terms of changes in Ca^{2+} fluxes is an exciting one. This is something that should be and could be explored in the mammalian system.

I've always been intrigued by the fact that the human suprachiasmatic nucleus, which drives the circadian rhythm of melatonin release by the pineal, contains receptors for melatonin. Those melatonin receptors are presumably responding to melatonin released by the pineal at a time determined by the suprachiasmatic nucleus itself. This seems reminiscent of some of the feedback loops that you described, occurring on an organ-system level rather than within the pace-making structure.

Block: I think we are going to find numerous feedback loops within organisms, some of which will have many of the properties of the 'central pacemaker loop'. For example, some of the processes of these loops will be rhythmic, and changing the level of some of these processes will lead to phase shifts of the rhythm. This raises the challenging question of when you can know that you really have an intrinsic part of the loop that times the oscillation, that plays a primary role in setting the circadian period. It's going to be like looking at an onion, peeling away layers of feedback from the behaviour right down to the intracellular processes, and not all of those layers will be involved in timing the circadian rhythm.

Czeisler: You are dealing with the eye in isolation from the rest of the organism. There may be other feedback loops within the whole organism that act in the same way.

Menaker: Dr Block, to what extent does your hyperpolarization–depolarization hypothesis require the two PRCs to be mirror images? We need to know this if we are going to use that relationship as a criterion for accepting or rejecting your hypothesis.

Block: All the model requires is that the PRCs for depolarization and hyperpolarization should be roughly 180° out of phase on the time axis—the model doesn't say anything about their shape. If you go to a more explicit model for the oscillator, as we have recently elaborated (Block et al 1993), you can begin to explain why the advances and delays occur where they do. I don't think the curves need to be true mirror images. Advances and delays, depending on when they occur in the pacemaker cycle, may be quite different mechanistically.

Moore: In the mammalian system glutamate is probably the retinohypothalmic transmitter. It's primary effect is to produce depolarization of SCN neurons. The other type of PRC is that obtained by activation of the intergeniculate leaflet–geniculohypothalamic tract system. This appears to be a GABAergic input to the SCN, which would cause hyperpolarization of SCN neurons. That would fit perfectly with your model. The interesting point is that there can be modifications of the responses depending on what these inputs interact with. The input neurons from the retina which excite the SCN neurons will in turn release GABA upon themselves. The input from the pacemaker itself probably should not change its period, so it must have a way of segregating the different kinds of inputs that come in from its afferents. That adds another interesting complication.

Rensing: In *Drosophila*, transcription and translation seem to be 180° out of phase, so you might also expect the PRCs produced in response to inhibitors of transcription and translation to be 180° out of phase.

Block: The PRCs are displaced, but are not 180° out of phase. I don't think an antiphasic relationship is a requirement for transcription and translation to be involved in the intracellular oscillation.

Rensing: Did you use cycloheximide to inhibit translation?

Block: Yes.

Rensing: In *Neurospora* cycloheximide almost completely stops protein degradation. You can never be sure exactly what you are doing when you apply that kind of inhibitor.

Takahashi: We have to be careful about how we refer to these two classes of PRC. They have been called mirror images, opposite, and 180° out of phase. The light and dark PRCs are *not* mirror images, and they are not opposite in the graphical sense. The two curves should, theoretically, be 180° out of phase if they have opposite effects on the same state variable of the oscillator. We should get the terminology straight.

As far as what their specific shape should be, we have delved into this informally using computer simulations with different classes of limit cycle oscillators. I was curious to see whether the simulation would produce symmetrical or identically shaped PRCs 180° out of phase. With a very simple limit cycle, such as a van der Pol oscillator, which is away from the axes, and is nice and round, you get a nice clean result in which the 'plus' and the 'minus' PRCs have the same qualitative shape and are just shifted by half a cycle. But, if you have an oscillator that looks like the Pavlidis oscillator, where the limit cycle is 'sitting' on one of the axes, you get a distorted shape when you go in the opposite direction. You don't get two clean classes of PRC. One looks like a light-type PRC but the dark one is complicated.

Roenneberg: Is there an agent that gives rise to a type 0 PRC in *Bulla*?

Block: No, I don't think there is. Almost every treatment we have tried that affects the membrane potential gives rise to a type 1 curve.

Miller: There is evidence that the intensity or duration of the zeitgeber may be responsible for the transition between type 1 and type 0. Perhaps if you just used a sufficiently high concentration of x, you would begin to see a type 0 curve.

Block: We have tried this. When one is trying to produce a phase shift through changes in membrane potential there are fairly strict limits on the magnitude of the phase shift. We believe that this is because of the mechanism by which phase shifts occur, which involves a transmembrane calcium flux. We are quite certain that in the *Bulla* retina, depolarizing pacemaker cells leads to a phase shift of the rhythm by increasing a transmembrane calcium flux. We suspect that when all of the voltage-sensitive calcium channels are open, which most probably occurs with a modest depolarization, further depolarization simply reduces the driving force for calcium. With a large enough depolarization there will be no calcium flux and consequently no phase shift. In other words, phase shift magnitude plotted as a function of depolarization will give a U-shaped curve.

Miller: Something else I wanted to address was the analogy between what you always see in *Bulla* and what we see in the SCN, particularly that the advance portions of the PRCs produced by cAMP and serotonin can be laid over each other. Either of those phase-shifting stimuli can be blocked with blockers of cAMP or protein kinase A. These phase shifts have a strong dependence on the opening of calcium-dependent potassium channels in the SCN. If you block those calcium-dependent potassium channels with agents such as apamin or charybdotoxin you prevent the zeitgeber action of both serotonin and cAMP analogues. Have you looked at more specific potassium channel blockers than tetraethylammonium?

Block: Not yet.

Do you see a reduction in spontaneous activity in the SCN when you apply cAMP or serotonin? That's what I predict should happen, that those drugs work by hyperpolarizing the SCN neurons.

Miller: Serotonin and 8-OH-DPAT (8-hydroxy-2-[di-*n*-propylamino]-tetralin) inhibit spontaneous activity when applied at CT 6. The big question is what serotonin does at CT 18, where it induces delays instead of advances.

Gillette: That question was directly addressed by Liou et al (1986) using a rat hypothalamic brain slice preparation. The majority of the neurons were inhibited by cAMP analogue applied ionophoretically in the daytime and excited by cGMP analogue applied at night.

Turek: I should like to explore further the question of how far we can draw analogies between Gene Block's preparation and the mammalian one. GABAergic activation can mimic the effects of an activity-inducing stimulus or a dark pulse in the hamster. Is there any indication that GABA is operative in the *Bulla* eye system?

Ralph: There was some indication that benzodiazepines act in a hyperpolarization-like direction. Midazolam, for example, reduced or eliminated compound action potentials when applied during the subjective day. However, phase-dependent effects on the rhythm were not apparent in these initial studies, and we did not look at GABA-specific agents such as muscimol or bicuculline.

Turek: We seem to have forgotten about the whole animal—we always get to the eye right away. What other things can cause phase shifts in *Bulla*? Does making them active when they normally would not be active induce phase shifts?

Block: We've tried to find out whether activity induces phase shifts by putting the animals into a turbulent stream; we managed to kill several animals doing that, so we stopped! I agree with you that these are interesting issues. Not everybody is familiar with the fact that the two eyes in *Bulla* are coupled to one another; if you induce a phase shift in one eye, the two eyes resynchronize one another. We don't know what role that coupling plays normally in the behaviour of the whole system.

Edmunds: We have to be careful not to classify all agents that cause phase shifts into just two categories, such as those treatments that result in either depolarization or hyperpolarization. It is amazing, in fact, that resetting treatments often do cluster! We have already mentioned the problem that dosage could have an effect on the PRC, and even determining where the peak is is often not precise, as it often does not occur at a sharp CT but rather covers a broad envelope. Thus, one must worry about the shape, or wave form, and curves that have different shapes may be different statistically. Also, if one is looking at other types of treatment which may fall in between the two, one might tend to push them to one camp or the other, with the danger of overlooking important differences. Nakashima (1985), for example, among many others, enjoyed drawing time maps of crossover points of PRCs for different agents in *Neurospora*. There were clusters, but I remember looking at those points in between and wondering to which cluster they belonged, if either, and found my own thoughts being influenced by my preconceptions about the significance of the apparent similarities.

Block: I hope I was careful to refer to membrane potential-based phase shifts, because there are a lot of other types of phase shift; those produced by protein synthesis inhibitors, for example, where the PRCs are very different from those generated by altering membrane potential.

Until we actually identify the pacemaker neurons in the SCN and can measure their membrane potential, it is not going to be useful to pursue this issue. You really have to know which cells you are recording from, because if you record from an SCN cell that's not a pacemaker cell but is driven by the pacemaker, its response may be very different. Only when the pacemaker cells are identified will we be able to test the various treatments to see whether they work through depolarization or hyperpolarization.

Edmunds: Suppose you have some compound that causes hyperpolarization, but then you build in some kind of a delay factor. How long does it take for something to happen? Might that not then affect where you place the displacement? The hyperpolarization does something, but certain downstream targets are affected more quickly than others, or may have to go through additional, secondary steps.

Block: That's possible, but one starts first with a simple hypothesis. If a drug causes hyperpolarization and causes a phase shift like that produced by hyperpolarization, you set up the hypothesis that hyperpolarization is a series element in the cascade. You then voltage clamp the cell and try the drug again. If the drug then has no effect you can go ahead with some confidence on the basis that it is working through hyperpolarization.

Mikkelsen: The basic question is, can you produce the two different PRCs in the same cell, or are you talking about one set of cells with one PRC and another set with another PRC?

Block: I can't answer that, because we have used whole retina, not single cells. There are 100 pacemaker cells in the base of the retina that are electrically coupled to one another. It's not impossible that there are two classes with different responses, but it's unlikely, because all the cells at the base of the retina seem to act in a similar fashion.

Roenneberg: There is always a possibility that two subgroups give rise to two distinct PRCs, but it's important to point out that light- and dark-induced PRCs produced by one cell, such as in unicellular organisms, can be out of phase and look different.

Mikkelsen: What is the evidence that a single cell can produce two different PRCs?

Roenneberg: In unicellular organisms, if there are two subpopulations, you should be able to isolate them, to get one responding to dark pulses and one to light pulses; there is absolutely no indication that such subpopulations exist.

Reppert: Dr Block, how do you think Ca^{2+} signalling brings about phase shifts? Do you think immediate early genes might be involved?

Block: We have not tested that yet, although it's one logical place to look. We have tried to find out whether calmodulin is involved; it seemed not to be. One problem we face is that transcription inhibitors and protein synthesis inhibitors, in our hands, do not block phase delays produced by light. We have no evidence that 'simultaneous' transcription and translation are involved in light-induced phase shifts.

Miller: You seemed to place transcription in the endogenous oscillator itself, but I'm wondering about the role of transcription in the incoming pathway, particularly in the case of hyperpolarization. Doesn't the fact that hyperpolarization-induced phase shifts are persistent suggest that genes are being transcribed under hyperpolarizing conditions? This has not been a major issue for most molecular biologists, because most transcription factors are studied under depolarizing conditions. There may be an entire family of genes that are transcribed under hyperpolarizing conditions.

Block: The work by Eskin et al (1984) has shown that protein synthesis is required for serotonin-induced phase shifts in *Aplysia*, and we've shown in *Bulla* that hyperpolarization-induced phase shifts and EGTA-induced phase shifts during the subjective day also require protein synthesis. This again raises the interesting issue that phase shifts induced by a stimulus given during the subjective day may have a mechanism quite different from that involved in phase shifts during subjective night. That doesn't bother me, because the pacemaker system is not static but constantly changing; different variables are active at different times of the cycle, and we are accessing the pacemaker at different phases in the cycle where different processes may be occurring at different rates. Light-induced phase delays do not require protein synthesis, whereas phase delays induced by EGTA during the subjective day do involve protein synthesis.

Unfortunately, we have to put some of these facts on the shelf for a while until we learn more about the basic process and then we can fit them back in.

References

Block GD, Khalsa SBS, McMahon DG, Michel S, Geusz ME 1993 Biological clocks in the retina: cellular mechanisms of biological timekeeping. Int Rev Cytol 146:83–144

Boulos Z, Rusak B 1982 Circadian phase response curves for dark pulses in the hamster. J Comp Physiol A Sens Neural Behav Physiol 146:411–417

Eskin A, Yeung SJ, Klass MR 1984 Requirement for protein synthesis in the regulation of a circadian rhythm by serotonin. Proc Natl Acad Sci USA 81:7637–7641

Liou SY, Shibata S, Shiratsuchi A, Ueki S 1986 Effects of dibutyryl cyclic adenosine-monophosphate and dibutyryl cyclic guanosine-monophosphate on neuron activity of suprachiasmatic nucleus of rat in hypothalamic slice preparation. Neurosci Lett 67:339–343

Nakashima H 1985 Biochemical and genetic aspects of the conidiation rhythm in *Neurospora crassa*: phase shifting by metabolic inhibitors. In: Hiroshige T, Honma K (eds) Circadian clocks and zeitgebers. Hokkaido University Press, Sapporo, p 35–44

Smith RD, Turek FW, Takahashi JS 1992 Two families of phase-response curves characterize the resetting of the hamster circadian clock. Am J Physiol 262:1149–1153

Circadian pacemakers in vertebrates

Martin R. Ralph and Mark W. Hurd

Department of Psychology, University of Toronto, 100 St George Street, Toronto, Ontario, Canada M5S 1A1

> *Abstract.* The identification and isolation of circadian pacemaker cells is of critical importance to studies of circadian clocks at all phylogenetic levels. In the vertebrate classes, a few structures of diencephalic origin have been implicated as potential sites but for only two, the avian pineal and the mammalian suprachiasmatic nucleus (SCN), has a pacemaker role in addition to oscillatory behaviour been demonstrated by the transfer of pacemaker properties from one organism to another. Studies of the mammalian system in particular have benefited from the ability to restore circadian function using transplantation of tissue from the SCN and from the availability of a hamster period mutant, *tau*, that allows donor-derived and host-derived rhythms to be distinguished easily. Initial cross-genotype transplantation studies and the subsequent creation of circadian chimeras expressing two phenotypes simultaneously demonstrated the pacemaker capability of the SCN, and demonstrated the relative autonomy of this nucleus as a pacemaking structure. Despite an abundance of information regarding the anatomy, physiology and pharmacology of these nuclei, the identity of the pacemaker cells and their methods of communication with each other and the organism remain obscure. None the less, it is possible under certain conditions to create chimeras with two clocks that interact. The behaviour of these animals provides a unique opportunity to study the nature and timing of pacemaker communication.
>
> *1995 Circadian clocks and their adjustment. Wiley, Chichester (Ciba Foundation Symposium 183) p 67–87*

The temporal organization of physiology and behaviour relies on endogenous circadian oscillations that are capable of being synchronized with the 24 h day. Central to this organization are pacemaker structures that generate circadian rhythms, respond and entrain to cyclic environmental factors, and convey timing information to the rest of the organism. The pacemaker thereby provides a temporal framework wherein physiological and behavioural processes may be prepared for and relatively restricted to times of day when conditions, both internal and environmental, are most likely to be optimal (Pittendrigh 1960).

Model circadian systems have been found and studied extensively at most phylogenetic levels. Pacemaker structure and identity are almost always addressed, and, depending on the specific example, this may include investigations from anatomical to cellular, biochemical and molecular genetic. For vertebrates, considerable effort has been directed at understanding the anatomical substrates of the pacemaker. These efforts have led to the identification of a handful of structures, derived embryologically from the diencephalon, that contain circadian oscillators—specifically, the pineal gland, eyes and suprachiasmatic nuclei. Among vertebrate classes, and even within non-mammalian classes, the degree to which the pacemaker role is distributed or assigned to a single oscillator structure may vary according to the species.

Although an intuitive sense of what a pacemaker comprises may be derived from the physical definition of a master or driving oscillator, it is not as easy to define a circadian pacemaker experimentally. This is due partly to the potential for multiple contributing oscillators and partly to the existence of complex feedback mechanisms that may contribute to determining period and may themselves be considered to be part of the pacemaker.

This issue may be more evident at the cellular or biochemical levels, as a functional dissection of each structure leads to similar questions of pacemaker identity. Hence, two general issues associated with all oscillator loci are the type of cell in which pacemaker capability resides and the cellular organization required to perform all pacemaker functions. Similarly, at subcellular levels the issue becomes one of whether relatively simple biochemical and membrane oscillations can account for rhythm generation, or whether a more complex interaction of biochemical loops is required.

Here, we examine evidence and features of mammalian pacemaker function at specific anatomical loci. To assign pacemaker function to a structure which produces circadian oscillations one must demonstrate unambiguously that a defined property of the pacemaker (phase, period, phase response curve) can be transferred from one individual to another by transplanting the organ or cells in question. Although evidence exists for oscillator activity in various organs, current results point to two structures—the suprachiasmatic nucleus (SCN) and the pineal gland—as sites of circadian pacemakers in vertebrates.

Transplantation of circadian phase

The phase of a circadian oscillation defines not only the overt behaviour and physiology of an organism, but also the biochemical and physiological state of its clock. Hence, it is possible to distinguish one oscillator from another on a temporal basis by examining the organism's behaviour. For transplantation studies, this means that, provided that the phase of a donor clock is not shifted significantly during the grafting procedure, the restored rhythm in the host should reflect the timing of the donor tissue. This would be the case if the pacemaker role were vested in a single site; in a more distributed system the resulting phase of the overt rhythm would depend on the relative strengths and phases of the components.

A convincing example of phase transfer was obtained by Zimmerman & Menaker (1979) following transplantation of the pineal gland in sparrows.

In these experiments, the host animal was pinealectomized before receiving graft tissue, rendering it arrhythmic in a constant environment (Gaston & Menaker 1968). Transplantation of a donor gland into the anterior chamber of the eye of the host restored the circadian rhythm of perch hopping with a phase that always reflected the lighting schedule to which the donor bird had been synchronized. The simplest explanation for this finding is that the pineal contains a clock that remained in synchrony with the prior light–dark schedule. Indeed, subsequent experiments with a variety of birds (e.g., Takahashi et al 1980), reptiles (e.g., Menaker & Wisner 1983) and fish (e.g., Falcon et al 1987, Kezuka et al 1989) have demonstrated independent oscillator functioning (the rhythmic release of melatonin) of the pineal gland *in vitro*.

However, the rhythmic release of melatonin is not shown by pineals of all species within a class, nor does pinealectomy produce arrhythmicity in all species (see Takahashi & Menaker 1979, Janik & Menaker 1990, Max & Menaker 1992). This indicates that the pineal does not occupy the same position in the circadian hierarchy in all species; and, even where removal of the pineal does result in arrhythmicity, there is some evidence for a second, damped, oscillator (Gaston & Menaker 1968), located in the SCN (Takahashi & Menaker 1982).

The organization of a generalized avian circadian system has been suggested to be a neuroendocrine loop (Cassone & Menaker 1984) with the pineal oscillator being influenced by periodic sympathetic input driven by the SCN, and the SCN in turn being influenced by the periodic melatonin signal from the pineal. If this is the arrangement of the circadian system in the sparrow, transplantation of an organized pineal oscillator into an arrhythmic, pinealectomized host could impose phase on the restored system even if the pineal were not the primary source of rhythmicity (see Takahashi & Menaker 1982). Phase transplantation, therefore, is not necessarily a conclusive demonstration of a pacemaker role, although phase control indicates that the transplanted structure is a critical component of the pacemaking system.

Circadian organization in mammals may be similar to that in birds, even though the pacemaker role may be vested entirely in the SCN. Current evidence indicates that rhythms in melatonin secretion from the pineal are driven by a clock in the SCN, but that in addition melatonin plays a role in photic entrainment (Quay 1970, Golombek & Cardinali 1993) and rhythm regulation (Redman et al 1982, Cassone et al 1986a,b, McArthur et al 1991, Cassone 1992, Aguilar-Roblero & Vega-Gonzalez 1993). The observation that periodic melatonin application reduces the latency to recovery of rhythmicity following fetal SCN engraftment in SCN-lesioned hamsters suggests that one role of the pineal during embryogenesis is to accelerate the organization of the pacemaker ensemble within the developing SCN (Romero & Silver 1989).

Transplantation of circadian period

The evidence that the SCN is a circadian pacemaker in mammals is substantial and has been reviewed extensively (see Ralph & Lehman 1991, Lehman & Ralph 1993). This evidence includes lesion studies (Moore & Eichler 1972, Stephan & Zucker 1972) demonstrating the loss of rhythmicity following ablation of the SCN; explantation studies (Groos & Hendricks 1982, Green & Gillette 1982, Earnest & Sladek 1986) showing the persistence of rhythmicity *in vitro*; and transplantation experiments in which rhythmicity is restored by fetal SCN grafts (see Ralph & Lehman 1991 for review).

In mammalian systems, the unambiguous demonstration of transplantation of phase is hindered by the fact that embryonic SCN tissue is used for grafting so that an *a priori* assessment of circadian phase is not possible. However, because the rhythms of fetuses and pups tend to be synchronized with those of the mother (Davis & Gorski 1985, Reppert & Schwartz 1983), it is possible to determine whether or not the donor tissue has restored a rhythm by extrapolating the time of onset of activity to the day of transplantation and comparing this phase with that of the mother on the same day. By this method phase has been shown to be conserved following transplantation and recovery (Romero & Silver 1989).

The pacemaker function of the mammalian SCN has, however, been demonstrated unambiguously with the transference of period rather than phase. These experiments have involved both cross-species transplants, where species-specific period is identifiable (Sollars & Kimble 1988), and cross-genotype transplants taking advantage of the period mutation, *tau*, in the golden hamster (*Mesocricetus auratus*) that shortens the period of the free-running rhythm from the wild-type value of about 24 h to about 22 h in heterozygotes and to about 20 h in homozygous mutants (Ralph & Menaker 1988; Fig. 1). For SCN transplantation studies, the *tau* mutation is an ideal behavioural marker because its inheritance follows simple Mendelian rules and the three phenotypic groups do not overlap (Ralph 1991).

Initial transplantation studies demonstrating the essential autonomy of the SCN in determining period involved reciprocal transplantation of SCN tissue among the three *tau* genotypes (e.g., Fig. 2; Ralph et al 1990). In these experiments, the period of restored rhythms always fell within the range expected of adult animals belonging to the donor genotype, suggesting that little influence on the basic period was derived from sources outside the SCN. Subsequent experiments in which circadian chimeras—animals with two functional circadian clocks—were produced by transplantation confirmed this finding: animals with two clocks of different *tau* genotypes may express, independently of each other, two behavioural rhythms with periods within the two ranges predicted by genotype (see below) (Ralph et al 1992, Vogelbaum & Menaker 1992).

Circadian pacemakers in vertebrates

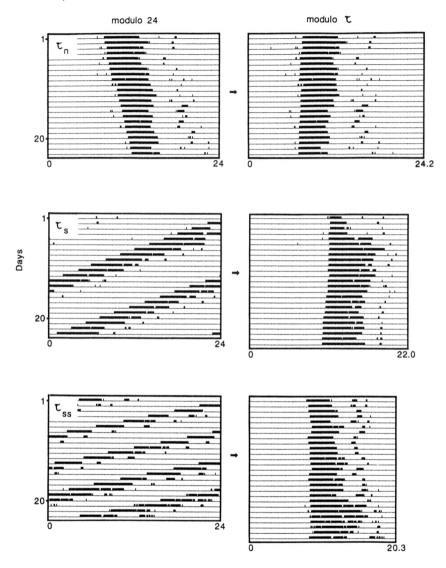

FIG. 1. Effects of the *tau* mutation on circadian period in golden hamsters (*Mesocricetus auratus*). Activity records (locomotor activity in constant darkness) are shown from a wild-type animal (*top panels*), a heterozygous animal (*centre panels*) and a homozygous *tau* mutant (*lower panels*). Actograms have been plotted at 24 h intervals (*left*) and re-plotted at intervals close to the free-running periods of the animals (*right*), so that the profile of activity is evident. (From Ralph 1991, with permission.)

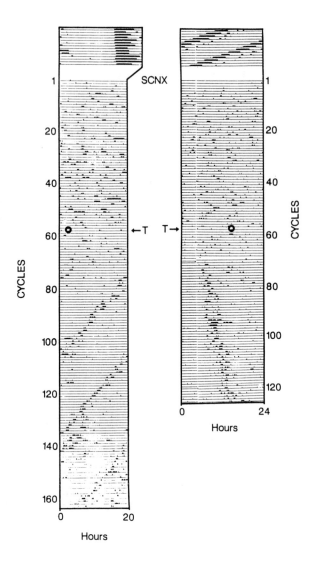

FIG. 2. Paradigm and examples of transfer of period by the transplantation of fetal SCN tissue from golden hamsters of known *tau* genotype. The left panel shows a wild-type host (period = 24 h, *top*), whose SCN is ablated on Day 1 (SCNX) and is given a graft of fetal SCN tissue from a homozygous *tau* mutant on Cycle 58 (T). The right panel shows a heterozygous host (period = 22 h) that received wild-type SCN on Day 58 (T). The animals were kept in constant darkness. Restored activity reflects the genotype of the donor. (From Ralph et al 1990, with permission.)

Pacemaker communication

The issue of communication is complicated somewhat by the fact that there are two types of communication in question: how the pacemaker (the SCN) influences the organism and how pacemaker cells (within the SCN) influence each other to produce coherent, overt rhythms. The first has received considerable attention in transplantation studies, but as yet it is not clear how the engrafted SCN convey temporal information to the organism, nor even whether these channels involve neural (synaptic) or humoral mechanisms. The second has received essentially no attention because the identity of the pacemaker cells within the SCN is unknown, and until recently there has been no indication that the grafted pacemaker has any influence on the intact pacemaker of the host.

If younger hamsters with robust locomotor rhythms are used as transplant hosts, the donor rhythm is not expressed unless the host SCN is at least partially ablated (Vogelbaum & Menaker 1992). The mechanism that enables expression under these circumstances is unknown. Partial ablation results in the fragmentation but not elimination of host rhythmicity, so it is possible that a reduction in the amplitude of the host rhythm is required for expression of the donor rhythm. Alternatively, the lesion may expose potential sites for neuronal connections from the graft to the host.

Both of these explanations raise the possibility that the quality of the host rhythm may be used to predict whether or not an SCN graft will be expressed in overt behaviour. If older animals, whose own locomotor rhythms have become fragmented, are used as transplant hosts, donor rhythmicity can be expressed without experimental destruction of the host SCN (Hurd & Ralph 1992, Hurd et al 1994). Under these conditions, not only is the donor rhythm expressed simultaneously with that of the host, but there is also evidence of pacemaker-pacemaker interactions.

Patterns of expression and interaction between the engrafted pacemaker and that of the aged host are highly variable, ranging from relative coordination between the two rhythms (Fig. 3), to complete dominance of the donor rhythm. In addition, a complex switching between the donor and host phenotype at unpredictable intervals has been observed (Fig. 4). The pattern displayed may depend upon the degree of fragmentation or disruption of the host rhythm, and a single animal may show multiple patterns following transplantation. An increase in the amplitude of the host rhythm is often observed as the restored rhythm becomes apparent (Fig. 4). It is possible that the fetal tissue, through its influence on host phase, tends to produce greater synchrony among host pacemakers which may be manifest as an increase in amplitude (Hurd et al 1994).

Together, these results suggest that the ageing process includes a progressive deterioration of organization within the SCN (Swaab et al 1985, Roozendaal

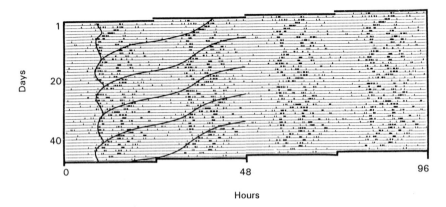

FIG. 3. Relative coordination between two rhythms in a circadian chimera. An aged, wild-type host was implanted with a fetal suprachiasmatic nucleus from a heterozygous *tau* donor. The segment of the record that is shown was taken two months after the transplantation operation. The figure has been quadruple-plotted at 24 h cycles so that the wild-type host rhythm appears vertically on the page. For both host and donor rhythms, a curve has been fitted by eye that connects each cycle to the next at CT 12 (left side of actogram). The periodic modulation of τ for one rhythm depends on the phase relationship between the two rhythms. (Hurd et al, unpublished observations.)

et al 1987, Chee et al 1988) that is mimicked by SCN lesions. However, the fact that interactions between pacemakers are observed when the SCN is grafted into aged hosts, but not when grafted into younger hosts with partial SCN lesions, indicates a fundamental difference between the disruptive effect of experimentally produced lesions and that which occurs with age. Perhaps the ageing process produces a more widespread diminution of pacemaker coupling and rhythm amplitude than lesions, which may spare islands of healthy SCN cells.

Pacemakers within the pacemaker

Although we have known the anatomical location of the mammalian circadian pacemaker for over 20 years, the cells within the SCN that generate rhythms have yet to be identified. Furthermore, while properties of the system have been studied extensively, it is not known whether these, including 24 h periodicity, are properties of single cells or an emergent feature of the pacemaker ensemble.

Evidence that there are multiple oscillators in the SCN comes from several sources. Partial lesions that destroy much of the SCN do not necessarily completely eliminate behavioural rhythmicity. Studies of phase-shifting, splitting and after-effects of entrainment have led to the conclusion that there are

Circadian pacemakers in vertebrates 75

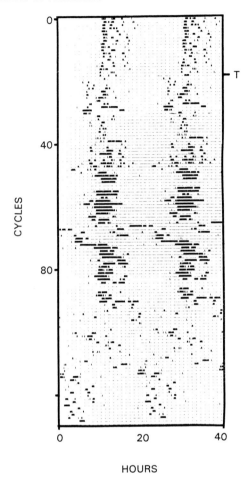

FIG. 4. Complex interactions between donor and host rhythms in a circadian chimera. An aged, homozygous *tau* mutant host received a graft of SCN tissue from a heterozygous donor (T). The presence of the donor clock produces three major changes in the record: (1) an increase in the amplitude of the host rhythm (about line 49); (2) expression of 22 h periodicity (lines 65–72 and 90–111); (3) relative coordination, suggested by the phase shifts of the host rhythm following overt expression of the donor clock. (Hurd et al, unpublished observations.)

at least two oscillators that time the transitions between subjective day and night (Pittendrigh 1974, Pittendrigh & Daan 1976). This is supported further by the finding that during entrainment of *tau* mutant hamsters to 24 h light cycles the daily onset of activity occurs three to five hours before normal, whereas the timing of the cessation of activity is the same as in the wild-type

(S. Osiel, unpublished results). In addition, during entrainment the onset is not simply moved earlier, but the entire behavioural profile is fragmented, suggesting a possible scattering of multiple oscillators under these conditions.

Even without knowing the identity of the pacemaker, it may still be possible to ascertain some of the cellular properties from a detailed analysis of circadian chimeras. Where relative coordination is observed following cross-genotype transplants into older hosts, the timing and direction of period modulation indicates that pacemaker cells may be either delayed or advanced by signals from other pacemakers. The maximum delay appears to occur when the responding pacemaker phase leads by about 90°. It is not yet possible to determine the timing of pacemaker communication, but the results suggest that coupling is phase dependent.

The question of whether a single cell is capable of producing a circadian rhythm may also be approached using chimeras. In this case, cells from two *tau* genotypes are mixed prior to transplantation, and the period of the rhythm produced by different ratios of normal and mutant cells is examined. The patterns of recovered rhythmicity obtained to date have been highly variable and are difficult to interpret. Although in some cases intermediate periods have been observed (Fig. 5), the same cell mixture has produced two distinct donor-specific rhythms in another host (Fig. 6). This, taken together with the fact that intermediate periods have never been reported for circadian chimeras produced under a number of different conditions, makes it seem unlikely that circadian rhythmicity is an emergent property of coupled ultradian oscillators, suggesting that the responsibility for generating the basic circadian period is vested in single pacemaker cells. Moreover, a corollary to this argument is that the *tau* mutation is itself an alteration of individual pacemaker period rather than a change in coupling among pacemaker cells.

Concluding remarks

Transplantation studies using golden hamsters carrying the *tau* mutation, particularly the production of circadian chimeras, have demonstrated the autonomy of the SCN in setting the basic circadian period in this species. Yet, it is still the case that other structures may influence either the phase or the period of the oscillation. One that is worthy of note is the intergeniculate leaflet (IGL), the ablation of which lengthens the period in constant dark and eliminates the period lengthening effect of constant light (Pickard et al 1987). The IGL, however, is usually not considered to be part of the pacemaker mechanism, but is thought rather to exert a regulatory influence on photic sensitivity or entrainment. Similarly, the pineal, although it has access to the pacemaker mechanism through its hormone, melatonin, is not a pacemaker in mammals. In non-mammalian species the situation may be different, not

Circadian pacemakers in vertebrates 77

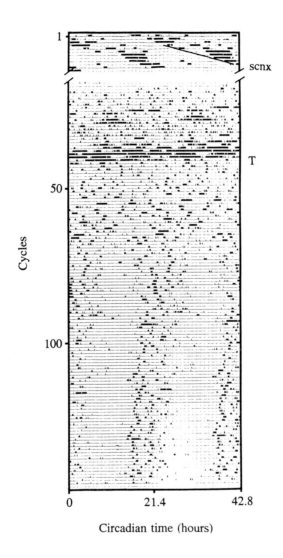

FIG. 5. Intermediate period produced by transplantation of cells from two *tau* genotypes. SCN tissue blocks from wild-type (24 h period) and homozygous mutant (22 h period) donors were dissociated and mixed thoroughly prior to transplantation into an arrhythmic host. Completeness of the dissociation was verified by microscopic examination of cell suspensions before mixing. Cells were pelleted, then about 50 000 cells were implanted in 2 μl Hank's balanced salt solution using established procedures (Ralph et al 1990, M. R. Ralph, E. R. Torre & M. N. Lehman, unpublished work). scnx, suprachiasmatic nucleus ablation.

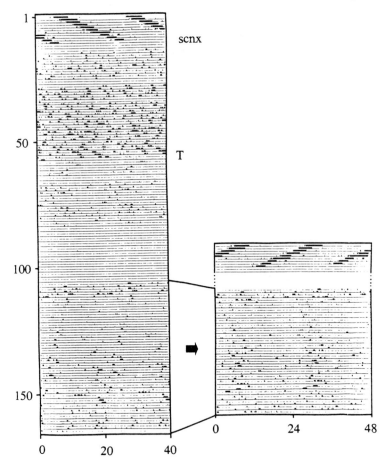

FIG. 6. Simultaneous expression of two rhythms restored using the dissociated cell preparation described in Fig. 5. The actogram has been plotted twice at intervals of 20 (*left*) and 24 (*right*) hours so that the two rhythms can be visualized. The two periods reflect the genotypes of the donor cells (M. R. Ralph, E. R. Torre & M. N. Lehman, unpublished work). scnx, suprachiasmatic nucleus ablation.

necessarily because the components of circadian organization are different, or even have different effects on each other, but because in some species, the non-SCN structures show oscillations. It remains to be seen, therefore, whether oscillators in the eyes (Besharse & Iuvone 1983) or pineal of non-vertebrates, the food-entrainable oscillator in rodents (see Rusak 1989 for review) or the visual oscillator in humans (Terman & Terman 1985) have partial pacemaker function.

The same problems of pacemaker identity may arise at other levels of organization. Within the SCN, the basic circadian period may be produced by a single type of cell, but it is not known how the other cell types within the nucleus may contribute. Moreover, within the pacemaker cells themselves, oscillating biochemical processes that may contribute to period determination may not be necessary for setting circadian period (e.g., calcium flux in *Bulla gouldiana*, Khalsa et al 1993). Hence, the difficulties encountered in rigorously defining the pacemaker at the systems level may recur at cellular and biochemical levels of organization.

Acknowledgements

This work is supported by NIH grant AG11084, AFOSR grant F49620-92-J-0517 and an Alfred P. Sloan Fellowship to M. R. R.

References

Aguilar-Roblero R, Vega-Gonzalez A 1993 Splitting of locomotor circadian rhythmicity in hamsters is facilitated by pinealectomy. Brain Res 605:229–236

Besharse JG, Iuvone PM 1983 Circadian clock in *Xenopus* eye controlling retinal serotonin *N*-acetyltransferase. Nature 305:133–135

Cassone VM 1992 The pineal gland influences rat circadian activity rhythms in constant light. J Biol Rhythms 7:27–40

Cassone VM, Menaker M 1984 Is the avian circadian system a neuroendocrine loop? J Exp Zool 232:539–549

Cassone VM, Chesworth MJ, Armstrong SM 1986a Dose-dependent entrainment of rat circadian rhythms by daily injection of melatonin. J Biol Rhythms 1:219–229

Cassone VM, Chesworth MJ, Armstrong SM 1986b Entrainment of rat circadian rhythms by daily injection of melatonin depends upon the hypothalamic suprachiasmatic nuclei. Physiol & Behav 36:1111–1121

Chee CA, Roozendaal B, Swaab DF, Goudsmit E 1988 Vasoactive intestinal polypeptide neuron changes in the senile rat suprachiasmatic nucleus. Neurobiol Aging 9:307–312

Davis FC, Gorski RA 1985 Development of hamster circadian rhythms. I. Within-litter synchrony of mother and pup activity rhythms at weaning. Biol Reprod 33:353–362

Earnest DJ, Sladek CD 1986 Circadian rhythms of vasopressin release from individual rat suprachiasmatic explants *in vitro*. Brain Res 382:129–133

Falcon J, Guerlette JF, Voisin P, Collin J-P 1987 Rhythmic melatonin biosynthesis in a photoreceptive pineal organ: a study in the pike. Neuroendocrinology 45:479–486

Gaston S, Menaker M 1968 Pineal function: the biological clock in the sparrow? Science 160:1125–1127

Golombek DA, Cardinali DP 1993 Melatonin accelerates reentrainment after phase advance of the light–dark cycle in Syrian hamsters: antagonism by flumazenil. Chronobiol Int 10:435–441

Green DJ, Gillette R 1982 Circadian rhythm of firing rate recorded from single cells in the rat suprachiasmatic brain slice. Brain Res 245:198–200

Groos GA, Hendricks J 1982 Circadian rhythms in electrical discharge of rat suprachiasmatic neurones recorded *in vitro*. Neurosci Lett 34:283–288

Hurd MW, Ralph MR 1992 Patterns of circadian locomotor rhythms in aged hamsters after suprachiasmatic nucleus transplant. Soc Res Biol Rhythms Abstr 3:67

Hurd MW, Golombek DA, Lehman MN, Ralph MR 1994 Pacemaker-pacemaker communication in hamsters with SCN transplants. Soc Res Biol Rhythms Abstr 4:123

Janik DS, Menaker M 1990 Circadian locomotor rhythms in the desert iguana. I. The role of the eyes and the pineal. J Comp Physiol A Sens Neural Behav Physiol 166:803-810

Kezuka H, Aida K, Hanyu I 1989 Melatonin secretion from goldfish pineal gland in organ culture. Gen Comp Endocrinol 75:217-221

Khalsa SBS, Ralph MR, Block GD 1993 The role of extracellular calcium in generating and phase-shifting the *Bulla* ocular circadian rhythm. J Biol Rhythms 8:125-139

Lehman MN, Ralph MR 1993 Modulation and restitution of circadian rhythms. In: Dunnett SB, Bjorklund A (eds) Functional neural transplantation. Raven Press, New York, p 467-487

Max M, Menaker M 1992 Regulation of melatonin production by light, darkness, and temperature in the trout pineal. J Comp Physiol A Sens Neural Behav Physiol 170:479-489

McArthur AJ, Gillette MV, Prosser RA 1991 Melatonin directly resets the rat suprachiasmatic circadian clock *in vitro*. Brain Res 565:158-161

Menaker M, Wisner S 1983 Temperature-compensated circadian clock in the pineal of *Anolis*. Proc Natl Acad Sci USA 80:6119-6121

Moore RY, Eichler VB 1972 Loss of a circadian adrenal corticosterone rhythm following suprachiasmatic lesions in the rat. Brain Res 42:201-206

Pickard GE, Ralph MR, Menaker M 1987 The intergeniculate leaflet partially mediates effects of light on circadian rhythms. J Biol Rhythms 2:35-56

Pittendrigh CS 1960 Circadian rhythms and circadian organization of living systems. Cold Spring Harbor Symp Quant Biol 25:159-184

Pittendrigh CS 1974 Circadian oscillations in cells and the circadian organization of multicellular systems. In: Schmitt FO, Worden FG (eds) The neurosciences: third study program. MIT Press, Cambridge, MA, p 437-458

Pittendrigh CS, Daan S 1976 A functional analysis of circadian pacemakers in nocturnal rodents. V. Pacemaker structure: a clock for all seasons. J Comp Physiol 106:291-331

Quay WB 1970 Precocious entrainment and associated characteristics of activity patterns following pinealectomy and reversal of photoperiod. Physiol Behav 3:109-118

Ralph MR 1991 Suprachiasmatic nucleus transplant studies using the *tau* mutation in golden hamsters. In: Klein DC, Moore RY, Reppert SM (eds) Suprachiasmatic nucleus: the mind's clock. Oxford University Press, New York, p 341-348

Ralph MR, Lehman MN 1991 Transplantation: a new tool in the analysis of the mammalian hypothalamic circadian pacemaker. Trends Neurosci 14:362-366

Ralph MR, Menaker M 1988 A mutation of the circadian system in golden hamsters. Science 241:1225-1227

Ralph MR, Foster RG, Davis FC, Menaker M 1990 Transplanted suprachiasmatic nucleus determines circadian period. Science 247:975-978

Ralph MR, Hurd MW, Lehman MN 1992 Culture and transplantation of a mammalian circadian pacemaker in rodents: circadian chimeras produced from mixed *tau* genotypes. Soc Res Biol Rhythms Abstr 3:68

Redman J, Armstrong SM, Ng KT 1982 Free-running activity rhythm in the rat: entrainment by melatonin. Science 219:1089-1091

Reppert SM, Schwartz WJ 1983 Maternal coordination of the fetal biological clock in utero. Science 220:969-971

Romero M-T, Silver R 1989 Control of phase and latency to recover circadian locomotor rhythmicity following transplantation of fetal SCN into lesioned adult hamsters. Soc Neurosci Abstr 15:725

Roozendaal B, van Gool WA, Swaab DF, Hoogendijk JE, Mirmira M 1987 Changes in vasopressin cells of the rat suprachiasmatic nucleus with aging. Brain Res 409:259–264

Rusak B 1989 The mammalian circadian system: models and physiology. J Biol Rhythms 4:121–134

Sollars PJ, Kimble DP 1988 Cross-species transplantation of fetal hypothalamic tissue restores circadian locomotor rhythm to SCN-lesioned hosts. Soc Neurosci Abstr 14:49

Stephan FK, Zucker I 1972 Circadian rhythms in drinking behavior and locomotor activity of rats are eliminated by hypothalamic lesions. Proc Natl Acad Sci USA 69:1583–1586

Swaab DF, Fliers E, Partiman TS 1985 The suprachiasmatic nucleus of the human brain in relation to sex, aging and senile dementia. Brain Res 342:37–44

Takahashi JS, Menaker M 1979 Brain mechanisms in avian circadian systems. In: Suda M, Hayaishi O, Nakagawa H (eds) Biological rhythms and their central mechanism. Elsevier Science Publishers, Amsterdam, p 95–111

Takahashi JS, Menaker M 1982 Role of the suprachiasmatic nuclei in the circadian system of the house sparrow, *Passer domesticus*. J Neurosci 2:815–828

Takahashi JS, Hamm H, Menaker M 1980 Circadian rhythms of melatonin release from individual superfused chicken pineal glands *in vitro*. Proc Natl Acad Sci USA 77:2319–2322

Terman M, Terman J 1985 A circadian pacemaker for visual sensitivity? Ann NY Acad Sci 453:147–161

Vogelbaum VA, Menaker M 1992 Temporal chimeras produced by hypothalamic transplants. J Neurosci 12:3619–3627

Zimmerman NH, Menaker M 1979 The pineal gland: a pacemaker within the circadian system of the house sparrow. Proc Natl Acad Sci USA 76:999–1003

DISCUSSION

Loros: You have shown us mixing of hamster genotypes in transplants of SCN material and the resulting periodicities in recipient animals. The two phases you get with the two different τs look consistent with an intracellular oscillator. They are each, when there are two different genomes, putting out information that controls the level of motor activity. I'm finding it difficult to figure out what's happening in the hamsters that ended up with an averaged τ. How can you have it both ways?

Ralph: The averaged τ showed up when we thoroughly mixed cells from the two genotypes together. We have never seen an averaged τ in the older animals with an intact SCN, or in any preparation where one or both nuclei are intact. We do see evidence in the latter preparations, however, that the two rhythms can interact. The degree to which cells of the two genotypes will interact with each other has something to do with the nature of how we put them together. If you thoroughly mix two types of cells together, the potential for interaction with the neighbouring cell may be a lot higher than when you take a piece of tissue, put it into a brain and let it settle where it may.

Loros: That is an important difference. You are still saying that the information that is being transferred from cell to cell can be averaged. Or, are you getting actual cell fusion to make chimeric cells?

Ralph: I think not, though I cannot be sure of that. What I am saying is that when you give single pacemaker cells from animals of different *tau* genotypes the opportunity to interact with each other, whether and how strongly they actually do interact may depend on how close they are together, how well mixed up they are and what other types of pacemaker are in contact. Averaging may result when there is the same extent of communication between genotypes as there is within a genotype. When you transplant a block of SCN cells into an intact organism the within-genotype communication may already be established by proximity and developmental history.

Turek: But in your mixed cultures most of the time you also got either one period or the other. Presumably you mixed your cultures up the same way.

Ralph: For the mixed cell transplants, all cells from both genotypes are mixed in the same tube.

Turek: So how can you sometimes get a combined intermediate period and at other times separate periods?

Ralph: The intermediate rhythm is rare. In the example I showed, with the same culture, one host developed an intermediate period while the other developed two distinct, genotype-specific periods. When all the cells are mixed up together, a pacemaker cell has an equal opportunity to interact with a cell of the same genotype or one of another genotype. So, the opportunity to form an ensemble with an average period is presented. However, any circumstance that might favour communication within one genotype population, such as incomplete dispersal or unequal cell numbers, might allow the formation of two oscillating groups.

Czeisler: How often do you see the intermediate period?

Ralph: We have only two examples out of 20 transplants. My interpretation is that the four-hour difference in period is at the limit of entrainment for one oscillator by another. For comparison, we find that the mutant (20 h) rhythm may be shifted by up to 12 h by short light pulses yet rarely can be entrained to a 24 h light–dark cycle. Perhaps this is true also of pacemaker–pacemaker influences. It may simply be easier for oscillators to synchronize with others of similar period, thus strengthening within-genotype coupling while reducing the likelihood that intermediate periods will emerge.

Edmunds: You seem to be saying that some cells are more equal than others under certain circumstances. There appear to be certain competing networks between cells of the same and of the other type. If you allow aggregates or clusters of interacting cells to exist, then, depending on whether they are coupled tightly or weakly, results can differ depending on the size of the initial aggregate. You can then invoke a threshold. An aggregate of a certain size could act as a block and overcome a competing cluster so as to impart its own phase. This is the kind of model that one has for *Dictyostelium* cells trying to come together, each giving out pulses of cAMP: they are all equal to begin with, but as soon as clustering occurs one mass eventually achieves dominance.

Ralph: In the case of the hamster SCN, though, there may be two types of oscillator with different periods, where type 1 and type 2 do not mutually entrain easily; perhaps their phase response curves to the incoming signal are of low amplitude, or the four-hour difference is at the limit of stable entrainment.

Edmunds: Their coupling strengths could be different.

Ralph: If their coupling strengths were different, I would expect amplitude in intact animals to vary with genotype, but we don't see this.

Kronauer: You are trying to make this too complicated. The two periods, 20 and 24 h, are really quite remote; you are looking for a large percentage change in the period. Rhythms separate principally because the coupling, whatever it is, is too weak to bring that period differential together most of the time. The interesting thing is that when they did come together, the compromise was almost exactly in the middle, which, according to a simple coupling model, says that interaction is the same in each direction. If it were stronger in one direction than the other you would expect the period to be off-centre.

Your results therefore say that the two types are essentially equivalent in the way they affect each other, but the difference in period, four hours, is so large that even though they are close to each other physically, they are unable to synchronize. In contrast, the population of cells of one type probably has a relatively small dispersion of period, so that a weak level of coupling existing between individual members is able to pull them together to express a single period. When there is a network with two distinct sets of elements all the type 1s can interact effectively with each other, as can all the type 2s, so that each separate type actually coalesces as a group with a single period because they aren't very far apart to start with.

Incidentally, four hours is the limit for spontaneous internal desynchrony in humans. When the sleep–wake rhythm differs from the intrinsic pacemaker rhythm by about four hours the two go their separate ways. Four hours is a large difference to expect these elements to bridge, and it's remarkable that they do it all.

Ralph: I agree; as our photic entrainment results indicate, the homozygous *tau* mutants rarely show stable entrainment to 24 h light–dark cycles.

Kronauer: It's also remarkable that when you have the two pacemakers, the transplant and the weaker endogenous SCN, each speaks to the other. That's an important finding: the interaction mechanisms work in both directions.

Cassone: Some years ago, Fred Davis and Mike Menaker published an infrequently cited paper entitled 'Hamsters through time's window', in which they wrote that it was common for hamsters' activity patterns to have a complex dynamic (Davis & Menaker 1980); there is a bout of activity of approximately 24.2 h, which has within it oscillations with quite different periods. Records like this don't seem to be appearing since the discovery of the *tau* mutant, as if those extra bout oscillators have been lost. Perhaps the *tau* mutant has lost one of these bout oscillators which originally contributed to the hamster's activity pattern. Their period was about 20 h, as I remember.

Menaker: No, the period was long, about 26 h.

Ralph: There are two ways of looking at those bout oscillators; one is that they are independent groups of cells whose output is expressed only during the subjective night, gated by the SCN, and the other is that within the ensemble of SCN pacemakers, every once in a while a subpopulation with an average period that is different from the whole gets together and the resultant output is sufficient to organize behaviour.

Cassone: There's a fundamental difference between putting cells into an adult animal and having an animal that develops a large number of genetically heterogeneous cells within its SCN. The expression of activity might be quite different in those two situations.

Hastings: If the averaging has something to do with the relative differences in period, the average period outcome should be more frequent when you mix heterozygotic cells with either of the homozygotic forms. Is that so?

Ralph: That will be our next experiment.

Kronauer: If you assemble the statistics of success rate (achievement of synchrony) for the different pairings (four-hour difference and two-hour difference), you might end up with a good measure of coupling strength appropriate to that particular preparation which could go into a model.

Hastings: The starting conditions would seem to be really important. If the initial pairing were random, and if the cells of the two types by accident started to beat together to give an intermediate, average period, then because of the minimal difference in period, it would be easier for individual cells of both types which are not yet paired to contribute to the emergent population rhythm. In contrast, if the first establishment of synchrony were between cells which all had a long or all had a short period, the individual cells with the opposite period would be excluded from the population oscillation.

I wondered whether you could stop the clock by some pharmacological or other means while the cells are forming their associations. That would prevent the two cell types expressing their different circadian periods and so they would not be able to tell the difference between each other. Consequently, the proportion of cells of the two periods in the population network that formed would reflect their relative abundance in the initial culture. When you allowed the clock to restart, would it always restart with an average period which was determined by this relative abundance?

Ralph: That's an interesting idea. You might be able to do that in culture, because eight-day-old cultures restore rhythms and you might be able to reduce coupling *in vitro*. However, cells may remain synchronous even though they are not communicating for a while, so it may still be difficult to produce random rewiring.

Meijer: Have you checked for periods shorter than 20 h or longer than 24 h after your transplantations? It's not necessarily true that the period of the two animals will produce some intermediate period; the coupling may drive them either to a very short period or to a very long period.

Ralph: It could, but, as I said, the only intermediate period that we've seen has been right in the middle. Our actograms are routinely analysed for periodicities between 18 and 26 h.

Meijer: I have been trying to relate your results to what we see at the onset of splitting. At the onset of splitting, in the first week, there is sometimes a big difference in the free-running period of the two components. Especially when the rhythms split quickly, the periods can differ temporarily by more than four hours. Nevertheless, this large difference in period does not prevent the two free-running periods from coupling. Dr Kronauer suggested that the two free-running periods might be too far apart to be coupled, but this may not be the case. One could argue that if the coupling between the two components is strong enough to drive two oscillator groups four hours apart, it may also be strong enough to drive your different oscillators together. What I found amazing in your results was the lack of coupling between the cells.

Ralph: In the case of splitting, the starting population is a group of oscillators which may have basic individual periods of 24 h. Subgroups within the population may move apart because they cause phase shifts in opposite directions. Still, within those groups, the individual cells may have 24 h periods. You do not see relative coordination after splitting because the cells have a propensity to come back to 24, and to re-couple with one another. In our chimeras, the intrinsic cellular periods may fall into two distinct groups. In splitting, the pacemakers are being forced apart, whereas in the chimeras they are being forced together.

Miller: If you assume that all the cells are intracellular circadian oscillators, you have to make some additional specific assumptions in order to explain the occasional occurrence of averaging. It seems to me that there is an alternative, that what you are really looking at is a network property, but that the network is extremely refractory to disruption, even in culture. On those occasions when you do manage to disrupt the network, it's possible to build up a new network with an averaged period. Your results haven't yet killed the network hypothesis.

Ralph: There are no cell clusters remaining after this process, as far as we can tell by microscopy. Physically, the coupling is not likely to be there.

Miller: Physically (or morphologically), true, but because we don't know the physical nature of the coupling that we are talking about anyway, who's to say that it isn't particularly difficult to disrupt? I have to admit that I'm disturbed by the fact that in monolayer culture, without three-dimensional structure, we can still see a 24 h rhythm in vasopressin secretion. But still, you cannot exclude the possibility that this network, whatever it is, is incredibly difficult to disrupt.

Arendt: Is it possible to select out from your cultures subpopulations of cells for transplant on the basis of cell surface receptors, for example? You could attach a ligand to an affinity column to select out subpopulations.

Ralph: We are looking for ways of separating subpopulations. We might be able to pull them out using flow cytometry.

Czeisler: Davis & Gorski (1984) and Pickard & Turek (1983, 1985) have reported that the free-running period of the activity rhythm is proportional to the volume of the SCN that's transplanted in the hamster. You were talking about one period being 24 h and the other period being 20 h, and yet, depending on the relative amount of tissue transplanted, there could presumably be differences in the periods observed.

Edgar et al (1993) have proposed that the SCN promote wakefulness, not inactivity or sleep. It seems from your results that there is a window, an entire zone of inactivity, with a near 24 h period. That zone of inactivity persists, even when it is coincident with the expected activity phase of the mutant SCN. Similarly, the zone of inactivity with a near 20 h oscillation persists, even when it is coincident with the expected activity phase of the near 24 h oscillation. Doesn't that argue that the SCN must actively promote sleep or inactivity at certain times as well as wakefulness at other times?

Ralph: This is really a semantic argument and will remain so until we have a better understanding of the neurochemical outputs of the clock. The promotion of one overt activity may be the result of the imposition of the other. For example, imposition of a period of inactivity (sleep) on an otherwise arrhythmic background may increase the total amount of activity (awake) in rebound from inhibition.

Basically, when we create a circadian chimera a periodic decrease in locomotor activity is imposed on the existing rhythm. We don't really add activity to the subjective day. Activity, at least in the hamster, may be a rebound from this inhibition.

Czeisler: The promotion of inactivity or sleep by the SCN would be quite different from what Edgar et al (1993) have proposed, where the primary role of the SCN is in the promotion of activity.

Moore: It is perfectly clear that lesions of the SCN reduce activity. In that context, one would say that the SCN promotes activity.

Ralph: An analogous situation may be the electrical activity in a single neuron. If you apply a hyperpolarizing pulse to a relatively quiet neuron and let it rebound, you will see an increase in electrical activity. In that way, your inhibitory pulse has promoted activity in that neuron. Thus, the total activity measured over time may increase in the presence of periodic hyperpolarization.

Menaker: How are your animals kept while you are holding them before starting experiments? Do you put them into constant darkness before you do the transplantation?

Ralph: Before the transplant they are kept in constant dim red light for about a month. When we began we kept them in dim red light for much less than that because we were worried about the older hosts dying. When the rhythm is broken down as much as we would like, the animals usually don't have long to live.

Menaker: So these animals presumably have large functional gonads and high levels of testosterone.

Ralph: Yes. They are not housed in the dark for a long time.

Menaker: Does the histology of transplanted tissue differ in old and young hosts?

Ralph: The grafted tissue, although we haven't done detailed cell counting, looks similar in young and old hosts. However, the age and health of the host often determines how long we can record following grafting, so we have not yet compared grafts that have been in young and old brains for the same length of time.

References

Davis FC, Menaker M 1980 Hamsters through time's window: temporal structure of hamster locomotor rhythmicity. Am J Physiol 239:149–155

Davis FC, Gorski RA 1984 Unilateral lesions of the hamster suprachiasmatic nuclei: evidence for redundant control of circadian rhythms. J Comp Physiol A Sens Neural Behav Physiol 154:221–232

Edgar DM, Dement WC, Fuller CA 1993 Effect of SCN lesions on sleep in squirrel monkeys: evidence for opponent processes in sleep–wake regulation. J Neurosci 13:1065–1079

Pickard GE, Turek FW 1983 The suprachiasmatic nuclei: two circadian clocks? Brain Res 268:201–210

Pickard GE, Turek FW 1985 Effects of partial destruction of the suprachiasmatic nuclei on two circadian parameters: wheel-running activity and short-day induced testicular regression. J Comp Physiol A Sens Neural Behav 156:803–815

Organization of the mammalian circadian system

Robert Y. Moore

Departments of Psychiatry, Neurology and Neuroscience, Center for Neuroscience and Alzheimer Disease Research Center, University of Pittsburgh, Pittsburgh, PA 15261, USA

Abstract. The mammalian circadian timing system is a set of related neural structures whose function is to provide a temporal organization for physiological processes and behaviour. The system has three major components, entrainment pathways, pacemakers and output pathways that couple the pacemakers to effector systems that express circadian functioning. The retinohypothalamic tract is a direct retinal projection to the circadian pacemakers, the suprachiasmatic nuclei. The retinohypothalamic tract arises from a discrete set of retinal ganglion cells that receive photic information from a unique population of retinal photoreceptors and it mediates photic entrainment of the suprachiasmatic nuclei. The geniculohypothalamic tract arises from neurons of a specialized subdivision of the lateral geniculate complex, the intergeniculate leaflet. The intergeniculate leaflet and geniculohypothalamic tract appear to provide integrated photic and non-photic input to the suprachiasmatic nuclei to modulate pacemaker function. The suprachiasmatic nuclei comprise individual neuronal oscillators coupled into a neural network. The output of the suprachiasmatic nuclei is quite restricted but becomes amplified by a set of downstream components of the system that appear to provide a widespread circadian signal.

1995 Circadian clocks and their adjustment. Wiley, Chichester (Ciba Foundation Symposium 183) p 88–106

The temporal organization of behaviour into circadian cycles of rest and activity is a fundamental feature of mammalian adaptation. This function is mediated by a specific neural system, the circadian timing system. Circadian biologists have used the term 'circadian system' for many years but it is only over the past 20 years that we have acquired a substantial body of detailed information about this unique brain system. Two sets of studies initiated this line of inquiry. The first was the demonstration of a direct retinohypothalamic tract (RHT) terminating in the suprachiasmatic nucleus (SCN) of the hypothalamus (Moore & Lenn 1972, Hendrickson et al 1972). The second was the demonstration that ablation of the suprachiasmatic nuclei results in loss of circadian functioning (Moore & Eichler 1972, Stephan & Zucker 1972), indicating that a discrete circadian pacemaker exists in the mammalian brain. From this beginning, a growing body of information has provided an extensive characterization of the mammalian circadian timing system, and it is the intent of this brief review to outline the current status of our understanding of this neural system.

Circadian rhythms have two principal properties: they are normally entrained to the light–dark cycle but, in the absence of photic cues, they 'free run' with a period approximating 24 h. From these properties, we can infer that the circadian timing system has three essential components, visual pathways mediating entrainment, circadian pacemakers and efferent pathways that couple the pacemakers to effector systems that express circadian functioning. Each of these will be considered in the sections that follow.

Entrainment pathways

Entrainment, in the simplest framework, is the setting of period and phase of the endogenous pacemakers to the external light–dark cycle. In this context, entrainment pathways must be visual pathways. It has become evident, however, that pacemaker operation may be affected by a series of non-photic variables; since these may also affect pacemaker period and phase, they also are mediated by what may be considered as entrainment pathways.

The retinohypothalamic tract

In early work, we found that sectioning all visual pathways leaving the optic chiasm produced animals that had lost the capacity for visually guided behaviour and visual reflexes, but not the capacity for visual entrainment of circadian function (Klein & Moore 1979). These findings indicated that there is a separation of function among visual pathways, with some mediating visual guidance of behaviour and visual reflexes and others mediating entrainment. The pathway mediating photic entrainment is the RHT. Early studies demonstrated that the RHT projects to the SCN of the hypothalamus (Fig. 1; Moore & Lenn 1972, Hendrickson et al 1972). With more sensitive methods, it has been shown that the RHT has several components; the largest projects to the SCN, particularly to the ventrolateral division, but there are additional projections to the lateral hypothalamic area, the anterior hypothalamic area and the retrochiasmatic area, a complex zone immediately caudal to the SCN (Johnson et al 1988a). The RHT is sufficient to maintain entrainment; as noted above, section of all visual pathways beyond the RHT does not affect stable entrainment (Klein & Moore 1979), whereas section of the RHT abolishes entrainment (Johnson et al 1988b).

The retinal ganglion cells projecting to the SCN also project to the intergeniculate leaflet (IGL) of the thalamus (Pickard 1985); these appear to be a separate set of ganglion cells that do not project significantly to other central retinal

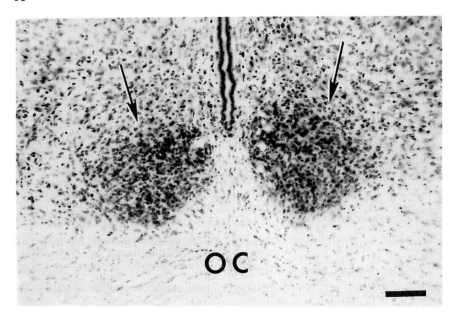

FIG. 1. Photomicrograph of a Nissl-stained coronal section from rat hypothalamus showing the suprachiasmatic nuclei (arrows) as two dense accumulations of cells lying above the optic chiasm (OC) and lateral to the third ventricle. Scale bar, 100 μm.

targets (Card et al 1991). In addition, the photoreceptors mediating entrainment appear to be separate from the classical photoreceptors (Foster et al 1991), indicating that the retinal components of the circadian timing system are distinct and specialized for circadian function.

Intergeniculate leaflet-geniculohypothalamic tract

The IGL is an anatomically and functionally distinct subdivision of the lateral geniculate complex that receives bilateral innervation from the retina (Moore & Card 1994). The IGL is characterized by a population of neuropeptide Y (NPY)-containing neurons that project to the SCN in a pattern overlapping the retinal input, and a population of enkephalin-containing neurons that project to the contralateral IGL (Card & Moore 1989). All IGL neurons also appear to contain GABA (γ-aminobutyric acid) (Moore & Speh 1993, Moore & Card 1994). The geniculohypothalamic tract (GHT) is the projection from the IGL to the SCN. The IGL receives input from the SCN, from brainstem noradrenaline, serotonin and acetylcholine nuclei and from the retrochiasmatic

Organization of the mammalian circadian system 91

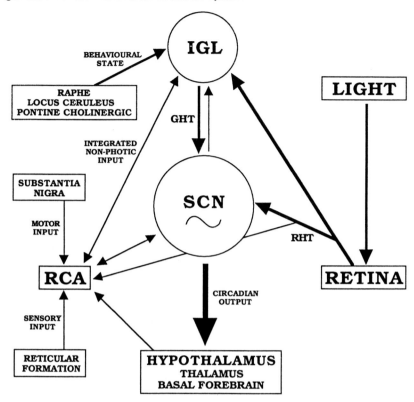

FIG. 2. Organization of afferents to the circadian timing system. The suprachiasmatic nucleus (SCN) receives photic input from the retina. It also receives input from the intergeniculate leaflet (IGL), which integrates photic input from the retina with non-photic input from the raphe serotonin neurons, locus ceruleus noradrenaline neurons, pontine cholinergic neurons and from the retrochiasmatic area (RCA). See text for further description. GHT, geniculohypothalamic tract; RHT, retinohypothalamic tract.

area (Moore & Card 1994). The retrochiasmatic area receives input from several brainstem and hypothalamic sources and can thus be viewed as a major source of non-photic input into the IGL (Fig. 2).

Stimulation of the IGL, or perfusion of NPY into the region of the SCN, produces changes in the phase of free-running rhythms with a phase response curve that differs markedly from that produced by light, having phase advances during subjective day and phase delays during subjective night (Moore & Card 1990, 1994). Stimuli that induce locomotor activity produce changes in the phase of free-running activity rhythms, with a phase response curve that is similar to that produced by stimulation of the GHT (see Moore 1992a and Mrosovsky 1994, this volume, for reviews). This effect appears to be mediated by the IGL-GHT (Johnson et al 1989). These observations indicate that the IGL-GHT system serves to integrate photic and non-photic information to modulate pacemaker function (Moore 1992b, Moore & Card 1994).

Other SCN afferents

The SCN receives inputs from several additional areas and all of these are likely to be involved in the modulation of pacemaker functioning. One of the largest inputs arises from the serotonin neurons of the midbrain raphe nuclei with a terminal plexus that largely overlaps the RHT and GHT projections (Moore et al 1978). Selective destruction of the raphe input to the SCN results in an alteration of the free-running period and in the phase angle of entrainment (Smale et al 1990). Application of serotonin to the SCN in slices *in vitro* changes the phase of the firing rate rhythm with a phase response curve similar to that for GHT stimulation (Shibata et al 1992). These results suggest that the function of the innervation of the SCN by raphe serotonin neurons is also to provide a feedback modulation of pacemaker functioning similar to that of the IGL-GHT projection. There are additional afferents to the SCN, including ones from areas receiving input from the SCN that are discussed below. Other afferents arising from the hypothalamus and basal forebrain (Pickard 1982) have not been studied functionally.

The pacemaker: the SCN

After the initial studies that demonstrated an RHT terminating in the SCN (Moore & Lenn 1972, Hendrickson et al 1972) and the studies showing that ablation of the SCN results in a loss of circadian function (Moore & Eichler 1972, Stephan & Zucker 1972), it was widely concluded that the SCN is a circadian pacemaker. However, the full evidence to support that conclusion was obtained only recently. Four lines of evidence are important. The first two are noted above, that the SCN is a major site of RHT input and that ablation of the SCN eliminates circadian functioning. The third is that circadian functioning is maintained in the isolated SCN, both *in vivo* (Inouye & Kawamura 1979) and *in vitro* (Shibata & Moore 1988; see Meijer & Rietveld 1989 for review). The fourth is that transplantation of fetal anterior hypothalamus containing the SCN into the third ventricle of arrhythmic, SCN-lesioned animals restores rhythmicity (Lehman et al 1987), with the period of the restored rhythm being determined by the graft (Ralph et al 1990).

Morphological organization of the SCN

Most studies of organization of the SCN have been done in rodents. The rat SCN contains about 8000 neurons on each side (van den Pol 1980). The neurons are small (8-12 μm diameter) with relatively sparse dendritic arbors that extend largely within the nucleus (van den Pol 1980). The neurons in the ventrolateral portion of the nucleus, the major retinorecipient zone, are larger than those in the dorsomedial portion, and this distinction between ventrolateral and dorsomedial SCN is quite evident in immunocytochemical studies of the organization of the nucleus (van den Pol & Tsujimoto 1985). The ventrolateral SCN is characterized by a large population of neurons that contain vasoactive intestinal polypeptide (VIP), whereas the major population of neurons in the dorsomedial SCN is vasopressin-containing. Recent results indicate that each of these peptides is co-localized with GABA (Moore & Speh 1993). The SCN is therefore a nucleus containing neurons producing an inhibitory transmitter. Studies of the organization of the SCN in other mammals (Cassone et al 1988) and in primates (Moore 1992a) indicate that it is quite similar in all species with respect to peptide content but the studies of GABA content have not been extended to species other than the rat.

Functional organization of the SCN

Both Golgi and immunocytochemical staining indicate that SCN neurons make extensive connections with other SCN neurons and, hence, that the nucleus is a neuronal network comprising interconnected neuronal units. The network exhibits a circadian rhythm in functioning with higher mean firing rate and higher glucose utilization during subjective day than subjective night (Shibata & Moore 1988). The question arises as to whether this circadian rhythm comes about as a consequence of the coupling of individual neuronal circadian oscillators, or whether it is an emergent property of the network. Several studies bear on this problem. In an important paper, Schwartz et al (1987) demonstrated that perfusion of the sodium channel blocker tetrodotoxin into the SCN region resulted in a loss of the circadian activity rhythm during the period of perfusion, when there would be no sodium-dependent action potentials. However, at the end of the treatment, circadian rhythmicity re-emerged at the phase at which it would have been expected to be if circadian functioning had been maintained through the tetrodotoxin administration. This indicates that maintenance of circadian functioning does not require an intact neuronal network. In other studies, Reppert and colleagues (see Reppert 1995, this volume) have shown that the circadian rhythm in glucose utilization in the rat begins in fetal life on Embryonic Day 19. At this time, shortly after the last neurons in the SCN have been generated, the neuropil is very immature and synaptogenesis has not yet been initiated (Moore & Bernstein 1989). Indeed, the total number of synapses in the SCN is less than one per neuron and these are widely dispersed, so that we can infer that circadian functioning is initiated before a neuronal network has been formed (Moore and Bernstein 1989). This can best be intrepreted by concluding that the SCN is made up of coupled circadian oscillators that function as a circadian pacemaker through formation of a neuronal network (Fig. 3).

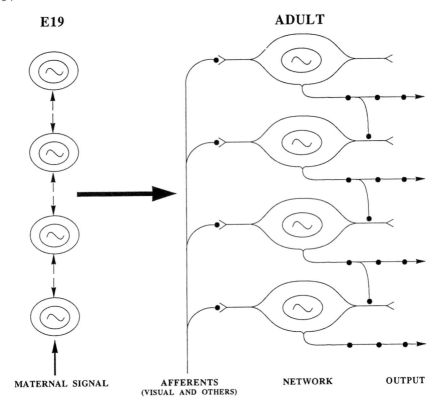

FIG. 3. Organization of the SCN pacemaker. At Embryonic Day 19 (E19), SCN neurons function as individual oscillators that are coordinated by maternal signals or by non-synaptic interactions. In the adult, the individual neurons are coupled by synaptic connections into a network that functions as a pacemaker.

Non-SCN circadian pacemakers

All of the evidence available at present indicates that the suprachiasmatic nuclei are the principal circadian pacemakers in the mammalian brain and are the pacemakers responsible for generating the rest–activity rhythm. There are studies, however, that demonstrate the existence of other circadian pacemakers in mammals. There is a rhythm in visual sensitivity and rod outer segment shedding that appears to be generated by an intrinsic retinal pacemaker (Reme et al 1991). In SCN-lesioned rats restriction of food produces anticipatory activity with a periodicity in the circadian range (Stephan et al 1979), and, similarly, giving methamphetamine to SCN-lesioned rats induces activity rhythms in the circadian range (Honma et al 1987); in neither of these instances is the location of the induced oscillator known, and it would appear to be a weak oscillator because circadian rhythmicity does not persist for a significant time after cessation of the inducing stimulus. It is not known whether the food-entrainable and methamphetamine-induced oscillators are the same, nor whether they participate in circadian functioning in the intact animal.

Output pathways: effector coupling

Early studies using the autoradiographic tracing method (see Watts 1991 for review) demonstrated that the SCN has few projections. Subsequent studies using anterograde transport of plant lectins have provided additional information but have largely confirmed the observations of the initial studies (Watts 1991). There are two major sites of projections originating in the SCN, within the SCN itself and into a region of the anterior hypothalamus dorsal and caudal to the SCN termed the subparaventricular zone (Watts 1991). From the projection into the subparaventricular zone, SCN axons extend caudally into the dorsomedial hypothalamic nucleus, into a zone lying between the ventromedial nucleus and the arcuate nucleus, the ventral tuberal area, and to a limited extent into the ventromedial nucleus. Some axons continue caudally from this projection into the posterior hypothalamic area and the rostral periaqueductal grey. In addition to these groups of SCN projections, a small group of fibres extends anteriorly into the anterior hypothalamic area and the medial preoptic area. Another group runs rostrally dorsal to these with a branch extending rostrally into the bed nucleus of the stria terminalis and lateral septum and a second branch dorsally into the anterior paraventricular thalamic nucleus. At the level of the SCN, there is a small group of fibres extending laterally into the lateral hypothalamic area with a small projection running along the optic tract to innervate the IGL. There is a larger group extending caudally into the retrochiasmatic area. With the exception of the projection into the subparaventricular zone, the remaining projections are relatively sparse. Watts and colleagues (Watts 1991) have emphasized, however, that the projections to the subparaventricular zone largely overlap those within the SCN but are more dense, and they conclude that the subparaventricular zone projections are likely to amplify and enhance SCN output.

Nevertheless, it is difficult to envision how these relatively limited patterns of projections will suffice to regulate the widespread and diverse functions controlled by the circadian timing system. To investigate this further, we have carried out a series of studies examining the projections of crucial areas that receive projections from the SCN and have reciprocal connections back to the SCN. Each of these areas receives in addition a direct retinal input, presumably from the ganglion cells that project to the SCN. Because of this organization of connections, we would propose that these areas, the IGL, retrochiasmatic area and paraventricular thalamic nucleus, be considered, along with the subparaventricular zone, as components of the circadian timing system. The projections from these areas provide a pattern of output that is in accord with

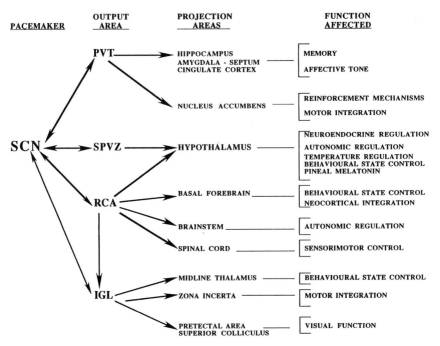

FIG. 4. Effector coupling by SCN connections. The SCN projects to four primary output areas, the paraventricular thalamic nucleus (PVT), the subparaventricular zone (SPVZ), the retrochiasmatic area (RCA) and the intergeniculate leaflet (IGL). Each of these projects to a series of areas through which the circadian signal is disseminated. The functions that might be under pacemaker control through these connections are indicated in the right-hand column.

the extensive set of functions under circadian regulation. A summary of the projections from these areas and the putative areas and functions that might be controlled is shown in Fig. 4.

The paraventricular thalamic nucleus projects rostrally and laterally to the forebrain to the septum, amygdala, hippocampus and cingulate cortex, areas certainly involved in memory, forebrain control of autonomic functioning and affective tone. It has other projections into the nucleus accumbens, a complex striatal region involved in diverse actions including reinforcement mechanisms and motor integration. The paraventricular thalamic nucleus also projects back to the SCN (Moga et al 1993). The projections of the subparaventricular zone are outlined above. As noted by Watts (1991), this area is ideally suited to directly regulate neuroendocrine functioning, autonomic control and temperature regulation. The rostral projections, not shown in Fig. 4, reinforce those of the paraventricular thalamic nucleus. The retrochiasmatic area, lying directly behind the SCN, projects quite widely to hypothalamic areas overlapping the subparaventricular zone projections, to basal forebrain areas that themselves project widely to the neocortex, to brainstem areas that are involved in autonomic regulation and to spinal cord grey (M. M. Moga & R. Y. Moore, unpublished work). Lastly, the IGL projects to midline thalamic nuclei involved in regulation of behavioural state, to subthalamic areas probably participating in motor integration and to pretectal and tectal zones involved in visual functioning (R. P. Weis, M. M. Moga & R. Y. Moore, unpublished work). Even with this widespread pattern of projections, it is not entirely clear how the SCN regulates the temporal organization of behavioural state. The areas most likely to exert this control are the hypothalamus and the midline thalamus, but this issue clearly requires further study.

Conclusions

The circadian timing system has three principal components, entrainment pathways, pacemakers and efferent pathways that couple the pacemaker to effector systems that exhibit circadian functioning. Light is the primary circadian zeitgeber. The pathway mediating direct effects of light on the pacemakers is the RHT, which arises from a distinctive set of retinal ganglion cells. These receive photic information from the circadian photoreceptors and project to the SCN and adjacent hypothalamic areas, particularly the subparaventricular zone and the retrochiasmatic area. Collaterals from one subset of the ganglion cells project to the IGL, and those from a second set to the olivary pretectal nucleus. The RHT is sufficient for entrainment. The GHT is another important entrainment pathway, which integrates photic and non-photic input to regulate SCN functioning. Other potential entraining pathways include the serotonergic input from the midbrain raphe, input from other hypothalamic nuclei and a humoral input, principally represented by melatonin. All these inputs to the SCN are probably involved in a continuing, precise regulation of the phase and period of the pacemaker.

The suprachiasmatic nuclei are the only circadian pacemakers as yet identified in the mammalian brain. Since the nuclei are paired and each can function independently to maintain circadian functioning, this is a multiple pacemaker system. Further, the suprachiasmatic nuclei themselves are composed of sets of independent oscillators, which are coupled into a network that operates as a pacemaker. The nuclei produce an apparently simple output expressed as a high firing rate during the day and a low firing rate at night.

The efferent projections of the SCN are largely to adjacent hypothalamic structures, with only limited projections to basal forebrain, thalamus and periaqueductal grey. The major projections are to areas that themselves receive retinal input and project reciprocally to the SCN—the IGL, retrochiasmatic area, subparaventricular zone and paraventricular thalamic nucleus. These areas provide a widely disseminated set of pathways that innervate critical areas expressing circadian functioning and are situated to provide the anatomical substrate for control by the SCN of effector functioning.

Thus, the circadian timing system is composed of three major divisions: a retinal component with photoreceptors and ganglion cells that are the origin of the RHT; the SCN, the circadian pacemaker; and a series of hypothalamic and thalamic nuclei that provide the circadian regulation of effector functioning.

Acknowledgement

The research in my laboratory that is cited in this review was supported by NIH grant NS-16304. I am grateful to J. Patrick Card and Joan Speh for their contributions to the work.

References

Card JP, Moore RY 1989 Organization of lateral geniculate–hypothalamic connections in the rat. J Comp Neurol 284:135–147

Card JP, Whealy ME, Robbins AK, Moore RY, Enquist LW 1991 Two alpha herpes virus strains are transported differentially in the rodent visual system. Neuron 6:957–969

Cassone VM, Speh JC, Card JP, Moore RY 1988 Comparative anatomy of mammalian suprachiasmatic nucleus. J Biol Rhythms 3:71–91

Foster RG, Provencio I, Hudson D, Fiske S, DeGrip W, Menaker M 1991 Circadian photoreception in the retinally degenerate mouse (rd/rd). J Comp Physiol A Sens Neural Behav Physiol 169:39–50

Hendrickson AE, Wagoner N, Cowan WN 1972 An autoradiographic and electron microscopic study of retino-hypothalamic connections. Z Zellforsch Mikrosk Anat 135:1–26

Honma KI, Honma S, Hiroshige T 1987 Activity rhythms in the circadian domain appear in suprachiasmatic nucleus lesioned rats given methamphetamine. Physiol & Behav 40:767–774

Inouye SIT, Kawamura H 1979 Persistence of circadian rhythmicity in a mammalian hypothalamic 'island' containing the suprachiasmatic nucleus. Proc Natl Acad Sci USA 76:5962–5966

Johnson RF, Moore RY, Morin LP 1988a Retinohypothalamic projections in the rat and hamster demonstrated using cholera toxin. Brain Res 462:301–312

Johnson RF, Moore RY, Morin LP 1988b Loss of entrainment and anatomical plasticity after lesions of the hamster retinohypothalamic tract. Brain Res 460:297–313

Johnson RF, Moore RY, Morin LP 1989 Lateral geniculate lesions alter circadian activity rhythms in the hamster. Brain Res Bull 22:411–422

Klein DC, Moore RY 1979 Pineal *N*-acetyltransferase and hydroxyindole-*o*-methyltransferase: control by the retinohypothalamic tract and the suprachiasmatic nucleus. Brain Res 174:245–262

Lehman MN, Silver R, Gladstone WR, Kahn MR, Gibson M, Brittman EL 1987 Circadian rhythmicity restored by neural transplant. Immunocytochemical characterization of the graft and its integration with the host brain. J Neurosci 7:1626–1638

Meijer JH, Rietveld WJ 1989 Neurophysiology of the suprachiasmatic circadian pacemaker in rodents. Physiol Rev 69:671–707

Moga MM, Weis RP, Moore RY 1993 Efferent projections of the thalamic paraventricular nucleus. Soc Neurosci Abstr 19:1615

Moore RY 1992a The organization of the human circadian timing system. Prog Brain Res 93:101–117

Moore RY 1992b The enigma of the geniculohypothalamic tract: why two visual entraining pathways? J Int Cycle Res 23:144–152

Moore RY, Bernstein ME 1989 Synaptogenesis in the rat suprachiasmatic nucleus demonstrated by electron microscopy and synapsin I immunoreactivity. J Neurosci 9:2151–2162

Moore RY, Card JP 1990 Neuropeptide Y in the circadian timing system. Ann NY Acad Sci 611:247–257

Moore RY, Card JP 1994 The intergeniculate leaflet: an anatomically and functionally distinct subdivision of the lateral geniculate complex. J Comp Neurol 344:403–430

Moore RY, Eichler VB 1972 Loss of a circadian adrenal corticosterone rhythm following suprachiasmatic lesions in the rat. Brain Res 42:201–206

Moore RY, Lenn NJ 1972 A retinohypothalamic projection in the rat. J Comp Neurol 146:1–14

Moore RY, Speh JC 1993 GABA is the principal neurotransmitter in the mammalian circadian system. Neurosci Lett 150:112–116

Moore RY, Halaris AE, Jones BE 1978 Serotonin neurons of the midbrain raphe: ascending projections. J Comp Neurol 180:417–438

Mrosovsky N 1995 A non-photic gateway to the circadian clock of hamsters. In: Circadian clocks and their adjustment. Wiley, Chichester (Ciba Found Symp 183) p 154–174

Pickard GE 1982 The afferent connections of the suprachiasmatic nucleus of the golden hamster with emphasis on the retinohypothalamic projection. J Comp Neurol 211:65–83

Pickard GE 1985 Bifurcating axons of retinal ganglion cells terminate in the hypothalamic suprachiasmatic nucleus and the intergeniculate leaflet of the thalamus. Neurosci Lett 55:211–217

Ralph MR, Foster RG, Davis FC, Menaker M 1990 Transplanted suprachiasmatic nucleus determines circadian rhythms. Science 247:975–978

Reme CE, Wirz-Justice A, Terman M 1991 The visual input stage of the mammalian circadian pacemaking system. I. Is there a clock in the mammalian eye? J Biol Rhythms 6:5–29

Schwartz WJ, Gross RA, Morton MT 1987 The suprachiasmatic nuclei contain a tetrodotoxin-resistant circadian pacemaker. Proc Natl Acad Sci USA 84:1694–1698

Shibata S, Moore RY 1988 Electrical and metabolic activity of suprachiasmatic nucleus neurons in hamster hypothalamic slices. Brain Res 438:374–378

Shibata S, Tsuneyoski A, Hamada T, Tominaga K, Watanabe S 1992 Phase-resetting effect of 8-OH-DPAT, a serotonin$_{1A}$ receptor agonist, on the circadian rhythm of firing rate in the rat suprachiasmatic nuclei *in vitro*. Brain Res 582:353–356

Smale L, Michels KM, Moore RY, Morin LP 1990 Destruction of the hamster serotonergic system by 5,7-DHT: effects on circadian rhythm phase, entrainment and response to triazolam. Brain Res 515:9–19

Stephan FK, Zucker I 1972 Circadian rhythms in drinking behavior and locomotor activity of rats are eliminated by hypothalamic lesions. Proc Natl Acad Sci USA 69:1583–1586

Stephan FK, Swann JM, Sisk CL 1979 Anticipation of 24-hr feeding schedules in rats with lesions of the suprachiasmatic nucleus. Behav Neural Biol 25:346–363

van den Pol AN 1980 The hypothalamic suprachiasmatic nucleus of the rat: intrinsic anatomy. J Comp Neurol 191:661–702

van den Pol AN, Tsujimoto KL 1985 Neurotransmitters of the hypothalamic suprachiasmatic nucleus: immunocytochemical analysis of 25 neuronal antigens. Neuroscience 15:1049–1086

Watts AG 1991 The efferent projections of the suprachiasmatic nucleus: anatomical insights into the control of circadian rhythms. In: Klein DC, Moore RY, Reppert SM (eds) The suprachiasmatic nucleus: the mind's clock. Oxford University Press, New York, p 77–106

DISCUSSION

Lewy: Are the anatomical pathways that mediate the acute suppressant effect of light on melatonin production the same as those that mediate entrainment?

Moore: I do not think there are two anatomically different pathways. The light input which is excitatory at night occurs on a baseline of an inactive SCN. Light increases the firing rate immediately. That information is passed directly into the paraventricular nucleus, which projects directly to the intermedial lateral cell column in the upper thoracic cord and thence to the pineal. The light input, because it is exciting an inhibitory neuron, inhibits the paraventricular output, and that inhibition turns off the pineal. The pineal requires a constant sympathetic input to maintain melatonin production.

Lewy: Is there no direct retinoparaventricular pathway that might mediate the suppressant effect of light?

Moore: No.

Roenneberg: Some of the output nuclei get both optic information and information from the SCN. Is that a good candidate for an anatomical pathway involved in masking effects?

Moore: There are several different kinds of masking, but certainly that would be one candidate mechanism which could produce a form of masking.

Menaker: Is there a projection from the lateral septum back to the SCN? You mentioned projections from all the other areas, but not from that one.

Moore: There are certainly projections from the SCN to the septal area (Watts et al 1987). The demonstration of septal–SCN projections has been more problematic.

Menaker: I asked the question because the lateral septal area is, at the moment, a hot candidate for the location of extraretinal photoreceptors in non-mammalian vertebrates (Foster et al 1993). If that turns out to be true, there's got to be a projection from those photoreceptors, direct or indirect, back to the SCN.

Moore: I don't mean to exclude any other potential components of the system; I can only talk about those for which we have done sufficient work to fulfil the criteria that I set out.

Redfern: What do we know about the neurotransmitters involved in the pathways from the SCN to the IGL?

Moore: Everything from the SCN is GABAergic with co-localized peptide. In general, the function of co-localized peptides is to enhance, prolong or otherwise slightly modify the effect of the small molecule transmitter.

Miller: In your work with pseudorabies virus, where you get the labelling of contralateral ganglion cells (Moore et al 1994), is there ever any retrograde transsynaptic transport into the bipolar layer, or does it stop at the ganglion cells?

Moore: That is a difficult question to answer. There are circumstances in which we have seen labelling of elements further up in the retina than the ganglion cell layer. We have not made any systematic attempt to characterize the labelled elements and it would be difficult to determine whether labelling of elements distal to the ganglion cell layer represented retrograde transynaptic transport or direct invasion by the virus.

Miller: What I am driving at is that if the ganglion cells are relatively specialized, W cells or whatever, and the photoreceptor for the circadian system is relatively specialized too, you might also expect the intermediate neuron in the bipolar layer to be a specialized component of the circadian visual system.

Moore: You would think that the entire retinal circuitry would be specialized. However, we do not know what the circadian receptor is. It could be in the ganglion cell layer, or somewhere else in the retina other than the receptor layer. This is a difficult problem.

Miller: From your maps of the efferents from the SCN can you find some combination of lesions in output structures that would reproduce the effects of an SCN lesion? You ought to be able to produce something like the arrhythmicity resulting from a direct SCN lesion.

Moore: We have just completed a set of studies in which we made highly selective lesions of the paraventricular nucleus of the hypothalamus. The rats with lesions have low nocturnal pineal melatonin content, but perfectly normal activity and temperature rhythms. Your suggestion is, however, perfectly reasonable.

Gillette: The issue of the effective agent in the pathway from the IGL to the SCN is an interesting one. You described an NPY-induced phase response curve, and of course you would expect GABA to be co-released from the NPY-containing neurons. Dr Mrosovsky has some interesting data suggesting that you absolutely have to have neuropeptide Y to get the non-photic phase shift. How would you interpret that result?

Moore: I was surprised that perfusion of the peptide into the SCN appears to have the same effect as stimulating the pathway, because we expect the primary message to be passed by the small molecule neurotransmitter.

Czeisler: When you transect the retinohypothalamic tract but still have visual input, do you get entrainment through the IGL?

Moore: No; those animals do not become entrained. The pathway to the IGL is intact, but we do not know whether or not the pathway back from the IGL to the SCN is intact. In all likelihood, the RHT lesions also transect the GHT.

Cassone: You have suggested that in the human SCN substance P may be a retinohypothalamic tract transmitter (Moore & Speh 1993). I usually think of substance P as an inhibitory neurotransmitter. Is it possible that glutamate and an inhibitory peptide, substance P, act at the level of the SCN? Is that a general finding in mammals?

Moore: Takatsuji & Tohyama (1993) described a substance P-containing set of retinal ganglion cells that project to the SCN in the rat. For a variety of reasons that I won't go into, we had several years ago stained human SCN for substance P, so I went back and looked at the material again. There is a dense substance P plexus that overlays the area of VIP-containing neurons, exactly where one would expect the human RHT to be. This does not prove, of course, that the substance P axons are of retinal origin but it is a suggestive observation. You are correct that substance P has generally been found in association with GABA, but there is no absolute association of a peptide with a particular small molecule transmitter.

Cassone: In the house sparrow SCN, one of the densest plexuses in the retinohypothalamic tract contains substance P (Cassone & Moore 1987).

Miller: The recently published phase response curve produced by substance P looks like that produced by light (Shibata et al 1992), and glutamate seems to produce a similar response curve, so this appears to be a second example of a small transmitter and a co-localized peptide doing more or less the same thing.

Moore: Yes, if that substance P phase response curve is entirely correct.

Redfern: There are plenty of examples in which substance P is excitatory—it's not necessarily inhibitory.

Mikkelsen: You didn't say anything about possible receptors in the mammalian SCN. Comparison of the immunohistochemical evidence about the content of various substances in afferent inputs of the SCN with that from autoradiographic and *in situ* hybridization studies of various receptors indicates a striking mismatch. The major transmitters in the afferents—glutamate, GABA, serotonin and neuropeptide Y—do not appear, perhaps except for glutamate, to have a corresponding receptor in the SCN. Also, a functional dopamine receptor (D_1) has been demonstrated in the rat SCN (Reppert 1995, this volume), but no dopamine-immunoreactive or tyrosine hydroxylase-immunoreactive nerve fibres have been described in the SCN.

Moore: Negative evidence is never entirely compelling. We know many examples of receptor mismatch—where there are receptors with no endogenous ligand, or endogenous ligand where there are no receptors. We can rely more on the methodology that demonstrates the presence of endogenous ligands. Evidence of ligand binding in an autoradiogram is not necessarily a demonstration of a receptor. Binding of antibodies against a known component of a receptor is more compelling.

Dunlap: My impression is that time information is generated in the SCN by a bunch of individual cells that are talking to one another, apparently not synaptically. How do they talk to one another? Time information from the SCN goes out to the rest of the brain, and it's an open question whether that is synaptic or not. You implied that the information is transmitted synaptically. What's the evidence, if any, that it doesn't go any other way?

Moore: There is only one instance in which there is fairly strong evidence that a humoral signal is involved, in the case of transplanted tissue. I know of no other evidence to support the view that information transfer within the SCN or from the SCN to other areas is anything other than synaptic.

Menaker: What is the evidence that there's humoral output from SCN transplants?

Moore: The variable location of the transplants and the relatively sparse connections that the transplanted tissue makes with adjacent brain tissue make it difficult to conceive that anything that even barely approximates the normal circuitry is restored.

Ralph: I agree. However, the connections may not have to be many in order to get function, as with gonadotropin-releasing hormone (GnRH) neuron transplants in the hypogonadal (*hpg*) mouse (Krieger et al 1982). We have seen coordination between the transplanted SCN tissue and the host indicative of a two-way interaction, where the graft itself is a long distance from the host SCN and there's little indication of neuronal outgrowth from the graft.

With the transplant work, everyone focuses on the big plug of tissue, but more than one bit of SCN will actually be transplanted—it's not all going to stay as a solid piece of tissue. You can't reach a firm conclusion from looking only at that one major piece of tissue; other pieces that you don't find may be better integrated.

Moore: Your comment about the hypogonadal mouse was misleading. Although it is true that six GnRH neurons can restore reproductive functioning in the hypogonadal mouse, they must make direct connections to the median eminence, to the zone where they can come into contact with the portal plexus.

Menaker: It is quite odd to imagine that a behaviour as complex as locomotor activity, which is normally driven synaptically, can be driven humorally by a graft. That doesn't make sense.

Ralph: The transplanted tissue doesn't restore the rhythm to its pre-lesion state. What usually is restored is the sinusoidal increase and decrease in the level of activity. There is occasionally good restoration, but we haven't found evidence for extensive neuronal outgrowth even in these cases.

Miller: There may be a subsidiary back-up system for neurotransmission. Bill Schwartz's results with tetrodotoxin *in vivo* show that as long as tetrodotoxin is pumped into the SCN so that synaptic transmission is blocked, the animal is arrhythmic (Schwartz et al 1987). The primary output of the SCN must be synaptic. However, we aren't forced to make any such assumptions about the endogenous clock mechanism itself. For a function as important as biological timing it would not be surprising if there were some sort of back-up paracrine system in addition to synaptic transmission for getting the signal out.

Gillette: For how many behaviours has restoration of function after SCN transplantation into arrhythmic lesioned hosts been examined?

Ralph: Those I'm aware of are rhythms of locomotor activity, sleep–wake, drinking and feeding and body temperature (for review, see Lehman & Ralph 1993).

Moore: Most people have not found restoration of body temperature rhythm.

Czeisler: Is that rhythm secondary to activity?

Menaker: It has to be secondary to activity, because when those animals run on wheels or are active in their cages their temperature goes up—there's just no way around that.

Moore: None of the hormonal rhythms has been restored.

Waterhouse: The point remains whether the temperature rhythm has been restored primarily and that this could influence the activity secondarily. The masking effect of activity might be comparatively trivial.

Ralph: What may be restored in a rhythmic fashion is general activation. All the rhythms that I mentioned are highly correlated in time with locomotor activity, so it's hard to look at causality. The animal has to perform all of these functions, physiological and behavioural, in a coordinated manner. I wouldn't be surprised if the rhythms were all restored by SCN grafts. Dr Moore is correct that the endocrine functions which clearly require neural connectivity, such as melatonin rhythms, are not restored.

Waterhouse: If a transplant makes some sort of neuronal outgrowths, are there problems in deciding whether the right kind of wiring up has resulted?

Moore: Yes; that's a terrific problem. There is a good marker for SCN efferents, the VIP neurons; there are not many other VIP neurons around so you can see what is coming out of the transplant by looking at the axonal plexus produced by the neurons. The number of fibres that go into the host brain in many of these transplants is vanishingly small. We have one in which the transplant was located at the junction between the lateral ventricle and the third ventricle in which we could see about three to five fibres going down, interestingly enough, into the paraventricular nucleus of the thalamus, but they were precious few.

Armstrong: The point about neuronal connections does not apply only to the SCN. Zimmerman & Menaker (1979) showed restoration of the activity rhythm of pinealectomized sparrows after transplantation of a donor pineal to the anterior eye chamber; rhythmicity was restored within a few cycles, in which time there could not have been any sympathetic innervation. Some sort of humoral/hormonal agent must have been involved rather than nervous reinnervation.

Cassone: There are demonstrated rhythms in noradrenaline in the anterior chamber of the eyes of several organisms, including chickens; there may not have been innervation, but there's clearly a noradrenaline rhythm.

Menaker: Not in a pinealectomized bird, though.

Armstrong: How does putting a donor pineal back in restore rhythmicity (a rhythm as distinct from a constant output)?

Cassone: It is true that in the pinealectomized sparrow pineal transplantation confers both the rhythmicity and circadian phase within one or two days. This strongly supports the view that the pineal *output* is humoral. However, it does not preclude the possibility that the transplanted pineal gland receives neurally derived information from the host. This information *may* arise from the release of sympathetic noradrenaline in the aqueous humour.

Menaker: We always felt strongly that the effect of a transplanted pineal was the result of a hormonal output, presumably melatonin, from the pineal gland (Chabot & Menaker 1992).

Armstrong: Would that be *rhythmic* melatonin release?

Menaker: Yes.

Redfern: What, if anything, does transplant experimentation tell us about possible mechanisms of feedback from the SCN to the pineal?

Ralph: There is not much information along those lines because the neural connection with the pineal isn't re-established. We are looking into the possibility that in circadian chimeras in which the graft and host SCN are able to influence each other, the grafted SCN could influence the pineal indirectly, through the intact nucleus of the host.

Reppert: What happens in pinealectomized host animals, which no longer have a melatonin rhythm that may influence any interactions?

Ralph: We have not done SCN transplants into pinealectomized animals. There is a suggestion that periodic melatonin can decrease the latency to recovery of overt rhythmicity after grafting.

References

Cassone VM, Moore RY 1987 Retinohypothalamic projection and suprachiasmatic nucleus of the house sparrow, *Passer domesticus*. J Comp Neurol 266:171–182

Chabot CC, Menaker M 1992 The effects of physiological cycles of infused melatonin on circadian rhythmicity in pigeons. J Comp Physiol A Sens Neural Behav Physiol 170:615–622

Foster RG, Garcia-Fernandez JM, Provencio I, DeGrip WJ 1993 Opsin localization and chromophore retinoids identified within the basal brain of the lizard *Anolis carolinensis*. J Comp Physiol A Sens Neural Behav Physiol 172:33–45

Lehman MN, Ralph MR 1993 Modulation and restitution of circadian rhythms. In: Dunnett SB, Bjorklund A (eds) Functional neural transplantation. Raven Press, New York, in press

Krieger DT, Perlow MJ, Gibson MJ et al 1982 Brain grafts reverse hypogonadism of gonadotropin-releasing hormone deficiency. Nature 298:468–471

Moore RY, Speh JC 1993 A putative retinohypothalamic tract (RHT) in the human demonstrated by substance P (SP) immunoreactivity. Soc Neurosci Abstr 19:1614

Moore RY, Speh JC, Card JP 1994 The retinohypothalamic tract originates from a distinct subset of retinal ganglion cells. J Comp Neurol, in press

Reppert SM 1995 Interaction between the circadian clocks of mother and fetus. In: Circadian clocks and their adjustment. Wiley, Chichester (Ciba Found Symp 183) p 198–211

Schwartz WJ, Gross RA, Morton MT 1987 The suprachiasmatic nuclei contain a tetrodotoxin-resistant pacemaker. Proc Natl Acad Sci USA 84:1694–1698

Shibata S, Tsuneyoshi A, Hamada T, Tominaga K, Watanabe S 1992 Effect of substance P on circadian rhythms of firing activity and the 2-deoxyglucose uptake in the rat suprachiasmatic nucleus *in vitro*. Brain Res 597:257–263

Takatsuji K, Tohyama M 1993 The development and innervation of two neuropeptides (substance P and neuropeptide Y) immunoreactive fibers in rat suprachiasmatic nucleus. In: Nakagawa H, Oomura Y, Nagai K (eds) New functional aspects of the suprachiasmatic nucleus of the hypothalamus. John Libbey, London, p 43–51

Watts AG, Swanson LW, Sanchez-Watts G 1987 Efferent projections of the suprachiasmatic nucleus. 1. Studies using anterograde transport of the *Phaseolus vulgaris* leucoagglutinin in the rat. J Comp Neurol 258:204–229

Zimmerman NH, Menaker M 1979 The pineal: a pacemaker within the circadian system of the house sparrow. Proc Natl Acad Sci USA 76:999–1003

General discussion I

Lewy: Dr Block, does the dark pulse phase response curve (PRC) seem anything other than the complement of the light PRC? Do light and dark pulses stimulate different chemical systems?

Block: In *Bulla*, we have never been able to generate a PRC in response to a dark pulse. If you leave the light on for 12 h the cells eventually adapt, so when you turn the light off you get only a small change in membrane potential.

Moore: The one clear example of dark pulse PRC is in the chicken pineal (Zatz & Mullen 1989).

Waterhouse: The question of how many types of PRCs there are was raised earlier, and we divided many of the PRCs into two groups, those produced by stimuli that hyperpolarize the membrane and those produced by stimuli that depolarize the membrane. What about the other PRCs? It's easy to see intuitively that hyperpolarizing and depolarizing membranes is the way the system might work, but then there appear to be these different types of PRCs. Are these an important category, which teaches us something about the system, or are they of no physiological relevance?

Turek: The vast majority of agents induce one of the two types of PRC, i.e., either a 'light pulse like' or a 'dark pulse like' PRC. Of course, there are slight differences between the curves generated in response to dark pulses on a running wheel, for example, but the differences are less important to me than the similarities. We are using such vastly different agents—running wheels, dark pulses or drugs—yet we see remarkable similarities in their phase shifting effects on the clock. Those agents which tend not to simulate one of these two types of PRC could still have physiological significance; perhaps I should make the point that the PRC for melatonin in rodents includes advances at a precise circadian time, somewhat outside the time of the normal rise in melatonin, and phase delays are not observed. There is a suggestion that exercise produces phase delays in humans but not advances. Protein synthesis inhibitors, at least in the hamsters that Dr Takahashi and I have looked at, produce a PRC that looks somewhat like a dark pulse PRC but its amplitude is different—we see small advances and large delays, the opposite to what we see in response to activity-inducing stimuli.

I would certainly like to think that the response to melatonin is 'physiological'. Does melatonin alter membrane conductivity?

Cassone: There's no evidence for melatonin receptors in the adult Syrian hamster suprachiasmatic nucleus (SCN). However, melatonin is inhibitory in several systems.

Gillette: In my paper, I shall suggest that the strongest behavioural stimuli dramatically depolarize or hyperpolarize the cells at the times of day when they act. However, the regulation of clocks, particularly the mammalian clock, is much more complex. There are many regulators coming into the SCN, some of which come in at the same time of day and can influence the clock in opposite ways. These photic and non-photic phase response relationships are the major types of response that could be made to a major environmental stimulus, be it light or arousal at different times of the day, but there is more complexity in the system than simple depolarization or hyperpolarization.

Waterhouse: The question still is whether these many influences that are coming in are acting through mechanisms other than hyperpolarization and depolarization. Hyperpolarization and depolarization could be produced in many different ways.

Gillette: That's true, and we are not yet in a position to address that question directly. Most of the receptors we are looking at are coupled to various second messenger pathways, so the complexity of the responses depends on which particular pathways are simultaneously activated or inhibited.

Edmunds: Dr Waterhouse asked whether PRCs can be divided into two physiological categories—hyperpolarizing and depolarizing—but I think you were also questioning whether the different types of PRC as categorized by Winfree (1975) have physiological significance. That's a more difficult question to answer, and Winfree (1980) himself in his book argues both sides. If one has an apparent type 0 or an apparent type 1 curve for a particular function in different organisms, is that significant? Does that change the category of a PRC? You can sometimes generate a family of PRCs going from type 1 to type 0 and vice versa depending on the conditions. Sometimes I am not convinced there is a physiological difference between a type 1 and a type 0 curve.

Hastings: When we plot PRCs we plot the time at which we apply the stimulus, not the time at which the stimulus ultimately affects the clock. We may apply the stimulus at circadian time (CT) 4 or CT 6, but we don't actually know, particularly with behavioural stimuli, e.g., increased wheel-running, which may have indirect actions on the clock when the clock is shifting, when the sensitive phase is. Another trap we might fall into is that it's easy to show what *can* shift the clock. Melatonin, for example, can shift the clock when injected into rats at CT 10, but when would that be physiological, since melatonin is synthesized and released only after CT 12 in rodents? Perhaps in the case of entrainment in the hamster fetus there is a true physiological action, but it is perhaps not physiological in the free-living adult where there would be no endogenously secreted melatonin at CT 10.

Lewy: In humans, exogenous melatonin can stimulate a phase shift in the pacemaker during at least the latter half of the endogenous melatonin secretory episode.

Cassone: Besides that, the time at which light is present (i.e., daytime) is also the time when most organisms are least sensitive to light. In contrast, the phase-shifting effects of light are greatest during subjective night, when light is unlikely to be encountered under natural conditions.

Hastings: That's not necessarily the case if we assume that entrainment occurs by either morning or evening effects, with just a short pulse of light inducing a daily delay or advance. The entraining responses to light may occur at either end of subjective night, depending on the individual period. The important point is that the oscillator will free-run relative to the light–dark cycle until light impinges on the photosensitive zone. The circadian clock of an animal can never free-run, i.e., hold a different period relative to its melatonin signal because they express identical periods. Therefore, endogenous melatonin can never be encountered at CT 10–11, the only sensitive phase we currently know of.

Cassone: The amplitude of the phase shift one obtains following illumination at either subjective dawn or dusk is small in comparison with the effects of light administered in the depths of night.

Hastings: But that may be all that's required under natural conditions.

Cassone: The same could be true of almost any of the other agents, including melatonin, that can induce phase shifts and are present anatomically and physiologically.

Miller: A morphological classification of PRCs ought to be topological. Small changes in form should be allowed—this is the essence of 'deformability' in topology—but radically different shapes probably represent radically different physiology. If you accept that much, you have to add to the list of PRCs those which are completely or almost completely composed of a phase advance or a phase delay region. There are numerous examples of both such monsters, and we have to treat them as separate categories because they cannot be topologically transformed into light-like or dark-like PRCs.

Block: It is possible to generate many types of PRC, but whether a particular PRC has any functional significance to the animal must always be determined. We know that in the *Bulla* ocular pacemaker the PRC produced in response to light pulses is generally similar to the phase-dependent responses of the animal's behaviour to light pulses. Thus, the PRC produced by light has functional significance in that it allows the animal to synchronize stably to a periodic environmental signal. In contrast, the PRC produced in response to membrane hyperpolarization may be an experimental oddity, something we do experimentally to the isolated eye, when in reality the circadian pacemaker neurons may not normally experience prolonged hyperpolarizations.

This is a complicated issue because we do not yet understand the interaction of various neurotransmitters and neuromodulators that individually can generate 'depolarization-like' or 'hyperpolarization-like' PRCs.

As a case in point, serotonin applied to the *Aplysia* eye at the same time as a light pulse modulates the size of the light-induced phase shift (Colwell 1990). It seems quite likely that serotonin in fact plays a minor role in modulating the amplitude of phase shifts induced by light. However, if you apply serotonin alone, without the concomitant light pulse, you obtain a PRC very similar to that obtained for hyperpolarization, 180° shifted on the time axis in relation to a light pulse-induced curve. This PRC may really be an epiphenomenon. If pacemaker neurons are entrained by periodic light-induced depolarizations leading to an increase in Ca^{2+} flux, serotonin-induced hyperpolarization of these same neurons will result in a very different PRC; but this PRC may have no relevance whatsoever for the animal, as the neuromodulator normally acts in concert with light and not as a separate agent.

Kronauer: If one lesions the hamster SCN and restores rhythm with a transplant, the phase response to light cannot be restored because the retinohypothalamic tract has been destroyed by the lesion. Do any of the effects of dark pulses survive? Might there be an input to the transplanted SCN of another variety?

Mrosovsky: Canbeyli et al (1991) found that triazolam, the effects of which may be mediated by activity or a correlate, and which produces a PRC like that produced by a dark pulse, could not cause a phase shift in hamsters with an engrafted SCN. They inferred that triazolam must normally act outside the SCN on some structure which sends information to the SCN. However, there was no quantitative check on the amount of activity that the triazolam produced. Perhaps there was too little activity induced in the operated animals to produce the phase shift. Even if there had been lots of activity, failure to find a phase shift in a transplant situation would tell you more about the ability of the transplant to function as a normal SCN than about the organization of the normal system.

Moore: We showed several years ago that the geniculohypothalamic tract is needed for triazolam to act (Johnson et al 1989). It is difficult to interpret the lack of an effect of triazolam in the animals carrying a transplanted SCN as it is unlikely that there would be innervation by the geniculohypothalamic tract.

Turek: Would we all agree that all our results are consistent with the inputs to SCN, from light and the induction of activity, being neuronally mediated? There is no evidence that those two inputs are humoral. Melatonin is in a separate class.

Takahashi: Rae Silver (personal communication) has told me that anisomycin does produce a phase shift in animals carrying transplanted SCN.

Czeisler: What about melatonin?

Takahashi: I don't know about that.

Miller: Melatonin should work, as should serotonin.

Cassone: The effects of melatonin depend on species and the age of the clock. Silver & Romero (1989) have reported acceleration of the restoration of the circadian rhythm in SCN-lesioned hamsters receiving fetal SCN transplants by daily melatonin injections, and have suggested that animals that received morning injections of melatonin had a phase different from those receiving evening injections. However, after the injections it was difficult really to tell where the phase was.

Armstrong: Are you all now agreed that the only circadian pacemaker in mammals is the SCN? In other words, are all the old reports of rhythms surviving in cultured hamster adrenal glands, isolated rat heart, human transplanted hearts, etc. (Armstrong 1989a) now discounted as non-oscillatory and non-pacemaker functions?

Miller: I don't think we have ever excluded slave oscillators.

Menaker: The food-entrainable oscillator persists, although not very well, after SCN lesion. Work by Terman et al (1993) suggests strongly that the circadian oscillations in the rat eye persist after SCN lesion. The work with isolated adrenals has not been replicated. I don't think this is a closed issue.

Moore: There's no reason to believe that there cannot be pacemakers in different tissues. By far the best characterized one is that in the rat eye; there is a rhythm in visual sensitivity that persists after SCN lesions (Terman & Terman 1985). For brain, the food-entrainable oscillator is the only well established non-SCN phenomenon.

Menaker: What about the methamphetamine-induced oscillation in SCN-lesioned rats (Honma et al 1987).

Miller: That's not an oscillator—that's an hour-glass.

Menaker: Do you say that because the animals drink and therefore are in a sense self-stimulating?

Moore: Yes, and they exhaust themselves then go to sleep.

Mrosovsky: But Honma et al (1987) say that methamphetamine restores the rhythm even when the drug is chronically administered by an osmotic minipump (Honma et al 1987).

Moore: In any event, none of the induced oscillators persists for any length of time in the absence of the stimulus that induces it. It is unlikely that they have any important function in the intact animal.

Armstrong: When I have suggested that melatonin sends signals to these other slave oscillators (Armstrong 1987, 1989b), people have usually jumped on me and said that they don't exist. That's the way the discussion seemed to be going here.

Mikkelsen: I agree, but there may be a reason why we have not moved far on this problem. On the basis of Dr Moore's comprehensive analysis of the neuroanatomy of the circadian timing system we may conclude that if there is another pacemaker(s) or slave oscillator(s) in the brain, the best candidates would be those structures which are anatomically connected to the SCN, e.g., the intergeniculate leaflet (IGL), the paraventricular thalamic nucleus or the septum. I think the reason why we have not yet seen pacemaker activity or slave oscillator capabilities in these regions is that they are very small (e.g., the IGL) or that the few neurons that might be responsible for this circadian function are spread over relatively large areas (e.g., the septum). That makes the techniques (e.g., 2-deoxyglucose utilization, transplantation, etc.) that have been used to locate the circadian pacemaker in the SCN inappropriate.

Ralph: In my presentation I emphasized the difference between oscillator and pacemaker. This discussion started with a question about pacemakers, not oscillators. There's plenty of evidence for oscillators, in the eyes for example, but the question was whether these are part of the pacemaker system, and whether or not they can be slave oscillators. I suspect that there are many oscillators yet to be discovered, but whether or not they are pacemakers is an entirely different question. It does seem reasonable to look for slave oscillators in anatomically connected structures. After all, overt rhythms are synchronized with the SCN pacemaker and communication pathways must exist.

Waterhouse: Dr Dunlap, to what extent do you see a link between the work in *Neurospora* and the mammalian work? Is there still a chasm between the genetics and the anatomical pathways and evidence for histochemical diversity in the SCN?

Dunlap: That of course is the $64 000 question. This question comes down to whether everybody really honestly in their heart of hearts thinks that clocks are assembled at the level of the cell, or whether in some cases a network is needed. If the clock is assembled within the cell, the chasm is not so great and the clock in mammals will probably work in a similar way to the clock in *Neurospora*. One thing that 15 years of molecular biology has shown us is that cells do the things they all do in much the same way—they divide in the same way, they interpret signals from heterotrimeric G proteins in the same ways, they make DNA, RNA and proteins in the same way and carry out glycolysis in the same way. If keeping time is something that most cells do, they will follow probably the same algorithm and have similar components, assembled in a similar kind of loop, talking to each other in the same sorts of ways. However, if timekeeping is a network phenomenon in the SCN but a cellular phenomena in *Gonyaulax*, *Neurospora* and *Drosophila*, the whole game is open.

Takahashi: We already have a few examples of the circadian clock being a cellular phenomenon in multicellular organisms. Dr Block's group has published a definitive paper showing that *Bulla* retinal neurons are circadian pacemakers at the single neuronal level (Michel et al 1993), and that there are a hundred of these which are electrotonically coupled. The network is still needed in a multicellular organism to allow coupling and a coherent output from the cellular oscillators. Network properties will be required in the SCN, the pineal and in the *Bulla* retina.

Dunlap: This brings us back to the distinction between an oscillator and a pacemaker. If you assemble a bunch of oscillators, you get a pacemaker; but if each unit part (i.e., each oscillator) is the same, and if each unit is a cell, that leads to one kind of answer to the question of how work on *Neurospora* relates to vertebrate clocks—the *Neurospora* clock should be similar to other eukaryotic clocks. However, if each unit oscillator, for instance, in the SCN consists of several cells communicating with each other in order to make a feedback loop, then that leads to a different type of answer—intracellular

General discussion I

feedback loop clocks such as *Neurospora*, pinealocytes and *Bulla* neurons may work differently from such (hypothetical) intercellular network clocks. You are right that, on balance, the evidence is tending strongly towards an intracellular basis for the feedback loops of all primary oscillator units. Such intracellular feedback loop clocks are consistent with the *Bulla* retinal neuron circadian oscillator (Michel et al 1993) and with the pineal and pinealocyte oscillator (Deguchi 1979, Robertson & Takahashi 1988), in addition of course to being consistent with the long tradition of unicellular oscillators such as those seen in *Gonyaulax*, *Paramecium*, *Euglena*, etc. (Edmunds 1988) and our own work in *Neurospora*. In all the systems where the analysis of a clock has been pursued to the cellular level, the clock has been found to reside within the cell, i.e., to be an intracellular regulatory phenomenon rather than an intercellular communication loop. The jury is still out on the SCN, although the preponderance of data seems to be leaning towards the SCN clock also being an intracellular feedback loop.

Miller: The only results supportive of a cellular basis that we have in the SCN come from work by Bos & Mirmiran (1990), which is not as definitive as Dr Block's because they were looking at SCN neurons which were not isolated but still hooked up to other neurons. However, their results force you to conclude that individual SCN neurons are sloppy pacemakers. The network would then be responsible for tightening up the pacemaker function.

Gillette: I agree with your ideas, but I don't think you can go that far with those results. Bos & Mirmiran (1990) showed that single cells oscillate, but a two-cell coupled oscillator with negative interactions could perform in the same way.

Miller: I'm pushing it, I know, but the dispersion of τ was large in that study.

Gillette: Eight of the 17 units Bos & Mirmiran (1990) sampled had periods from 16 to 23 h, another eight had 25–30 h periods and one fell at 24 h. This was the behaviour of single neurons in single organotypic slices cultured for several weeks where the behaviour of the ensemble is unknown. If the ensemble still has a near-24 h rhythm such as we see in acute slices, there must be cellular interactions different from those in *Bulla* that orchestrate the circadian output.

Waterhouse: Perhaps this would be the moment to focus our thoughts and consider where we think we stand with the concept of clock proteins, and how we know when we are within a central pacemaker. What properties should we expect of such a pacemaker and its component protein or proteins?

Loros: How do we know when we are, at a molecular level, within the central pacemaker? From my perspective, the first thing I expect, and think is necessary, is that disruption, by mutation for example, should affect all of the properties that are driven by this pacemaker. For an individual pacemaker, if you are within the clock and have disrupted it in some way, every output of that clock, everything that's driven by it, must be affected.

You might also expect to see pleiotropic effects. This is perhaps not necessary, but is an expectation. In other words, you would expect to affect more than one canonical clock property, be it τ, compensation, either temperature compensation or culture and feeding regime compensation, or the ability of stimuli such as light to affect the clock.

So, how do you test this? At a molecular level a clock component must change in functional activity over the cycle. If you actually have a protein in hand, and can turn on its expression constitutively, it must be in an active form. Proteins can be modified, so even if a particular protein is expressed constitutively, it is not necessarily active at all times of the subjective day. If you can provide that protein, or whatever moiety you think is involved, in an active form constantly around the clock cycle, you *must* disrupt the clock. Also, removal of the active form of the component should set the clock to the specific phase associated with the component's inactivity.

Block: I agree with what you said, but would restrict the discussion even further to those proteins that are actually involved in the causal loop that generates the oscillations. Other proteins will play roles other than as rhythmic elements in the timing loop. I can see three criteria that would need to be fulfilled for a protein to be part of the causal loop: (1) the concentration of the protein should oscillate, with a period the same as that of the overall system; (2) if you change its concentration you should produce a stable phase shift; (3) if you hold that protein at a constant level, the oscillation must stop. The third is the tricky one. The oscillation should stop if you hold the protein at *any* constant level, not just at the highest or the lowest level, outside the physiological range. There may be a number of proteins closely associated with this causal loop which might stop or reset the oscillation if held at high or low concentrations though they may not be part of the causal loop. In the cardiac oscillator, if you open up chloride channels in the membrane, even though this conductance is not part of the causal loop generating the oscillation, you will stop the oscillation by clamping the membrane potential.

Rensing: Apart from its being a part of the causal mechanism, would you assume that the protein also has other functions in the cell, for example as an activator or repressor of other genes?

Loros: That idea has been discussed. Must disruption be specific to the clock, or is the clock tied in with other cellular metabolic processes? With *per* and *frq*, and perhaps *tau*, it seems that metabolism is not particularly affected by mutational alteration of the clock. Certainly, something within a central pacemaker transfers information that results in the control of things other than the clock itself.

Kronauer: A chemical oscillator has to have at least two constituents for it to oscillate. The indications are that there may have to be at least three or four components to get temperature compensation. The problem is, of course, that chemical reactions all increase their rate with temperature. It is necessary to

General discussion I

put them together in a special arrangement so that some rate increases will actually cause the period to decrease. This implies the need for a strong oscillator, usually with many components. If you hold one component constant, it wouldn't necessarily have to stop the clock, but it ought to have a major effect on the period. Oscillations might still continue.

Block: I was thinking of a simple single loop oscillator.

Kronauer: It's highly unlikely that a chemical oscillator will be that simple.

Takahashi: Perhaps we could, as a group, attempt to establish some criteria for a clock component. I can can think of five criteria which will cover all the possibilities, if I started with a mutation in a clock gene such as *per* or *frq*. (1) Mutations should affect the period of the rhythm. There's a caveat here— there should be a significant change in period. We would like to see both alleles that shorten period and alleles that lengthen period by a large magnitude, more than a couple of hours, because you can imagine a lot of secondary effects that could cause small changes in period. (2) If you delete the candidate gene product, for example by creating a null mutation, you must abolish the rhythm. The component needs to be necessary, which differs a little from what Dr Kronauer just said. This is the case for both *frq* and *per*. (3) I would restate what Dr Block said, that the component must oscillate, if it is a state variable in the feedback loop; and (4) this oscillation must be necessary. Simply clamping the concentration at a low or high level is not sufficient; you have to express the clock gene constitutively at all levels within the dynamic range of the normal process to show that it is a component. As I understand it, Dr Dunlap has done this experiment in *Neurospora*, using the inducible promoter to drive *frq* expression in a null background, not a wild-type background, where you can titrate the amount of inducer and produce different constant levels of FRQ. If the rhythm is not re-established in the null background at any level of the gene product, which I believe Dr Dunlap has shown with *frq*, this criterion is fulfilled. (5) Finally, as Dr Block said, you must be able to reset the phase by perturbing the level of the clock gene product. You can imagine several different experiments that could be done to address this. You could see if pulse induction of the gene product resets the phase, or you could do the type of experiment that Dr Dunlap described, overexpress the product and then switch it off; then the phase should be locked to the time at which you switched off the induced gene product.

These five criteria cover most of the possible arguments against something being a clock component. To address what Ludger Rensing said, I can think of two further considerations. We have to explain how entraining inputs have access to the oscillator loop. Finally, we have to explain how the loop can 'talk' to the rest of the organism—what is the output signal? But again, these are requirements for coupling, not for the loop itself.

Czeisler: What do you think about the pleiotropic effects?

Takahashi: I agree with Dr Loros that one would expect to see them, but I wouldn't view pleiotropism as a necessary criterion.

Loros: I agree that doesn't need to be on this list, but I do think that the sixth criterion is that all the outputs of the clock must be affected by a mutation or disruption of the clock.

References

Armstrong SM 1987 Melatonin: a link between the environment and behavior. Integr Psychiatry 5:19–21
Armstrong SM 1989a Melatonin: the internal zeitgeber of mammals? Pineal Res Rev 7:157–202
Armstrong SM 1989b Melatonin and circadian control in mammals. Experientia 45:932–938
Bos NPA, Mirmiran M 1990 Circadian rhythms in spontaneous neuronal discharges of the cultured suprachiasmatic nucleus. Brain Res 511:158–162
Canbeyli RS, Romero M-T, Silver R 1991 Neither triazolam nor activity phase advance circadian locomotor activity in SCN-lesioned hamster bearing fetal SCN transplants. Brain Res 566:40–45
Colwell CS 1990 Light and serotonin interact in affecting the circadian system in the eye of *Aplysia*. J Comp Physiol 167:841–845
Deguchi T 1979 A circadian oscillator in cultured cells of the chicken pineal gland. Nature 282:94–96
Edmunds LN JR 1988 Cellular and molecular bases of biological clocks. Springer-Verlag, New York
Honma K-I, Honma S, Hiroshige T 1987 Activity rhythms in the circadian domain appear in suprachiasmatic nuclei lesioned rats given methamphetamine. Physiol Behav 40:767–774
Johnson RF, Moore RY, Morin LP 1989 Lateral geniculate lesions alter circadian activity rhythm in the hamster. Brain Res Bull 22:411–422
Michel S, Geusz M, Zanitsky J, Block G 1993 Circadian rhythm in membrane conductance expressed in isolated neurons. Science 259:239–241
Robertson L, Takahashi J 1988 Circadian clock in cell culture. I. Oscillation of melatonin release from dissociated chick pineal cells in flow-through microcarrier culture. J Neurosci 8:12–21
Romero MT, Silver R 1989 Control of phase and latency to recover circadian locomotor rhythmicity following transplantation of fetal SCN into lesioned adult hamsters. Soc Neurosci Abstr 15:725
Terman M, Terman JS 1985 A circadian pacemaker for visual sensitivity? Ann NY Acad Sci 453:147–161
Terman JS, Remé CE, Terman M 1993 Rod outer segment disk shedding in rats with lesions of the suprachiasmatic nucleus. Brain Res 605:256–264
Winfree AT 1975 Unclocklike behaviour of biological clocks. Nature 253:315–319
Winfree AT 1980 The geometry of biological time. Springer-Verlag, New York
Zatz M, Mullen DA 1989 Photoendocrine transduction in cultured chick pineal cells. 3. Ouabain (or dark pulses) can block, overcome, or alter the phase response of the melatonin rhythm to light pulses. Brain Res 501:46–57

The effects of light on the *Gonyaulax* circadian system

Till Roenneberg

Institut für Medizinische Psychologie, Ludwig-Maximilians-Universität, Goethestrasse 31, D-80366 München, Germany

Abstract. The circadian system of the marine unicellular alga *Gonyaulax polyedra* consists of at least two separate circadian oscillators. One of these controls the rhythm of bioluminescence, the other the rhythm of swimming behaviour. These two oscillators have separate light input mechanisms. The bioluminescence oscillator responds mainly to blue light whereas the aggregation oscillator is also sensitive to red light. Therefore, one of the chlorophylls is a likely candidate for the light receptor of the aggregation oscillator. Owing to their differences in spectral sensitivity, the two oscillators can be internally desynchronized when frequent dark pulses (e.g., five minutes every 20 min) are given in otherwise constant red light. Single bright red light pulses interrupting a constant dim blue background shift the bioluminescence oscillator similarly to dark pulses. They also lead to after-effects in the period of the bioluminescence rhythm, indicating that the aggregation oscillator has a different phase response to red light pulses. In contrast, blue light pulses interrupting a dim red background shift both oscillators in a similar way and do not significantly alter the circadian period following the light pulse. The mammalian phosphagen creatine shortens the period of the bioluminescence rhythm significantly in blue light but not in red. Because it also increases the sensitivity of the phase response of the bioluminescence oscillator, we propose that creatine acts on its blue-sensitive light input mechanism.

1995 Circadian clocks and their adjustment. Wiley, Chichester (Ciba Foundation Symposium 183) p 117–133

Light is the most important environmental signal for circadian systems, and its effects have been extensively investigated. In spite of the importance of light as a circadian effector, little is known about the transduction mechanisms by which it produces changes in phase and period. We have studied the effects of light on the circadian system of the marine unicellar alga *Gonyaulax polyedra* and have found that its input involves complex mechanisms beyond the time-dependent changes in the responsiveness of the oscillator itself.

In this dinoflagellate, many cellular functions are under circadian control. Bioluminescence is emitted at night as brief flashes and a sustained low intensity glow (Fritz et al 1990, Hastings & Sweeney 1958). Photosynthetic activity

(Hastings et al 1961, Samuelsson et al 1983) and the activity of the enzyme superoxide dismutase (Colepicolo 1991) are greater during the day, and cell division is restricted to the early hours of the morning (Homma & Hastings 1989, Sweeney & Hastings 1958). During the day, *Gonyaulax* cells form aggregations near the water surface, whereas they sink during the night to form a loose 'carpet' on the bottom of the container (Roenneberg et al 1989, Roenneberg & Hastings 1992). All these daily changes continue unabated under constant conditions. The fact that several of these circadian rhythms can be recorded automatically for many weeks (Roenneberg & Hastings 1992, Taylor et al 1982) makes *Gonyaulax* an excellent model with which to investigate the circadian system in unicellular organisms.

A circadian system with two light receptors

Both phasic and tonic light influences contribute to entrainment, and the two effects have been suggested to occur via the same mechanism. In most organisms, the period length under constant conditions (τ) depends on light intensity (Aschoff 1979), with τ either shortening or lengthening as the light intensity is increased. The specific adjustments of τ to different intensity levels have been attributed to the phase response characteristics of the pacemaker of the respective organism (velocity response) (Daan & Pittendrigh 1976). In *Gonyaulax*, τ can be either shortened or lengthened by increasing light intensity, depending on the spectral quality of the light (Roenneberg & Hastings 1988). Increasing the intensity of blue or cool white light (<400 nm) shortens τ, whereas τ lengthens in higher intensities of red and yellow light (Fig. 1). Thus, the circadian system of *Gonyaulax* perceives light via at least two receptors. The effects of blue and red light appear to be additive, because changes in the intensity of spectrally balanced white light or simultaneously presented blue and red light do not alter τ significantly. The same is true for changes in intensity of green light (unpublished results). The red light effect shows no far-red reversibility and therefore cannot be attributed to the action of a phytochrome.

An important part of the elucidation of the light transduction mechanisms of any system is the identification of the pigments and receptors involved. The magnitude of phase responses to light pulses of different spectral qualities can be used to determine the absorption characteristics of the circadian light receptors. Hastings & Sweeney (1960) have constructed an action spectrum for phase shifting in the *Gonyaulax* circadian system in which light pulses are given in otherwise constant darkness. They found a large maximum at 475 nm and a smaller one at 650 nm, a pattern corresponding roughly to the absorption spectrum of chlorophyll. A possible involvement of photosynthesis in circadian light effects is supported by experiments with inhibitors of photosynthesis such as dichlorophenyldimethylurea (DCMU). At high concentrations (in the micromolar range), DCMU can suppress phase shifting elicited by incremental

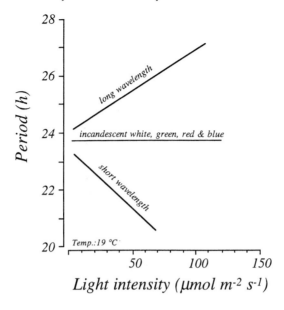

FIG. 1. The period of the bioluminescence rhythm in *Gonyaulax polyedra* is a function of the intensity of the constant light. The sign of this correlation depends on the spectral composition of the light. In long wavelengths, the period lengthens with increasing intensity, whereas the oscillator is accelerated in short wavelengths. The two opposing responses are additive, because increasing the intensity of (a) a combination of red and blue light, (b) spectrally balanced white light or (c) green light has no significant effect on period length.

light pulses (Johnson & Hastings 1989). Much lower concentrations (< 100 nM) can shorten the period in short and long wavelength light, and pulses of DCMU (100 nM removed from the medium after one hour) generate phase shifts similar to dark pulses (Roenneberg 1993).

Because *Gonyaulax* is an obligatory photoautotroph and cannot survive long periods of darkness, responses to light are difficult to study in darkness or very dim light. The determination of the phase response to light must therefore be conducted in constant light. Usually, dim red light is used as constant background and bright white or blue light for the pulse. Under these conditions, the circadian system of *Gonyaulax* responds with strong (type 0) phase shifting (Fig. 2A), with a break-point around CT (circadian time) 15 (Johnson & Hastings 1989, Roenneberg & Taylor 1993). The phase response curve (PRC) of *Gonyaulax* is somewhat unusual in comparison with the type 0 light-generated PRCs of other organisms (Pittendrigh 1981) in that the delay portion in the early subjective night is greatly reduced or even completely absent, which gives it an asymmetrical shape. The break-point of the PRC produced by light is fixed at

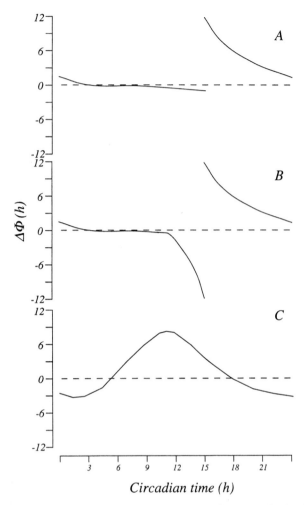

FIG. 2. Phase response curves (PRCs) (for the bioluminesence rhythm in *Gonyaulax polyedra*) resulting from single light and dark pulses. When bright white or blue pulses replace a dim red background, the resulting type 0 PRC lacks the delay region commonly found in other circadian systems (A). This delay region is, however, present if creatine is added to the cells two days before the light pulse is given (B). When bright red light pulses replace a dim blue background, the resulting PRC resembles that normally found with dark pulses (C). Δφ, phase shift.

CT 15, regardless of the intensity of the applied light; even with low intensity pulses, which give rise to a weak (type 1) PRC, the transition between delays and advances is at CT 15.

The shape of the PRC produced by incremental light pulses could explain the shortening of the period in short wavelength light. Because increasing the intensity of constant red light lengthens the period, we have investigated the phase-shifting capacity of pulses using long wavelength light (Roenneberg & Hastings 1991). If pulses of bright red light replace a dim blue light background, the resulting PRC resembles, with a lower amplitude, that found with dark pulses replacing a white light background (Fig. 2C) (Broda et al 1986). Because the magnitude of the phase shift decreases when the intensity of the red light pulse is increased, red light must be perceived by the phase-shifting mechanism as a lack of blue light. In contrast, the fluence (intensity) response curve produced by blue light pulses on a dim red background shows a positive correlation (Johnson & Hastings 1989).

Synergistic effects of light and creatine

The discovery that creatine shortens the period of the *Gonyaulax* bioluminescence rhythm (Roenneberg et al 1987) (Fig. 3) has also indicated that there are two different light input mechanisms because creatine shortens τ in short wavelength light but is almost ineffective in long wavelengths (Roenneberg et al 1988). Thus, creatine appears to act on the light transduction mechanism responsible for the circadian responses to short wavelength light. The involvement of light transduction in the effects of creatine is supported by the fact that this substance also affects the cells' phototactic response. The distance from a light source at which daytime aggregations form depends on its intensity. The cells show a preference for specific light intensities: the brighter the light the further they move away from it. The same reaction can also be elicited by adding creatine to the culture medium (Roenneberg & Taylor 1993).

In search of the mechanisms by which creatine shortens τ in constant white or blue light, we investigated its effects on the blue-sensitive phase-shifting mechanism by adding creatine to the cultures two days before blue light pulses were given in otherwise constant dim red light (Roenneberg & Taylor 1993). In the advance region of the PRC, the extent of phase shifting induced by light depended on the creatine dose. Creatine appears to act analogously to an intensity increase of the pulse, i.e., the light intensity required to produce a saturating pulse is decreased as creatine is added to the cells. The similarity between creatine dose and pulse intensity is also evident in the change of the waveform of the glow rhythm. At the transition from type 1 to type 0 resetting (caused by increasing either the pulse intensity or the creatine dose), the glow rhythm is flattened and shows a bimodal distribution (Johnson & Hastings 1989, Roenneberg & Hastings 1991).

In the delay portion of the PRC (early subjective night), when there were no phase shifts in the control cultures even at intensities which were saturating in the advance portion, negative phase shifts occur under the influence of creatine (Fig. 2B). Thus, creatine has two effects on the phase-shifting mechanism: first, it increases the amplitude of phase shifts in the late subjective night, and second,

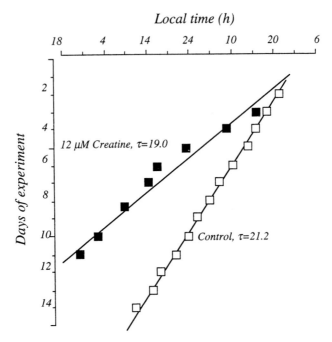

FIG. 3. The mammalian phosphagen creatine (added to the cells at the beginning of the experiment) shortens the period of circadian rhythms in *Gonyaulax polyedra*. The squares represent the times of the daily glow peaks; periods (τ) were calculated by linear regression. This effect requires the presence of short wavelength light (in this case about 30 μmol m^{-2} s^{-1} cool white fluorescent light at a temperature of 19 °C). The shortening of the period might be mediated by an increase in the light sensitivity of the phase response mechanism because creatine also increases the amplitude of phase shifts produced by pulses of white and blue light of subsaturating intensity.

it disinhibits the responsiveness of the system before CT15, such that the generic symmetrical shape of a type 0 light-produced PRC results.

The mechanisms by which creatine shortens the period and increases the sensitivity to pulses of blue and white light remain a mystery. Measurements of oxygen evolution at different times during the circadian cycle and with spectrally different forms of light have shown that the rate of photosynthesis is not influenced by creatine (Fig. 4), excluding the involvement of the chloroplast in these effects. These results show that the effects of photosynthesis inhibitors on phase shifting in response to light are not caused by a changed rate of oxygen evolution. This was already predicted by the fact that DCMU affects the circadian bioluminescence rhythm only at concentrations much higher than those necessary to completely inhibit photosynthesis (Johnson & Hastings 1989).

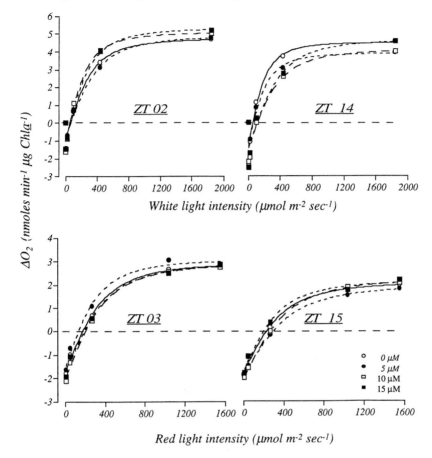

FIG. 4. The effect of creatine and light on O_2 evolution by *Gonyaulax polyedra*. In contrast to its effects on the circadian system, creatine does not affect the rate of O_2 evolution in either white (*top pair of curves*) or red (*bottom pair of curves*) light. Different concentrations of creatine (0–15 μM) were added to the cells two days before the measurements were made. 40 ml cells was taken from the culture room (white light : darkness, 12 h : 12 h) at different times of the circadian cycle, placed for 30 min in the light to be used during the O_2 evolution measurements which were made in a Clarke-type oxygen electrode (Roenneberg & Carpenter 1993) at different light intensities. Curves were fitted according to the algorithm of Smith (1936). Chl$\underline{a}$, chlorophyll *a*; ZT, zeitgeber time, with reference to the lighting conditions in the culture rooms, with ZT 0 corresponding to lights on and ZT 12 to lights off.

Differential effects of dark pulses

Perturbations of the free-running system by a single pulse of light can change not only the phase of the rhythm, but also the subsequent period for several cycles. If changes in τ following light or dark pulses gradually decrease until the original period is restored, they are generally interpreted as transient shifts of the observed rhythm towards the new phase of the oscillator. If the changes in τ are stable for several days, however, the phenomenon is classified as a period after-effect (Boulos & Rusak 1982) and is attributed to interactions between several circadian oscillators.

In *Gonyaulax*, incremental light pulses such as those described above do not lead to significant after-effects (Johnson & Hastings 1989). However, dark pulses, and especially pulses of bright red light, can lengthen the period of the glow rhythm for five to seven days (Fig. 5). The magnitude of these changes in τ correlates with the intensity of the red pulse; a systematic relationship with the time of the circadian phase at which the pulse was given is, however, not obvious. The fact that the extent of the phase shift decreases with increasing intensities of the red pulse, whereas the after-effects on τ increase indicates that phase and period changes have different mechanisms, suggesting the involvement of two oscillators.

Earlier studies showed that dark pulses in otherwise constant white light shifted the phase of various rhythms in *Gonyaulax* by an equal amount and in the same direction (McMurry & Hastings 1972), consistent with a single circadian oscillator driving the different circadian rhythms. However, simultaneous recording of

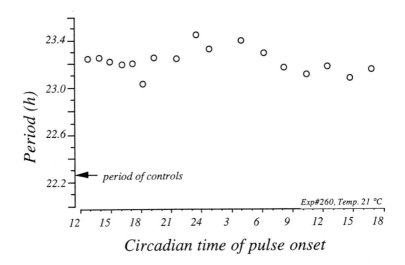

FIG. 5. When single red light pulses (4 h, 35 μmol m^{-2} s^{-1}) replace a dim blue background (12 μmol m^{-2} s^{-1}, 21 °C) the subsequent period is lengthened relative to the control cultures regardless of the phase at which the pulse is given. The lengthening of the period correlates with the intensity of the red light pulse and can be as much as two hours. The altered period remains stable for five to seven days.

the bioluminescence rhythms and the aggregation rhythm in constant red light showed that different rhythms can adopt different periods in *Gonyaulax* (Roenneberg & Morse 1993). Similar cases of internal desynchronization have been reported in higher organisms including humans (Aschoff et al 1967, Ebihara & Gwinner 1992, Meijer et al 1990). The desynchronizing effect of dark pulses in constant red light could be explained if the oscillator controlling bioluminescence was sensitive to only short wavelength light (i.e., could not distinguish between red light and darkness) but the oscillator controlling aggregation was also sensitive to red light (perhaps by using chlorophyll as the receptor). Indeed, unpublished results show that three-hour dark pulses shift the phase of the aggregation oscillator by as much as three hours, while the bioluminescence oscillator responds with only small phase shifts (<30 min).

A different receptor for each oscillator?

Comparison of the PRC produced by incremental light (Fig. 2A) and that produced by decremental light (Fig. 2C) shows that the two are not complementary or reversed, as might be suspected from the fact that light and dark pulses involve the same light changes but in reverse order (on–off versus off–on). Unlike the PRC produced by incremental light, that produced by decremental light has no dead zone, suggesting that information about the absence and presence of light reaches the circadian system via different mechanisms. Different transduction mechanisms might be involved or different state variables of the oscillator. Similarly, phase advances and delays may also be mediated by different mechanisms. Generally, for a symmetrical type 0 PRC, the distinction between delays and advances is somewhat arbitrary, because pulses reset the oscillator to a certain phase of the circadian cycle regardless of when they are applied. The peculiarities of the phase response to light in *Gonyaulax* make this distinction less arbitrary. In mammals, this distinction can even be made on the basis of chemicals which specifically inhibit either delay or advance phase shifts in response to light (Menaker & Ralph 1986, Ralph & Menaker 1985). The differential effect of creatine on different regions of the PRC supports the notion that specific reactions lead to either phase advances or delays. The disinhibition of the response to pulses of light between CT 10 and CT 15 suggests that the light input pathway may be actively inhibited during this part of the circadian cycle, perhaps by feedback control by the oscillator of its own light transduction pathway. Such feedback loops have been documented clearly in arthropods (Fleissner & Fleissner 1988).

The investigation of the circadian system in *Gonyaulax* has provided evidence for two coupled oscillators and for light perception via two receptors. It is possible that the two light transduction pathways are connected separately to the two oscillators. On the one hand, results from several independent experiments indicate that the bioluminescence oscillator is mainly blue sensitive and reacts to red light as the absence of blue. On the other hand, the *Gonyaulax* circadian system can also be actively affected by red light (Hastings & Sweeney 1960, Roenneberg & Hastings 1991) (an effect not mediated by phytochrome), and inhibition of photosynthesis affects the circadian phase, period and light perception (Johnson & Hastings 1989, Roenneberg 1994). Chlorophyll is, therefore, a likely candidate for the receptor that supplies information to the oscillator controlling the aggregation rhythm.

Acknowledgements

This work was supported by the Deutsche Forschungsgemeinschaft and the Friedrich-Bauer-Stiftung. I thank Iris Neher for her excellent laboratory assistance and David Morse for his critical comments on this manuscript.

References

Aschoff J 1979 Circadian rhythms: influences of internal and external factors on the period measured under constant conditions. Z Tierpsychol 49:225–249

Aschoff J, Gerecke U, Wever R 1967 Desynchronization of human circadian rhythms. Jpn J Physiol 17:450–457

Boulos Z, Rusak B 1982 Circadian phase response curves for dark pulses in the hamster. J Comp Physiol A Sens Neural Behav Physiol 146:411–417

Broda H, Gooch VD, Taylor WR, Aiuto N, Hastings JW 1986 Acquisition of circadian bioluminescence data in *Gonyaulax* and an effect of the measurement procedure on the period of the rhythm. J Biol Rhythms 1:251–263

Colepicolo P 1991 A circadian rhythm in the activity of superoxide dismutase in the photosynthetic alga *Gonyaulax polyedra*. Chronobiol Int 9:266–268

Daan S, Pittendrigh CS 1976 A functional analysis of circadian pacemakers in nocturnal rodents. III. Heavy water and constant light: homeostasis of frequency? J Comp Physiol A Sens Neural Behav Physiol 106:267–290

Ebihara S, Gwinner E 1992 Different circadian pacemakers control feeding and locomotor activity in European starlings. J Comp Physiol A Sens Neural Behav Physiol 171: 63–67

Fleissner G, Fleissner G 1988 Efferent control of visual sensitivity in arthropod eyes: with emphasis on circadian rhythms. In: Lindauer M (ed) Information processing in animals. Gustav Fischer, Stuttgart, vol 5:1–67

Fritz L, Morse D, Hastings JW 1990 The circadian bioluminescence rhythm of *Gonyaulax* is related to daily variations in the number of light-emitting organelles. J Cell Sci 95:321–328

Hastings JW, Sweeney BM 1958 A persistent diurnal rhythm of luminescence in *Gonyaulax polyedra*. Biol Bull (Woods Hole) 115:440–458

Hastings JW, Sweeney BM 1960 The action spectrum for shifting the phase of the rhythm of luminescence in *Gonyaulax polyedra*. J Gen Physiol 43:697-706

Hastings JW, Astrachan L, Sweeney BM 1961 A persistent daily rhythm in photosynthesis. J Gen Physiol 45:69-76

Honma K, Hastings JW 1989 The S-phase is discrete and is controlled by the circadian clock in the dinoflagellate *Gonyaulax polyedra*. Exp Cell Res 182:635-644

Johnson CH, Hastings JW 1989 Circadian phototransduction: phase resetting and frequency of the circadian clock of *Gonyaulax* cells in red light. J Biol Rhythms 4:417-437

McMurry L, Hastings JW 1972 No desynchronization among the four circadian rhythms in the unicellular alga, *Gonyaulax polyedra*. Science 175:1137-1139

Meijer JH, Daan S, Overkamp GJF, Hermann PM 1990 The two-oscillator circadian system of tree shrews (*Tupaia belangeri*) and its response to light and dark pulses. J Biol Rhythms 5:1-16

Menaker M, Ralph MR 1986 Effects of diazepam on circadian phase advances and delays. Brain Res 372:405-408

Pittendrigh CS 1981 Circadian systems: entrainment. In: Aschoff J (ed) Biological rhythms. Plenum, New York, vol 4:95-124

Ralph MR, Menaker M 1985 Bicuculline blocks circadian phase delays but not advances. Brain Res 325:362-365

Roenneberg 1994 The *Gonyaulax* circadian system: evidence for two input pathways and two oscillators. In: Hiroshige T, Honma K (eds) Evolution of circadian clocks. Hokkaido University Press, Sapporo, in press

Roenneberg T, Carpenter E 1993 Daily rhythm of O_2-evolution in the cyanobacterium *Trichodesmium thiebautii* under natural and constant conditions. Mar Biol (Berl) 117:693-697

Roenneberg T, Hastings JW 1988 Two photoreceptors influence the circadian clock of a unicellular alga. Naturwissenschaften 75:206-207

Roenneberg T, Hastings JW 1991 Are the effects of light on the phase and period of the *Gonyaulax* clock mediated by different pathways? Photochem Photobiol 53:525-533

Roenneberg T, Hastings JW 1992 Cell movement and pattern formation in *Gonyaulax polyedra*. In: Rensing L (ed) Oscillations and morphogenesis. Marcel Dekker, New York, vol 5:399-412

Roenneberg T, Morse D 1993 Two circadian oscillators in one cell. Nature 362:362-364

Roenneberg T, Taylor W 1993 Light-induced phase responses in *Gonyaulax* are drastically altered by creatine. J Biol Rhythms 9:1-12

Roenneberg T, Hastings JW, Krieger NR 1987 A period shortening substance in *Gonyaulax polyedra*. Chronobiologia 14:228

Roenneberg T, Nakamura H, Hastings JW 1988 Creatine accelerates the circadian clock in a unicellular alga. Nature 334:432-434

Roenneberg T, Colfax GN, Hastings JW 1989 A circadian rhythm of population behavior in *Gonyaulax polyedra*. J Biol Rhythms 4:201-216

Samuelsson G, Sweeney BM, Matlick HA, Prezelin BB 1983 Changes in photosystem II account for the circadian rhythm in photosynthesis in *Gonyaulax polyedra*. Plant Physiol 73:329-331

Smith EL 1936 Photosynthesis in relation to light and carbon dioxide. Proc Natl Acad Sci USA 22:504-511

Sweeney BM, Hastings JW 1958 Rhythmic cell division in populations of *Gonyaulax polyedra*. J Protozool 5:217-224

Taylor W, Wilson S, Presswood RP, Hastings JW 1982 Circadian rhythm data collection with the Apple II microcomputer. J Interdiscipl Cycle Res 13:71-79

DISCUSSION

Edmunds: Because I work with *Euglena* I have no problem with the notion of there being more than one clock in a cell. There's preliminary evidence (reviewed in Edmunds 1988, p 372–373) from a number of systems for this. For example, in *Euglena*, the circadian rhythms of chlorophyll content and photosynthetic capacity sometimes display different free-running periods and varying phase angles, suggesting the possibility of internal desynchronization in a multi-oscillator, unicellular 'clockshop' (Edmunds & Laval-Martin 1981).

Roenneberg: Our results on internal desynchronization and differential phase responses are the strongest evidence yet.

Edmunds: A lot of earlier work has been leading gradually to this conclusion, but has not been as conclusive as yours. I understand that you consider cell division to be a possible output. Do you have evidence against there being a separate oscillator for control of cell division?

Roenneberg: As yet, we can only speculate on which rhythms other than bioluminescence and aggregation are coupled to which oscillator. When we try to associate the different rhythms with different oscillators, we might find rhythms that are coupled to a third, as yet unknown, oscillator.

Waterhouse: When there is more than one rhythm simultaneously present, do the rhythms have any simple relationship to one another with respect to period?

Roenneberg: It's not simple, but we see relative and phase-dependent coordination. The day-to-day τ depends on the phase relationship between the two oscillators. At certain phase relationships one can see transient coupling. We can also see amplitude modulations when the two rhythms cross. When we force decoupling in the dark pulse regime, or when we use dark pulses in T cycle experiments where the cycle length lies outside the range of entrainment for one of the oscillators but within the range of entrainment of the other, we can even see complete suppression of the rhythms when they cross each other, and they pick up when they have crossed.

Miller: In humans, spontaneous internal desynchronization, and on occasion external desynchronization, has been related to various types of psychopathology (e.g., Wehr et al 1979). Does this desynchronization condition have any implication for *Gonyaulax* with respect to long-term viability?

Roenneberg: Although we have only anecdotal evidence, it is apparent that they don't like internal desynchronization and die earlier.

Kronauer: With a PRC containing a strong discontinuity, the suggestion is that in the vicinity of the discontinuity the amplitude has been greatly reduced. Have you attempted to quantify the amplitudes?

Roenneberg: I was waiting for you to come to Munich to do that. We talk about you a lot when we look at these data.

Waterhouse: What would be the advantage of investigating this?

Effects of light on the *Gonyaulax* circadian system 129

Kronauer: If for a very small change of stimulus phase you get a large difference in the response, the suggestion is that you have in fact reduced the amplitude so that you are close to a critical point in some sense, where a small change can make a big difference in the resultant phase. We ought, if possible, to quantify amplitudes of rhythms in order to elucidate this effect. At PRC transitions between delays and advances phase shifts may be small, but in humans and other species one finds that amplitude reductions can be very large, thereby bringing the pacemaker close to the critical region.

Roenneberg: Creatine apparently reduces the amplitude of the output. Also, we have the remarkable result that with addition of creatine you go from no phase response suddenly to a strong phase response at the same CT. We have a crazy hypothesis that this might have something to do with the two clocks. One of the two clocks has an earlier break-point in its PRC curve than the other. The bioluminescence clock has a CT 15 break-point, whereas that of the aggregation clock is apparently around CT 12. Thus, under synchronized conditions you could get a phase advance in one clock and a phase delay in the other. We know that creatine acts differently on the two clocks, so we might be pulling these two clocks apart such that they do not give opposing phase responses to the same light pulse; we might therefore suddenly see the true type 0 PRC.

Kronauer: You don't need such a complicated explanation if in fact the amplitude in that regime is reduced. Any secondary influence could bring about a large change in the phase shifts.

Daan: I gather that you plan to test the idea that the effect on τ of constant light of different intensities is somehow an artefact of the light sensitivity expressed in the PRC. Creatine shortens τ in blue light, but you have found that it increases the phase delay portion of the PRC while having no effect on the advance portion. How can an increase in the delay portion explain a shortening of τ? Wouldn't you expect τ to lengthen?

Roenneberg: When we test the effect of creatine with pulses of lower light intensities we get a type 1 PRC. In these cases, creatine increases the amplitude in the advance portion. We get large phase delays at circadian times at which the system normally does not respond only when creatine is given together with light pulses of very high intensity. We haven't tested the effect of such high intensities of blue light under constant conditions. The shortening of τ by increasing light intensities may turn around at a certain point.

Daan: So at this point you are not quite clear why the relationship is in this direction.

Roenneberg: At lower light intensities the sensitivity hypothesis fits, but at higher light intensities it may not. The question is whether in type 0 PRCs the delay is really a delay. It could just as well be a huge advance.

Czeisler: What you describe as a huge advance could be viewed as very small delay. What is the difference between a 12 h advance and a 12 h delay?

Roenneberg: From all the PRCs I've looked at in *Gonyaulax*, I would say that phase shifts at this subjective CT are more likely to be advances.

Czeisler: This is a problem of definition: it's hard to say which is an advance and which is a delay in these kinds of regions.

Roenneberg: This is an issue of semantics because a 16 h advance is identical to an eight-hour delay.

Waterhouse: But they might imply different mechanisms.

Loros: Can't you distinguish an advance from a delay by using different intensities of light?

Menaker: You can, but then you have to assume that it continues in the same direction. We have the same problem with the *tau* mutant hamster.

Rensing: Have you tried keeping the cells under a different regime of blue light and red light, perhaps with red light in the day and blue light at night? Are the two oscillators then entrained by the different qualities of the light? Do you get a dissociation of the two oscillators?

Roenneberg: You get a slight phase angle difference, because one of the oscillators is more sensitive to red light than the other, but they are both very sensitive to blue. Therefore, the blue light phase in a blue/red regime will always fall into the subjective day of both oscillators.

Rensing: So you can't separate them by such a regime.

Roenneberg: You could perhaps if you played around with the intensity levels. A very bright red light might mean subjective day for the aggregation oscillator when the blue light part of the cycle is very dim, but it is difficult to do such experiments with red filters in the laboratory without having problems with temperature.

Menaker: Your interpretation of the decreasing effect of red light of increasing intensity depends on the idea that more blue light gets through as you increase the amount of red. You could test that directly using monochromatic light.

Roenneberg: We are planning to do this. The absorption spectra of both oscillators now have to be tested again with monochromatic light. Unfortunately, with *Gonyaulax* this is not easy because one has to get a lot of photons through the narrow band filter.

Menaker: You could use lasers.

Roenneberg: That would be too expensive for me.

Takahashi: If it's true that when you increase the intensity of red light you get leakage of the red light into the blue absorbing region, you should get the same result with monochromatic light, not the opposite result. Your implication was that with monochromatic red light, you wouldn't see a decreasing phase shift as you increase the intensity.

Roenneberg: I'm not so sure. If the narrow band width filter lies outside of the region where the absorption spectra of the two pigments overlap, we might not find that the phase response gets smaller with higher intensities of red light pulses.

Takahashi: You would come to that conclusion if you plot the absorbance on a linear scale. If you plotted it on a logarithmic scale, you would see that the photopigment continued to absorb red photons but at a very low level. Even with monochromatic light, you would get the same result.

Roenneberg: I agree, but that doesn't contradict our interpretation.

Menaker: The effect would be in the same *direction*, but would not be the same *quantitatively*.

Takahashi: You would still see a decreasing phase shift with increasing red light intensity, assuming the red light is acting on the blue light photoreceptor at very long wavelengths. A photon has the same effect no matter what its wavelength is.

Roenneberg: And the pigment can't distinguish between intensity changes and spectral changes.

Waterhouse: You speculated that the red receptor might be connected in some way with chlorophyll.

Roenneberg: It's not a red receptor—it's a blue and red receptor, which is why I think it might be chlorophyll.

Waterhouse: What about the other receptor? Are there inclusion bodies with an appropriate spectral absorption?

Roenneberg: No.

Kronauer: In the spectral response postulated for the resetting mechanism, do you think there is an insensitive region between the two sensitive peaks?

Roenneberg: In a plot of τ versus intensity (Fig. 1), the red light gives a positive correlation and the blue light gives a negative correlation, but green light has no effect on the period of the bioluminescence rhythm.

Takahashi: Also, in Hastings & Sweeney's (1960) action spectrum measured in *Gonyaulax* there is an insensitive region like that you have suggested. The action spectrum is similar to the absorption spectrum for chlorophyll.

Roenneberg: This action spectrum was done in constant darkness, by the way, but we don't know what our two clocks do in darkness because we can't test that anymore. I would speculate that the bioluminescence responses to red light that were found by Hastings & Sweeney (1960) are due to strong coupling between the two clocks.

Menaker: Why can't you test this anymore? Has the strain changed?

Roenneberg: The strain has become pampered! It dies after 30–36 h in darkness. We could adopt the regime used by Kondo et al (1991) with *Chlamydomonas*, where they put the cells into darkness, give them monochromatic light pulses and then release them into dim white light. There are some difficulties with that protocol because if you have a second pulse it's difficult to know what's really going on.

Edmunds: When we worked with baker's yeast some years ago (Edmunds 1980), we found direct damaging effects, on amino acid transport and membrane transport, of white light at high intensities that wouldn't normally affect *Euglena* or probably *Gonyaulax* because of all their shielding, accessory pigments. This damaging effect of white light was due mainly to the blue light component, and we identified a flavin, cytochrome-type receptor using point mutations (Ulaszewski et al 1979). Do you find any damaging effects at higher intensities of blue light?

Roenneberg: We haven't done any direct measurements that would answer that question. In very bright light *Gonyaulax* becomes bleached and the cells die. Bright white light also has strong effects on the expression of the bioluminescence rhythm, and it can apparently stop the bioluminescence oscillator (Hastings 1964).

Edmunds: You are probably working at upper intensity levels, and it is difficult to translate this to the effect attributable to the blue light component.

Roenneberg: To increase the blue light even further is difficult. We already use slide projectors.

Menaker: Is it possible that in the course of being pampered the coupling between these two oscillators has weakened, so that you can now see things in the strain that would have been harder to see 20 or 30 years ago?

Roenneberg: That's possible, but it's not necessary. The experiments by McMurray & Hastings (1972) to see whether different rhythms have different phase responses were all done in white light, where one wouldn't see any coupling.

Menaker: It would still be interesting to do some of these experiments again with re-isolated *Gonyaulax*. Would that be difficult?

Roenneberg: No.

Menaker: That would be important with regard to what your results mean to the organism in the context of the real world.

References

Edmunds LN Jr 1980 Blue-light photoreception in the inhibition and synchronization of growth and transport in the yeast *Saccharomyces*. In: Senger H (ed) The blue light syndrome. Springer-Verlag, Berlin, p 584–596

Edmunds LN Jr 1988 Cellular and molecular bases of biological clocks. Springer-Verlag, New York

Edmunds LN Jr, Laval-Martin DL 1981 'Free-running' circadian rhythms of photosynthesis elicited by short-period cycles of light and darkness in synchronously dividing and nondividing *Euglena*. In: Photosynthesis. Akoyunoglou G (ed) Proc 5th Int Congr Photosynthesis, Halkidiki, Greece, 1980. Balaban International Science Services, Philadelphia, PA, vol VI:313–322

Hastings JW 1964 The role of light in persistent daily rhythms. In: Giese AC (ed) Photophysiology: action of light on living materials. Academic Press, New York, vol 1:333–361

Hastings JW, Sweeney BM 1960 The action spectrum for shifting the phase of the rhythm of luminescence in *Gonyaulax polyedra*. J Gen Physiol 43:697–706

Kondo T, Johnson CH, Hastings JW 1991 Action spectrum for resetting the circadian phototaxis rhythm in the CW15 strain of *Chlamydomonas*. 1. Cells in darkness. Plant Physiol 95:197–205

McMurray L, Hastings JW 1972 No desynchronization among the four circadian rhythms in the unicellular alga, *Gonyaulax polyedra*. Science 175:1137–1139

Ulaszewski S, Mamouneas T, Shen W-K et al 1979 Light effects in yeast: evidence for participation of cytochromes in photoinhibition of growth and transport in *Saccharomyces cerevisiae* cultured at low temperatures. J Bacteriol 138:523–529

Wehr TA, Wirz-Justice A, Goodwin SK, Duncan W, Gillin JC 1979 Phase advance of the circadian sleep–wake cycle as an antidepressant. Science 206:710–713

Intrinsic neuronal rhythms in the suprachiasmatic nuclei and their adjustment

Martha U. Gillette*†‡°, Marija Medanic†, Angela J. McArthur†, Chen Liu‡, Jian M. Ding*‡, Lia E. Faiman*, E. Todd Weber†, Thomas K. Tcheng‡ and Eve A. Gallman*‡°

Department of Cell & Structural Biology, *Department of Physiology*†, *Neuroscience Program*‡ *and the College of Medicine°, University of Illinois, 506 Morrill Hall, 505 S Goodwin Avenue, Urbana, IL 61801, USA*

> *Abstract.* The central role of the suprachiasmatic nuclei in regulating mammalian circadian rhythms is well established. We study the temporal organization of neuronal properties in the suprachiasmatic nucleus (SCN) using a rat hypothalamic brain slice preparation. Electrical properties of single neurons are monitored by extracellular and whole-cell patch recording techniques. The ensemble of neurons in the SCN undergoes circadian changes in spontaneous activity, membrane properties and sensitivity to phase adjustment. At any point in this cycle, diversity is observed in individual neurons' electrical properties, including firing rate, firing pattern and response to injected current. Nevertheless, the SCN generate stable, near 24 h oscillations in ensemble neuronal firing rate for at least three days *in vitro*. The rhythm is sinusoidal, with peak activity, a marker of phase, appearing near midday. In addition to these electrophysiological changes, the SCN undergoes sequential changes *in vitro* in sensitivities to adjustment. During subjective day, the SCN progresses through periods of sensitivity to cyclic AMP, serotonin, neuropeptide Y, and then to melatonin at dusk. During the subjective night, sensitivities to glutamate, cyclic GMP and then neuropeptide Y are followed by a second period of sensitivity to melatonin at dawn. Because the SCN, when maintained *in vitro*, is under constant conditions and isolated from afferents, these changes must be generated within the clock in the SCN. The changing sensitivities reflect underlying temporal domains that are characterized by specific sets of biochemical and molecular relationships which occur in an ordered sequence over the circadian cycle.
>
> *1995 Circadian clocks and their adjustment. Wiley, Chichester (Ciba Foundation Symposium 183) p 134–153*

The paired suprachiasmatic nuclei of the hypothalamus are the seat of the primary circadian timekeeping mechanism of mammals. They mark the passage of time in near 24 h cycles. It is here that entraining signals, initiated by environmental change, are integrated so as to adjust the phase of the pacemaker.

Efferent signals from the suprachiasmatic nucleus (SCN) organize and regulate metabolic, physiological and behavioural functions that occur in circadian patterns. Other chapters in this volume address aspects of the mammalian circadian system studied in whole organisms, whereas we shall examine the neuronal and pacemaker properties of the SCN that are expressed in a hypothalamic brain slice. Because our studies evaluate the SCN isolated *in vitro*, we directly assess intrinsic properties. We have found that each SCN is composed of an electrophysiologically diverse group of neurons that function remarkably autonomously: the SCN generates stable, near 24 h rhythms of ensemble neuronal activity *in vitro* and undergoes an orderly sequence of changes in sensitivity to stimuli that adjust phase. The pattern of responsiveness to neurochemicals known to be contained in SCN afferents demonstrates that the organization of the clock is more complex than simply night-versus-day processes; rather, our findings suggest there is a continuously changing series of sensitivities to multiple phase-adjusting stimuli. This sequence must reflect changes in underlying cellular processes, or 'temporal domains'. These domains are functional epochs, characterized by specific biochemical or molecular substrates and their interactions, which are linked together to generate the daily order of the circadian clock's cycle.

The experimental system that we use is a hypothalamic slice from inbred Long-Evans rats 5–10 weeks old. It differs from whole organism studies of clock function in several important ways.

First, we have dissected the SCN out of the central nervous system into a 500 μm coronal slice of hypothalamus. Therefore, the activities and circadian properties that we measure must be generated spontaneously within the tissue slice. The slice contains less than the 800 μm rostrocaudal extent of each nucleus; we study properties of the medial portion, primarily. Any circadian patterns that we observe therefore are the result of activity in less than the entire nucleus and would indicate that there is redundancy in pacemaker organization.

Second, the suprachiasmatic slice is studied in a defined, constant environment with minimal supplementation. The fresh slice is placed in a brain slice chamber at the interface of a moist, 95% O_2/5% CO_2 atmosphere and the glucose–bicarbonate-supplemented medium (Earle's balanced salt solution, 37 °C), maintained at pH 7.4 and perfused at 34 ml/h (see Prosser & Gillette 1989 for complete methods). The two suprachiasmatic nuclei are clearly visible in the slice, which gives the investigator precise control over the sites of measurement and drug treatment. Thus, with the *in vitro* preparation we have the advantage of a high degree of control over the chemical and physical environment while making manipulations under clear visual guidance.

Third, the suprachiasmatic nuclei that we study within the brain slice are removed from the influence of other brain regions and humoral factors. After the suprachiasmatic slice has been cut, the base of the hypothalamus is surgically reduced to exclude the supraoptic nuclei, although part of the paraventricular nuclei may be included in the slice because the third ventricle is intact dorsally. Thus, in these studies, the suprachiasmatic nuclei are free of the primary brain regions that project to and regulate them, as well as most feedback loops. Electrical properties, spontaneous activities and sensitivities to chemical perturbations are assessed directly in the free-running clock with little, if any, contamination by signals from other sites. Our measure of phase is the peak of the circadian rhythm of firing activity of the neuronal ensemble, which most probably represents both intrinsic signalling and the primary neural output of the clock. Any circadian changes that we measure *in vitro* are those generated spontaneously within the SCN, that is, those which are components of or are driven by the timekeeping mechanism.

A primary property attributed to the SCN from organismic studies is that of endogenous oscillation with a period about that of the day, i.e. circadian. This behaviour can be studied in our brain slices because they survive under the conditions we use for at least three days *in vitro*. At any one time point within the daily cycle, SCN neurons show diversity in rates and patterns of spontaneous activity. Circadian measurements made with whole-cell patch recording methods in the brain slice (Blanton et al 1989) demonstrate that, electrophysiologically, the SCN may be composed of considerably more cell types than first reported (Wheal & Thompson 1984). Preliminary determinations found that 75% of over 50 neurons sampled were spontaneously active, with firing patterns ranging from very regular to irregular random and burst patterns (Gallman & Gillette 1992). Neurons of the various types could be found across the circadian cycle, but circadian patterns of firing frequency, membrane potential and specific membrane conductance were apparent (Gallman & Gillette 1993). Circadian changes in firing frequency of the sample measured with whole-cell patch recording resemble the sinusoidal oscillation recorded extracellularly for the neuronal ensemble (shown in Fig. 1; Gillette 1991): mean firing rate peaks at mid subjective day, alternating with a trough during subjective night.

The pattern of oscillation in spontaneous activity of the ensemble of single units monitored extracellularly is remarkably stable over consecutive 24 h cycles, so that the time of peak activity is a reliable marker of phase (Prosser & Gillette 1989). Despite the fact that the tissue contains significantly less than the entire SCN and is maintained in glucose-supplemented minimal salt solution, the period of the daily oscillation is near 24 h. These observations reveal that isolated suprachiasmatic nuclei are able to regenerate circadian rhythmic neuronal activity cycles spontaneously given only an exogenous energy source. Further, because this occurs regularly in coronal slices which inevitably have been cut through some portion of the SCN, there must be redundancy of the pacemaker within the entire rostrocaudal extent of the SCN (Gillette 1991).

A second property attributed to circadian clocks from the study of organisms' behaviour is differential sensitivity to phase adjustment. This was tested in the SCN *in vitro* with structural analogues of the ubiquitous second messenger cyclic

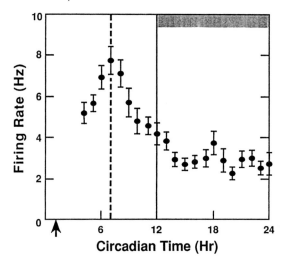

FIG. 1. The circadian oscillation in spontaneous firing rate of the ensemble of neurons of the suprachiasmatic nucleus in a rat hypothalamic brain slice. Mean firing rate (Hz) for all units encountered with an extracellular electrode and sampled within a two-hour interval is plotted against circadian time (CT) of the lighting cycle to which the rat had been entrained, where CT 0 is the time of the dark–light transition. The slice was prepared at CT 1 (arrow) from an 8-week-old rat reared in a 12 h light:12 h dark lighting cycle and placed in constant light in the interface brain slice chamber; recording commenced at CT 4. Activity peaked at CT 7 (dashed line), seven hours into the entrained light cycle, and was generally low at night (shaded horizontal bar). Error bars, SEM; $n = 6-15$ neurons per two-hour bin (each two-hour bin was offset by 15 min to generate a running average).

AMP (cAMP). Several 8-substituted cAMP analogues, including 8-benzylamino-cAMP (BA-cAMP), 8-bromo-cAMP (Br-cAMP) and 8-chlorophenylthio-cAMP, when applied in a bath for up to one hour during subjective daytime, permanently advanced the time of peak neuronal activity; non-cyclic 8-BA-5′-AMP was ineffective (Gillette & Prosser 1988, Prosser & Gillette 1989). When the SCN was treated at midday, peak firing appeared 4.5 h before the expected time in the two cycles monitored thereafter. These analogues did not produce phase shifts when given during subjective night (Fig. 2). Likewise, agents that increase endogenous cAMP, such as forskolin or RO20-1724, were effective only in the daytime. Conversely, Br-cGMP, a structurally related purine cyclic nucleotide analogue, was effective in antiphase to Br-cAMP and BA-cAMP; it adjusted phase in the SCN only when applied at night-time (Fig. 2, Prosser et al 1989). These were the first demonstrations not only of phase adjustment in the isolated SCN, but also of spontaneous waxing and waning of one sensitivity followed by another. These findings established that SCN clock properties persist *in vitro*.

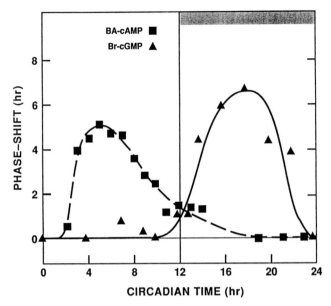

FIG. 2. The sensitivity of the rat SCN to cAMP is in antiphase to that to cGMP. Phase shifts in the time of the peak in the electrical activity rhythm are plotted against the time at which the SCN was treated *in vitro* for one hour with analogues, 8-benzylamino-cAMP (BA-cAMP) or 8-bromo-cGMP (Br-cGMP), applied in a bath. Bromo-5′-AMP was without effect. The rats were entrained to the same light–dark cycle as in Fig. 1. Shaded horizontal bar represents subjective night. Adapted from Prosser et al (1989).

Both cAMP and cGMP are potent regulators of cell state. Their 8′-analogues are exceptional activators of the protein kinases regulated by these cyclic nucleotides and are degraded only slowly (Meyer & Miller 1974). Differential sensitivity of the clock to each nucleotide over the circadian cycle may be modulated at many levels. Although the concentrations of the kinases do not appear to change, the concentrations of their regulatory cyclic nucleotides, as well as the phosphorylation states of the kinases and some substrates, oscillate over the 24 h cycle (Prosser & Gillette 1991, Faiman & Gillette 1991, Weber & Gillette 1991). The critical point at which these changes adjust the clock mechanism has yet to be determined.

The identity of the synaptic neuromodulators mediating phase adjustment, including those activating these second messengers, is not yet known; however, candidates are neuroactive substances known to be localized in afferent fibres and those with ligand-binding sites within the SCN. Major projections to the SCN include those from: (1) the dorsal raphe, containing serotonin (5-hydroxytryptamine, 5-HT); (2) the intergeniculate leaflet of the lateral geniculate nucleus, containing neuropeptide Y (NPY) and γ-aminobutyric acid (GABA);

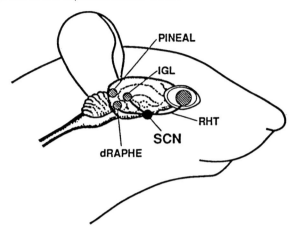

FIG. 3. Schematic of the head of a rat showing major brain sites that regulate the SCN. The suprachiasmatic nuclei are positioned at the base of the hypothalamus, directly over the optic chiasm. They receive serotonergic projections from the dorsal raphe (dRAPHE), NPY/GABAergic projections from the intergeniculate leaflet (IGL), and glutamate-liberating axons from the retina via the retinohypothalamic tract (RHT). Additionally, the pineal is the source of circulating melatonin.

and (3) the retina, via the retinohypothalamic tract (RHT), containing a glutamate precursor (Fig. 3). Additionally, the SCN is one of the few brain regions that bind significant amounts of melatonin, the indoleamine produced nocturnally by the pineal. We have explored the sensitivities of the SCN to candidate modulators.

Neurons of the ventrolateral SCN (vlSCN) of the rat receive serotonergic fibres from the dorsal raphe (Azmitia & Segal 1978, Moore et al 1978, van de Kar & Lorens 1979) as well as NPY- and glutamate-containing terminals from the IGL and retina, respectively (Card & Moore 1989, Mikkelsen & O'Hare 1991, Castel et al 1993). This indicates that the vlSCN in the rat is a major site of signal integration; furthermore, because these fibre types may synapse both on each other and on the same SCN neuron (van den Pol & Gorcs 1986, Guy et al 1987), this integration may have both presynaptic and postsynaptic components. Cognizant of the localized nature of the termination sites of these projections, we designed these experiments to examine the effects of localized application of test substances in microdrops applied directly to the vlSCN.

Both 5-HT and NPY adjust phasing in the SCN when applied *in vitro* during subjective daytime. The period of sensitivities to these two neuromodulators encompasses the period of phase advance induced by non-photic stimuli (Mrosovsky et al 1989; see Mrosovsky 1995, this volume). 10^{-6} M 5-HT applied to the vlSCN in a 10^{-11} ml droplet for five minutes between CT 2 and CT 10 caused significant phase adjustments (> 1 h) (Fig. 4; Medanic & Gillette 1992).

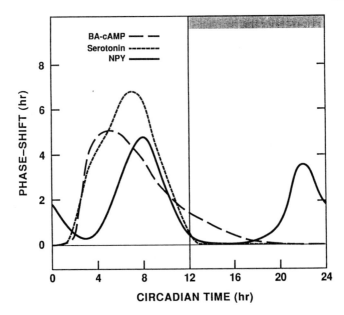

FIG. 4. The family of phase response curves generated in response to three daytime phase-shifting agents, 8-benzylamino-cAMP (BA-cAMP), 5-HT and NPY. BA-cAMP was applied in a bath for one hour at 5×10^{-5} M, whereas 5-HT and NPY were applied at 10^{-6} M in microdrops to the vlSCN for five minutes. Notice the similarities and differences within this family of curves. Curves were fitted by eye to results from the following: BA-cAMP, 25 experiments at 16 CTs from Prosser & Gillette (1989); serotonin, 20 experiments at 7 CTs from Medanic & Gillette (1992); NPY, 30 experiments at 11 CTs from Medanic & Gillette (1993). The rats were entrained as in Fig. 1. Subjective night is represented by the shaded horizontal bar.

The maximal response occurred to application of 5-HT at CT 7 which induced a 7.0 ± 0.1 h phase advance. This period of sensitivity to 5-HT applied as a microdrop overlaps with but is not identical to that for cAMP applied in the bath (Fig. 4). Although the differences in the details of the phase responses to 5-HT and cAMP could be a result of the differing extent of exposure to the stimulus between microdrop and bath application, the similarities, especially between CT 2 and CT 5, are striking and suggestive of a common pathway. Studies with agonists and antagonists have implicated a 5-HT_{1A}-like receptor (Medanic & Gillette 1992, Shibata et al 1992, Prosser et al 1993). A new cAMP-coupled 5-HT receptor (5-HT_7) has recently been cloned from rat brain (Lovenberg et al 1993); although it has not been localized to the SCN, should this receptor reside either in vlSCN neurons or in presynaptic terminals, it could modulate the early phase of the 5-HT-induced shift by a cAMP-dependent mechanism. The disparity between the cAMP- and 5-HT-induced phase response curves after this point would suggest that the latter response, from CT 6 to CT 11 and including the maximal effect, is mediated at least partially by a non-cAMP-dependent mechanism.

Daytime sensitivity to NPY microdrops applied to the vlSCN significantly lags behind the sensitivities to cAMP and 5-HT (Fig. 4; Medanic & Gillette 1993). It first appears at CT 5 and peaks at CT 8 with a maximal phase shift of 4.5 h—about 2.5 h less than the 5-HT peak—and then dissipates in synchrony with the waning 5-HT sensitivity. This temporal pattern of sensitivity of the SCN to NPY in subjective day makes it unlikely that NPY utilizes a cAMP-stimulating signal transduction mechanism. A second period of sensitivity to NPY anticipates dawn, peaking at CT 22, a time when the suprachiasmatic nuclei are insensitive to regulation by both cAMP and cGMP. This bimodal pattern of sensitivity of the free-running SCN to NPY raises questions about the mechanism(s) regulating sensitivity. Is the receptor itself disappearing, then reappearing, or is regulation at the level of intracellular signal transduction elements or at the level of cellular substrates of the transduction cascade? Our results regarding changing sensitivity to cAMP stimulation would suggest that regulation of cellular substrates is the most likely. Whatever the level(s) of regulation, these changes must be close to or driven by the clock, and thus understanding them should lead us toward a better understanding of the timekeeping elements.

Nocturnal sensitivity of the SCN to phase adjustment must include the neural substrates of photic entrainment. Because many lines of evidence support a role for glutamate in mediating light signals from the environment at RHT synapses at the vlSCN (Rusak & Bina 1990, Kim & Dudek 1991), we examined temporal changes in the effect of focal application of glutamate to the SCN. One μl drops, which effectively covered one SCN with 10 mM glutamate for 10 min, produced phase delays and phase advances which are remarkably like those produced by light in time, shape and amplitude, with a maximal delay at CT 14, and maximal advance at CT 19 (Ding & Gillette 1993). Interestingly, although this period of sensitivity spans that to cGMP, the biphasic shape and lesser amplitude of the response to glutamate suggest that activation of cGMP pathways cannot wholly explain the result with glutamate.

Interspersed between the daytime and night-time sequences of sensitivities of the SCN to these neuromodulators are periods that represent the entrained light–dark and dark–light transition zones. The first of these, which surrounds environmental dusk, coincides with a period when injections of the pineal hormone melatonin have been shown to entrain rats and alter energy utilization in the SCN (Redman et al 1983, Cassone et al 1986, 1988). Not surprisingly, bath application of 10^{-9} M melatonin for 60 min adjusts the subsequent cycles of firing rate in the SCN (McArthur et al 1991) by advancing the clock by up to 4.5 h. Careful examination of sensitivity surrounding dawn revealed a second, narrow period of sensitivity to melatonin that appeared sharply at CT 23 and then decayed over the next two hours (McArthur & Gillette 1992). Both of these sensitive periods closely follow those to NPY. In this respect, melatonin resembles NPY: both show two windows of sensitivity, separated by many hours, which appear spontaneously in constant conditions *in vitro* without exposure to the phase-adjusting agent. In the case of melatonin, these sensitive periods occur at times when seasonal changes in night-length can be expected to stretch the duration of pineal melatonin synthesis into these temporal domains of the SCN.

Overall, we have accumulated evidence that the properties of a circadian clock persist in the SCN *in vitro*. The suprachiasmatic nuclei generate multiple, near 24 h cycles of ensemble neuronal firing rate and show differential sensitivity to phase-adjusting stimuli. The patterns of membrane properties and neuronal activity are circadian, even though less than the entire SCN is present in the brain slice. The SCN *in vitro* continues to generate a stable circadian oscillation in the ensemble electrical activity, even though component neurons show diversity in electrophysiological properties. The contribution of the various electrophysiological types of neurons to integrative or timekeeping functions of the SCN is presently unknown. Further, the isolated SCN proceeds through an orderly sequence of sensitivities to resetting stimuli. This occurs without prior exposure to the stimuli during that circadian cycle. The onset of each sensitive period is generally more rapid than its decay. The neuromodulatory stimuli identified so far act through different signal transduction pathways which are insensitive to direct activation during non-sensitive periods. This suggests that the processes that regulate sensitivity operate beyond the level of the receptor, at least one step into the cell. The cellular substrates that determine the sensitivity and the characteristics of the response to extracellular regulators define separate temporal domains, which appear as an orderly sequence of biochemical and molecular relationships that together make up the circadian cycle. Understanding the components of each domain, its positive and negative regulators and the linkages between successive domains, should permit fine control of phase adjustment.

Acknowledgements

The generous research support of the US Public Health Service (NINDS NS22155) and the US Air Force Office of Scientific Research (90-0205 and F49620-93-1-0413) is gratefully acknowledged.

References

Azmitia EC, Segal M 1978 An autoradiographic analysis of the differential ascending projections of the dorsal and median raphe nuclei in the rat. J Comp Neurol 179: 641–668

Blanton MG, Lo Turko JJ, Kreigstein AR 1989 Whole cell recording from neurons in slices of reptilian and mammalian cerebral cortex. J Neurosci Methods 30:203–210

Card JP, Moore RY 1989 Organization of lateral geniculate–hypothalamic connections in the rat. J Comp Neurol 284:135–147

Cassone VM, Chesworth MJ, Armstrong SM 1986 Entrainment of rat circadian rhythms by daily injection of melatonin depends upon the hypothalamic suprachiasmatic nuclei. Physiol & Behav 36:1111–1121

Cassone VM, Roberts MH, Moore RY 1988 Effects of melatonin on 2-deoxy-[1-^{14}C]-glucose uptake within rat suprachiasmatic nucleus. Am J Physiol 255:R332–R337

Castel M, Belensky M, Cohen S, Ottersen OP, Storm-Mathisen J 1993 Glutamate-like immunoreactivity in retinal terminals of the mouse suprachiasmatic nucleus. Eur J Neurosci 5:368–381

Ding JM, Gillette MU 1993 Glutamate induces light-like phase shifts in the rat SCN in brain slice. Soc Neurosci Abstr 19:1815

Faiman L, Gillette MU 1991 Temporal changes in protein kinase A substrates in the rat suprachiasmatic nucleus. Soc Neurosci Abstr 17:671

Gallman EA, Gillette MU 1992 Whole cell recording of neurons of the suprachiasmatic nucleus (SCN) studied in rat brain slice. Soc Res Biol Rhythms Abstr 3:29

Gallman EA, Gillette MU 1993 Circadian modulation of membrane properties of SCN neurons in rat brain slice. Soc Neurosci Abstr 19:1703

Gillette MU 1991 SCN electrophysiology *in vitro*: rhythmic activity and endogenous clock properties. In: Klein DC, Moore RY, Reppert SM (eds) Suprachiasmatic nucleus: the mind's clock. Oxford University Press, New York

Gillette MU, Prosser RA 1988 Circadian rhythm of rat suprachiasmatic brain slice is rapidly reset by daytime application of cAMP analogs. Brain Res 474:348–352

Guy J, Bosler O, Dusticier G, Pelletier G, Calas A 1987 Morphological correlates of serotonin-neuropeptide Y interactions in the rat suprachiasmatic nucleus: combined radioautographic and immunocytochemical data. Cell Tissue Res 250:657–662

Kim YI, Dudek FE 1991 Intracellular electrophysiological study of suprachiasmatic nucleus neurons in rodents: excitatory synaptic mechanisms. J Physiol 444:269–287

Lovenberg TW, Baron BM, de Lecea L et al 1993 A novel adenylyl cyclase-activating serotonin receptor (5-HT$_7$) implicated in regulation of mammalian circadian rhythms. Neuron 11:449–458

McArthur AJ, Gillette MU 1992 Melatonin resets the SCN circadian clock *in vitro* within a narrow window of sensitivity near dawn. Soc Neurosci Abstr 18:879

McArthur AJ, Prosser RA, Gillette MU 1991 Melatonin directly resets the suprachiasmatic circadian pacemaker. Brain Res 565:158–161

Medanic M, Gillette MU 1992 Serotonin regulates the phase of the rat suprachiasmatic circadian pacemaker *in vitro* during the daytime. J Physiol 450:629–642

Medanic M, Gillette MU 1993 Suprachiasmatic circadian pacemaker of rat shows two windows of sensitivity to neuropeptide Y *in vitro*. Brain Res 620:281–286

Meyer RB, Miller JP 1974 Analogs of cyclic AMP and cyclic GMP: general methods of synthesis and the relationship of structure to enzymatic activity. Life Sci 14:1019–1040

Mikkelsen JD, O'Hare MMT 1991 An immunohistochemical and chromatographic analysis of the distribution and processing of neuropeptide Y in the rat suprachiasmatic nucleus. Peptides 12:177–185

Moore RY, Halaris AE, Jones BE 1978 Serotonin neurons of the midbrain raphe: ascending projections. J Comp Neurol 180:417–438

Mrosovsky N 1995 A non-photic gateway to the circadian clock of hamsters. In: Circadian clocks and their adjustment. Wiley, Chichester (Ciba Found Symp 183) p 154–174

Mrosovsky N, Reebs SG, Honrado GI, Salmon PA 1989 Behavioural entrainment of circadian rhythms. Experientia 45:696–702

Prosser RA, Gillette MU 1989 The mammalian circadian clock in the suprachiasmatic nuclei is reset *in vitro* by cAMP. J Neurosci 9:1073–1081

Prosser RA, Gillette MU 1991 Cyclic changes in cAMP concentration and phosphodiesterase activity in a mammalian circadian clock studied *in vitro*. Brain Res 568:185–192

Prosser RA, McArthur AJ, Gillette MU 1989 cGMP induces phase shifts of a mammalian circadian pacemaker at night, in antiphase to cAMP effects. Proc Natl Acad Sci USA 86:6812–6815

Prosser RA, Dean RR, Edgar DM, Heller HC, Miller JD 1993 Serotonin and the mammalian circadian system. I. *In vitro* phase shifts by serotonergic agonists and antagonists. J Biol Rhythms 8:1–16

Redman JR, Armstrong SM, Ng KT 1983 Free-running activity rhythms in the rat: entrainment by melatonin. Science 219:1089–1091

Rusak B, Bina KG 1990 Neurotransmitters in the mammalian circadian system. Annu Rev Neurosci 13:387–401

Shibata S, Tsuneyoshi A, Hamada T, Tominaga K, Watanabe S 1992 Phase-resetting effect of 8-OH-DPAT, a serotonin$_{1A}$ receptor agonist, on the circadian rhythm of firing rate in the rat suprachiasmatic nuclei *in vitro*. Brain Res 582:353–356

van der Kar LD, Lorens SA 1979 Differential serotonergic innervation of individual hypothalamic nuclei and other forebrain regions by the dorsal and median raphe nuclei. Brain Res 163:45–54

van den Pol AN, Gorcs T 1986 Synaptic relationships between neurons containing vasopressin, gastrin-releasing peptide, vasoactive intestinal polypeptide, and glutamate decarboxylase immunoreactivity in the suprachiasmatic nucleus: dual ultrastructural immunocytochemistry with gold-substituted silver peroxidase. J Comp Neurol 252:507–521

Weber ET, Gillette MU 1991 Temporal changes in protein kinase G substrates in the rat suprachiasmatic nucleus. Soc Neurosci Abstr 17:671

Wheal HV, Thompson AM 1984 The electrical properties of neurones of the rat suprachiasmatic nucleus recorded intracellularly *in vitro*. Neuroscience 13:97–104

DISCUSSION

Armstrong: Although we have looked very carefully, we find that melatonin injections given around dawn to freely moving rats have no effects on locomotor activity (Armstrong et al 1989). We find just a narrow window of sensitivity between CT 9 and CT 12. Recently, we even tried giving injections on three consecutive days at the same CT and this did not work either (unpublished work). So, although the SCN in your preparation seems to be sensitive to melatonin at subjective dawn, we cannot pick up that sensitivity in overt locomotor behaviour.

Gillette: All I would ever claim is that we are showing the sensitivities the SCN is capable of when it is treated with one stimulus in an isolated situation. It will be interesting to start treating the SCN with different stimuli simultaneously, because that's undoubtedly what would happen normally. We've begun by treating the SCN simultaneously with serotonin and NPY. Much to our surprise, we don't get an intermediate response. The response to NPY predominates, unless its concentration is low. All we have is a skeleton of response capabilities. At dawn there may normally be a number of stimuli coming in regulating the sensitivity to melatonin. It would be interesting to put the animal on different photoperiods and see if the sensitivity changed.

Cassone: Of course, in the real world, there are changes according to the time of year. Equally important is the fact that the light cycle of the real world is not a square wave made up of 300 lux or nothing. The various neural and endocrine inputs to the SCN might modulate pacemaker functioning in addition to the effects demonstrated on phase. They may also regulate the response to the ambiguous light intensities and colours that occur in the natural world.

Gillette: I would agree with that. We've described the pathways which have been best studied, but they are by no means the only pathways. Dr Moore spoke earlier about substance P as a potential input, and Cote & Harrington (1993) have evidence that the SCN can respond to histamine. The suprachiasmatic nuclei express many other receptors, which may or may not be functional. This is going to turn out to be a system whose regulation is complex.

Reppert: I gather that you think melatonin is activating protein kinase C. How do you think it does this?

Gillette: We think it's working through the classic phospholipase C pathway (McArthur & Gillette 1992), although we haven't been able to measure changes in inositol 1,4,5-trisphosphate ($InsP_3$). This pathway usually involves a receptor interacting through a heterotrimeric G protein with phospholipase C, which then cleaves membrane phospholipids into diacylglycerol, which activates protein kinase C in the membrane, and $InsP_3$, which liberates intracellular Ca^{2+}. Often, $InsP_3$ has a short half-life in the cell and is moved on into other phosphoinositol intermediates, so you need to look not only for $InsP_3$, but also all its various breakdown products.

Reppert: We too have not found melatonin-induced $InsP_3$ changes in the rat SCN (L. L. Carlson & S. M. Reppert, unpublished work). You could also look for changes in intracellular Ca^{2+} by imaging.

Gillette: We haven't done that.

Reppert: Vanecek & Klein (1992) have shown in another system, the neonatal rat pituitary gland, that melatonin actually *inhibits* the gonadotropin-releasing hormone-induced increase in intracellular Ca^{2+}.

Hastings: In the ovine pars tuberalis we've been unable to find any effects of melatonin on phospholipase C, phospholipase D or phospholipase A, but that's a different tissue (McNulty et al 1994).

Gillette: We have looked at this very carefully, to see if we can show the same effects as your group (Hazelrigg et al 1991) and Dr Reppert's (Carlson et al 1989) with melatonin blocking forskolin-stimulated cAMP production. We could not. I would suggest the signal transduction pathway in the rat SCN may be different from that in the median eminence/pars tuberalis.

Reppert: I disagree with your assessment of the results with cAMP. The concentration of melatonin receptors in the SCN is low. Thus, an SCN slice or explant will have only a small proportion of the cAMP pool coupled to melatonin receptors. When you activate all the cAMP in SCN-containing tissue with forskolin, you would not expect to see a decreased cAMP response with melatonin. In the hypophysial pars tuberalis, the situation is different. The melatonin receptor concentration is high in that tissue such that melatonin inhibits most of the forskolin-stimulated cAMP (Carlson et al 1989).

Gillette: Our results with TPA (12-*O*-tetradecanoylphorbol 13-acetate) and staurosporine correlate surprisingly closely with those with melatonin in the magnitude of the phase advance at both dusk and dawn (McArthur & Gillette 1992). Furthermore, the protein kinase C inhibitor staurosporine completely blocks the phase shifts induced at dawn and dusk by melatonin and TPA. The SCN is sensitive to neither earlier in the day. Thus, protein kinase C is implicated by agonists and antagonists at both windows of sensitivity to melatonin.

Turek: A few years ago there was some interest in the fact that the SCN has functionally distinct regions. You mentioned redundancy in the SCN, but most of your results are from the ventrolateral SCN. Do you think there are differences between the dorsomedial and the ventrolateral SCN?

Gillette: The redundancy lies, I believe, anterioposteriorially. We and others have been saying that for several years. Before we were good at making brain slices, we were happy to see half of the SCN in our slice, but we always, even new students, got the same pattern of activity. That suggests to me that when you are sampling in a coronal slice, no matter whether you have the rostral or caudal half or the medial portion, you get the same pattern. Recordings in the intact SCN slice, whether dorsomedial or ventrolateral, gave a nice oscillation (Tcheng & Gillette 1991). The difference came when we then took that coronal slice and cut it at an angle of 45°. The ventrolateral SCN behaved as though we hadn't done anything to it, as though it were in the intact slice. We were left with much less than half of the SCN yet we got a strong 24 h rhythm. However, there was a change in the dorsomedial portion. If we made a high dorsomedial cut, we could not see a clear oscillation. If we moved the cut down towards the ventrolateral portion, the high amplitude, 24 h oscillation returned in the dorsomedial piece (Gillette et al 1992). This suggests there is something unusual about the intermediate zone that can confer rhythmicity on the dorsomedial region. What surprised us was that the dorsomedial cells were not primarily the pacemaker region, which was our original hypothesis. The ventrolateral portion not only gets the signals and integrates them, but can also generate the rhythm.

Miller: The effect of melatonin in late subjective night is interesting in the light of Kilduff et al's (1992) report that c-*fos* is expressed by the SCN in rats injected subcutaneously with melatonin at that time. We have no idea what behavioural relevance that has.

Gillette: This may be involved in shaping entrainment patterns at the end of long nights in the annual solar cycle.

Miller: There is at least one indole, serotonin, that does seem to mobilize Ca^{2+} in the SCN, at least in fluorescence studies (van den Pol et al 1992).

One problem here is that you are looking at a whole mishmash of SCN cells, within which there is a population of clock cells. We have strong evidence that the 5-HT receptor of relevance here increases the concentration of cAMP yet we have never been able to demonstrate an increase in cAMP concentration by biochemical methodology. Perhaps because only certain cells, a small proportion, are involved, we cannot visualize the increase. Similarly, you may be losing the $InsP_3$ change in the noise.

You have to be phenomenally careful in talking about time domains when you're also mixing paradigms. Part of your conclusions are based on the microdrop methodology and part on bath administration. As you know, our results with serotonin are very different from yours, with respect to the phase delay we see with bath-administered 5-HT or quipazine in the SCN slice at CT 18 (Prosser et al 1993).

Gillette: You have to be even more careful about whether you are using serotonin itself or an agonist which can act as an antagonist at some receptors.

Miller: We have used both quipazine and 5-HT at CT 18, though we haven't generated a full phase response curve (PRC) for serotonin.

Gillette: My point is that you cannot equate serotonin with quipazine, because they can act differently. We have found that agonists of different receptor subtypes induce phase shifts of different magnitudes (Medanic & Gillette 1992).

Miller: The effects of serotonin were identical to the effects of quipazine, and we blocked both effects with metergoline. Although the serotonergic innervation of the SCN is primarily in the ventrolateral division, there is no doubt that there is serotonergic innervation above and beyond the ventrolateral division, so it is quite possible that bath application and microdrop application only to the ventrolateral portion could produce different responses. One of us has to do the corresponding study to straighten this out. The lack of a delay portion in the PRC may not represent a difference in time domain but a difference in technique.

Gillette: Perhaps we should clarify this for the benefit of others. With bath application of serotonin or quipazine in the early portion of the night Dr Miller gets a phase delay. He does not get this with 8-OH-DPAT (8-hydroxy-2-[di-*n*-propylamino]tetralin), so the event is not mediated by a $5-HT_{1A}$-like receptor, which is interesting. With microdrop application of serotonin or 8-OH-DPAT to the ventrolateral SCN, we do not observe this phase delay at night. We each used different methods of drug application and different agents, and, provocatively, have slightly different results. Interestingly, Mike Rea's group has found that serotonergic agonists can block the optic nerve-stimulated volley of depolarization in the SCN *in vitro* and light-stimulated phase shifts *in vivo* (without themselves stimulating phase shifts) (Rea et al 1994). These findings suggest that there is an interaction between the serotonin and glutamate systems in the SCN, as in other brain regions. The pharmacology suggests a 5-HT$_1$-like receptor mediates this effect. Yet, there are other interesting serotonin receptor subtypes found in the SCN which apparently don't mediate the phase advance late in the day or the blocking effect on photic stimuli at night. The serotonin story is going to be complex. This is not surprising considering the diversity of the serotonin system, in terms of both receptor subtypes and signal transduction pathways, and subsequent intersection with other signalling systems.

Waterhouse: This discussion has raised a more general point, that the exact method by which the study is done will influence the kind of result which is obtained. The larger problem to me is the question of how the *in vitro* work relates to the *in vivo* situation. The SCN has many inputs from many brain areas, yet what, essentially, you have described is a similarity between what is found in the total system and what is found in your isolated system. That similarity seems to decrease the importance of what's coming in to the SCN.

Moore: We always hope that the *in vitro* system will provide information about the *in vivo* system, but it's clear that it provides information which is flawed in various ways. Dr Gillette gets nice phase advances but rarely gets phase delays, whereas NPY given *in vivo* produces phase delays, as do dark pulses. The *in vitro* situation is not giving us the same kind of information that the *in vivo* situation is. This system is so tremendously regulated *in vivo* that the *in vitro* system will never approximate it.

Gillette: I agree that there are differences, but we should learn something from them. It's difficult to know that one treatment *in vivo* is really affecting the sites that would normally be stimulated as a consequence of handling or cage changing, for example, and to know that the injected substance is itself the primary player.

Meijer: We have applied glutamate in the *in vivo* situation through cannulae implanted into the SCN. Injection of glutamate produces a PRC like that produced by a dark pulse, whereas you found a light pulse-like PRC *in vitro*. We have always been afraid that glutamate injections in the SCN stimulate the cells at non-physiological sites, and that this was why we obtained this PRC which we did not expect. A possible explanation for our unexpected PRC is that by injecting glutamate into the SCN *in vivo* we also stimulate pathways to the intergeniculate leaflet, for example, and in turn stimulate the projection from the IGL to the SCN which would enhance the release of NPY and GABA, which may produce a dark pulse-like PRC. *In vivo*, you stimulate all these pathways, which in turn stimulate the SCN.

Mrosovsky: Another difference between the *in vivo* and *in vitro* experiments is that the amplitudes you get from slices tend to be large. However, the glutamate-induced phase shifts *in vitro* seem to be an exception. What is the amplitude of the PRC produced by a light pulse in the rat, and how closely does this correspond to what you are getting in the slice with glutamate?

Gillette: We noticed from the beginning that we got phase shifts of large amplitude. One of the first things I did was to apply depolarizing stimuli which induced phase shifts of huge amplitude. We thought this was because when we made the slice we were cutting away the feedback loops which would normally dampen any change. This makes adaptive sense, because there's no reason for an animal suddenly to undergo a phase shift of seven hours. It was quite a surprise to find that dropping glutamate onto the SCN gave us something that looked so much like the photic response. The results suggest that this phase shift is processed primarily within the SCN. Light is one of the strongest entraining stimuli, so perhaps this makes sense. It's really interesting that we see a delay that looks like that produced in response to light and an advance that also has an amplitude of the same order of magnitude as that produced by light.

Mrosovsky: My other point relates to the phase response to NPY, and the apparent differences between Albers & Ferris's (1984) PRC and your bimodal curve. Albers and Ferris's work was a pioneering effort, but their PRC was based on rather few points. In the delay area, near where you found your second advance area, there was one advance point and two delay points, with the average as a delay. That is not enough really for comparisons between different preparations.

Gillette: This finding of two sensitive periods was an unexpected result. This must be interpreted as being what the SCN is capable of when it experiences only NPY in the ventrolateral region. The sensitivity to NPY changes *in vitro*, so it must be regulated by whatever is ongoing and changing over 24 h *within* the SCN.

Shinohara et al (1992) showed that there is increased NPY release in the SCN just after light–dark and dark–light transitions *in vivo*. For either of these NPY releases to affect SCN phase, the photic transitions must intersect the SCN's sensitive periods. Because NPY injected into the SCN *in vivo* causes advances in subjective day, like the *in vitro* response, but at night causes delays, the opposite phase changes from *in vitro* responses, additional regulators must participate nocturnally *in vivo*.

Miller: It is definitely the case that there are conditions in which the *in vivo* situation reflects the *in vitro* situation. Injection of a serotonin agonist into the third ventricle induces phase advances at CT 6 and delays at CT 18 *in vivo* (Edgar et al 1993). It is true that the magnitude of those phase shifts was smaller than we find in SCN slices, but the time course was qualitatively the same.

Daan: Another aspect which is not fully appreciated is that these measurements are made on the first day after the pulse—these are immediate responses. One does not see phase advances in the activity of the whole animal on the first day after a pulse. This is an important difference, which suggests that the pacemaker is reset much more rapidly then the whole system. Do you see differences between Day 1 and Day 2, Dr Gillette, or can you say that there are no transient changes in phase at all in these slices?

Gillette: We have not been able to see what I would call transients. We are cautious about interpreting any change of less than an hour as a significant change. Everything that we have used to induce a phase shift we have checked in two subsequent cycles; in each case, the peak appears 24 h after the first shifted peak, suggesting the system has shifted fully in the first day after the treatment.

Menaker: A classic observation made with *Drosophila* years ago was that whereas phase shifts produced by light are 'permanent', and reset the whole system, others, such as those produced by temperature steps, are transient and decay rapidly (Pittendrigh et al 1958). In your paradigm it is difficult, if not impossible, to distinguish permanent and transient phase shifts and therefore it will be extremely difficult to make the correlation between *in vitro* and *in vivo* results.

Gillette: I don't know to what extent mechanisms could be evaluated from the results with temperature steps, but you might suggest that a permanent resetting involves a fundamental change in the biochemical machinery, whereas a shift in temperature might change the kinetics of something which is not tweaking the clock mechanism, whatever that is.

Menaker: The results in *Drosophila* were explained convincingly and elegantly in terms of the action of a temperature change on one of two coupled oscillators.

Gillette: Was a second treatment given? In Pittendrigh's classical study (1974), the speed of phase resetting was assessed through the effect of double pulses of light on *Drosophila* eclosion rhythms. When two pulses of light were given within one cycle, the response was the sum of the individual responses, suggesting that the pacemaker's state was rapidly shifted to a new time point after the first treatment and it was on this new time state that the second pulse acted. Similarly, Dr Hastings has shown that handling of the animal can change within one hour the ability of light to induce *c-fos* (Mead et al 1992).

Menaker: Two-pulse experiments have been done with *Drosophila*.

Gillette: With the temperature steps?

Edmunds: No, but experiments with temperature pulses were done, which gave steady-state results quite different from those obtained with temperature steps (Zimmerman et al 1968). The results with the pulses could be predicted from the algebraic addition of the separate effects yielded by the steps down and steps up of which the pulses were formally composed.

Lewy: It's impressive that the second day's phase shift is so much less than that after the first day. This suggests that much of the shift occurs within one day. Have you done, or are you considering doing, a two-pulse experiment within one day, to see how instantaneous the phase shift is?

Gillette: That would be a good experiment.

When I gave my first talk on the SCN at the Society for Neuroscience meeting in 1985, Dr Block asked me whether the phase shift *in vitro* was instantaneous; since I was new to the field, I asked what an instant was for a clock. For the

Neuronal rhythms in the suprachiasmatic nuclei

first molecular interaction, is it five minutes? We know that our five-minute treatments can produce long-lasting change, but we are setting into motion a cascade of events.

Edmunds: The possible differences between PRCs based on data from Day 1 and data from later on are important, although in your system you are limited, of course, by the time the preparation will survive *in vitro*. Do you find lasting, stable, phase shifts with pulses of cAMP and its agonists or antagonists?

Gillette: They're stable over at least two cycles.

Edmunds: Do you also apply forskolin as a pulse?

Gillette: Yes.

Edmunds: I ask because we have done the same thing in *Euglena*, but we also tried an experiment in which we added saturating concentrations of forskolin to build up the concentration of cAMP (Carré & Edmunds 1993). We reasoned that if we wiped out the normal bimodal cAMP signal, the *rhythm* downstream would be wiped out if it were mediated by the bimodal cAMP signal. Cell division suddenly became essentially arrhythmic, though not stopped. Have you done any experiments like that, blocking not the function but the rhythm?

Gillette: None of the agents I spoke about in my presentation wipes out the rhythm, but some things do. All I can say is that with some treatments we can't measure a rhythm of electrical activity, which is our measure of clock functioning. Several treatments given at night make it impossible to determine phase in the next cycle as there is no clear peak. Since we don't maintain the slices for more than two cycles after treatment, we haven't pursued those agents.

Hastings: Apart from glutamate, most of your treatments seem to have advancing effects. If these influences are operational in the animal's life, are you not surprised that when you take the SCN out and isolate the slice from those influences that the period doesn't lengthen?

Gillette: You might be surprised by that. All I can say is that we haven't been able to observe a lengthened period. The real surprise is that the period we see is so close to 24 h even though we have variable amounts of SCN tissue. Longitudinally in time, the rhythm is running at slightly less than 24 h. We would like to look at this more carefully.

Hastings: Do you know the individual periods of your rats before you prepare the slices?

Gillette: No. It would be useful to know that.

Block: I understand that you have found conductance rhythms (Gallman & Gillette 1993), like those we have observed in *Bulla*, with conductance being low when impulse activity is high, and high when impulse activity is low. Perhaps K^+ channels are closing, causing an increase in activity. The only caveat is that there may be inhibitory synapses that are active at night, which would also give an increase in conductance. Nevertheless, it's reassuring these two systems show the same changes in conductance.

Gillette: There is even more reason to think that conductance changes are basic clock features. A recent study on a plant organ, the monkey pod leaf, has shown circadian conductance rhythms in K$^+$ channels in opposing regions of the leaf-movement organ (Kim et al 1993).

References

Albers HE, Ferris CF 1984 Neuropeptide Y: role in light–dark cycle entrainment of hamster circadian rhythms. Neurosci Lett 50:163–168

Armstrong SM, Thomas EMV, Chesworth MJ 1989 Melatonin induced phase-shifts of rat circadian rhythms. In: Reiter RJ, Pang SF (eds) 1989 Advances in pineal research. J Libbey, London, vol 3:265–270

Carré IA, Edmunds LN Jr 1993 Oscillator control of cell division in *Euglena*: cyclic AMP oscillations mediate the phasing of the cell division cycle by the circadian clock. J Cell Sci 104:1163–1173

Carlson LL, Weaver DR, Reppert SM 1989 Melatonin signal transduction in hamster brain: inhibition of adenylyl cyclase by a pertussis toxin-sensitive G protein. Endocrinology 125:2670–2676

Cote NK, Harrington ME 1993 Histamine phase shifts the circadian clock in a manner similar to light. Brain Res 613:149–151

Edgar DM, Miller JD, Prosser RA, Dean RR, Dement WC 1993 Serotonin and the mammalian circadian system. II. Phase shifting rat behavioral rhythms with serotonergic agonists. J Biol Rhythms 8:17–31

Gallman EA, Gillette MU 1993 Circadian modulation of membrane properties of SCN neurons in rat brain slice. Soc Neurosci Abstr 19:1703

Gillette MU, McArthur AJ, Liu C, Medanic M 1992 Intrinsic organization of the SCN circadian pacemaker studied by long term extracellular recording *in vitro*. Soc Res Biol Rhythms Abstr 3:28

Hazelrigg DG, Morgan PJ, Lawson W, Hastings MH 1991 Melatonin inhibits the activation of cyclic AMP-dependent protein kinase in cultured pars tuberalis cells of ovine pituitary. J Neuroendocrinol 3:597–603

Kilduff TS, Landel HB, Nagy GS, Sutin EL, Dement WC, Heller HC 1992 Melatonin influences Fos expression in the rat suprachiasmatic nucleus. Mol Brain Res 16:47–56

Kim HY, Coté GG, Crain RC 1993 Potassium channels in *Samanea saman* protoplasts controlled by phytochrome and the biological clock. Science 260:960–962

McArthur AJ, Gillette MU 1992 Melatonin resets the SCN circadian clock *in vitro* within a narrow window of sensitivity near dawn. Soc Neurosci Abstr 18:879

McNulty S, Morgan PJ, Thompson M, Davidson G, Lawson W, Hastings MH 1994 Phospholipases and melatonin signal transduction in the ovine pars tuberalis. Mol Cell Endocrinol 99:73–79

Mead S, Ebling FJP, Maywood ES, Humby T, Herbert J, Hastings MH 1992 A nonphotic stimulus causes instantaneous phase advances of the light-entrainable circadian oscillator of the Syrian hamster but does not induce c-*fos* expression in the suprachiasmatic nucleus. J Neurosci 12:2516–2522

Medanic M, Gillette MU 1992 Serotonin regulates the phase of the rat suprachiasmatic circadian pacemaker *in vitro* during the daytime. J Physiol 450:629–642

Pittendrigh CS 1974 Circadian oscillations in cells and the circadian organization of multicellular systems. In: Schmitt FO, Worden FG (eds) The neurosciences, third study program. MIT Press, Cambridge, MA, p 437–458

Pittendrigh CS, Bruce V, Kaus P 1958 On the significance of transients in daily rhythms. Proc Natl Acad Sci USA 44:965–973

Prosser RA, Edgar DM, Dean R, Heller HC, Miller JD 1993 Serotonin and the mammalian circadian system. I. *In vitro* phase shifting by serotonergic agonists and antagonists. J Biol Rhythms 8:1–16

Rea MA, Glass JD, Colwell CS 1994 Serotonin modulates photic responses in the hamster suprachiasmatic nuclei. J Neurosci 14:3635–3642

Shinohara K, Tominaga K, Isobe Y, Inouye S-I 1992 Photic regulation of peptides located in the ventrolateral subdivision of the suprachiasmatic nucleus of the rat: daily variations in vasoactive intestinal polypeptide, gastrin-releasing peptide and neuropeptide Y. J Neurosci 13:793–800

Tcheng TK, Gillette MU 1991 Characterization of regional neuronal activity in the suprachiasmatic nucleus using a curve-fitting program. Soc Neurosci Abstr 17:672

Vanecek J, Klein DC 1992 Melatonin inhibits gonadotropin-releasing hormone-induced elevation of intracellular Ca^{2+} in neonatal rat pituitary cells. Endocrinology 130:701–707

van den Pol AN, Finkbeiner SM, Cornell-Bell AH 1992 Calcium excitability and oscillations in suprachiasmatic nucleus neurons and glia *in vitro*. J Neurosci 12:2648–2664

Zimmerman WF, Pittendrigh CS, Pavlidis T 1968 Temperature compensation of the circadian oscillation in *Drosophila pseudoobscura* and its entrainment by temperature cycles. J Insect Physiol 14:669–684

A non-photic gateway to the circadian clock of hamsters

N. Mrosovsky

Departments of Zoology, Physiology and Psychology, University of Toronto, Toronto, Ontario, Canada M5S 1A1

> *Abstract.* This paper considers the neural mechanisms underlying a particular kind of non-photic phase shifting, that produced by novelty-induced wheel running in the hamster. The projection from the intergeniculate leaflet (IGL) to the suprachiasmatic nucleus (SCN) appears to be an important part of the mechanism mediating such phase shifts. A number of experiments support this view. First, expression of immediate-early genes in the IGL is induced by non-photic phase-shifting stimuli. Second, Fos-like immunoreactivity in the IGL co-localizes with neuropeptide Y (NPY) immunoreactivity. Third, direct application of NPY to the SCN produces phase shifts which do not depend on the hamsters becoming active following the injections. Fourth, blocking the normal actions of NPY at the SCN blocks or greatly attenuates the phase shifting that is normally produced by novelty-induced wheel running. Progress on the physiological basis of phase shifts associated with activity, or a correlate, depends on understanding the behavioural aspects of this phenomenon. The activity–shift response curve is especially useful.
>
> *1995 Circadian clocks and their adjustment. Wiley, Chichester (Ciba Foundation Symposium 183) p 154–174*

There are many different phase-shifting stimuli and entraining situations that can be classified as non-photic; for example, social zeitgebers, pharmacological treatments and entrainment by periodic feeding. Most have received little study. We have no idea whether or not they depend on the same neural mechanisms. It is definitely premature to attempt to review the mechanisms of non-photic entrainment in general. Instead, I shall concentrate on a particular type of non-photic phase-shifting, that produced by confinement of hamsters (*Mesocricetus auratus*) to novel running wheels for a period of a few hours in the middle of their subjective day. The procedure is simple (Mrosovsky & Salmon 1987, Reebs & Mrosovsky 1989a). For most of a test, a hamster is kept in its home cage which usually contains its own wheel for monitoring circadian activity rhythms. At the desired time, the animal is placed in a nearby clean running wheel from which there is no exit (Fig. 1). Its main options are to run or to rest. In our laboratory conditions most hamsters choose to run. After a standard period,

FIG. 1. A simple apparatus for producing non-photic phase shifts in hamsters. The procedure is also simple: the hamster is removed from its home cage and put in the novel wheel for a specified time, usually three hours in our experiments. The animal is confined to the wheel by the plastic sides. The number of wheel revolutions produces a measure of activity.

three hours usually, the animal is returned to its home cage. Novelty-induced running is not so esoteric a starting point as it may appear and has a number of considerations in its favour.

(a) It is a manipulation capable of producing large phase shifts, as large in fact as those produced by light in hamsters (Fig. 2).

(b) Novelty-induced wheel running shares some features with other non-photic phase-shifting stimuli. It is probably just one way of tapping into a phase-shifting system that requires arousal or activity to become engaged. For example, the phase response curve (PRC) for triazolam injections is similar in shape to that for novelty-induced running, and it has been pointed out that triazolam often makes hamsters hyperactive (Mrosovsky & Salmon 1987). It has now been shown that phase-shifting effects of triazolam are greatly attenuated by restricting the

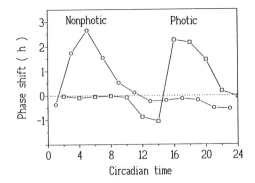

FIG. 2. Non-photic (○) and photic (□) phase response curves (PRCs) for hamsters kept in the dark. The non-photic PRC is for three-hour pulses of novelty-induced wheel running (replotted from Mrosovsky et al 1992). The photic PRC is for one-hour pulses of saturating light (replotted from Takahashi et al 1984). Points show mean values for two-hour bins.

hamster's activity after injection (Van Reeth & Turek 1989, Mrosovsky & Salmon 1990). Phase response curves for pulses of social interaction, for cage changing and for neuropeptide Y (NPY) injections, as well as those for application of a number of substances *in vitro* to suprachiasmatic nucleus (SCN) slices (Gillette 1991, Prosser et al 1993), have some similarities to those for activity pulses or triazolam. Some authors classify PRCs into two main families, one with shapes similar to the classic PRC for light pulses and the other with shapes similar to those for dark pulses and various non-photic stimuli (Smith et al 1992). The latter family has also been designated Y-type because 'in all probability, the phasically active peptide in hamster brain is neuropeptide Y' (Morin 1991). This might turn out to be correct, but grouping PRCs into families without explanation of the considerable differences within a family could be superficial. PRCs within a family can have peaks differing by as much as five hours (Smith et al 1992). Also, the delay portions of the curves within a family often do not match up well; for example, NPY produces phase delays when given in the early subjective night (Albers & Ferris 1984) but triazolam produces delays mainly when given in the late subjective night. Nevertheless, despite many differences in detail, there could be some underlying features common to PRCs with advance portions somewhere in the subjective day.

(c) When studies on activity pulses (Mrosovsky et al 1989, Janik & Mrosovsky 1993) are taken together with those on triazolam (Turek 1988), they constitute the beginnings of a database.

(d) Finally, there is some recent information about the physiological substrates for novelty-induced wheel running.

For these reasons, it may be profitable at this stage to try to understand one particular kind of non-photic phase shifting, and to go on from that base to ask

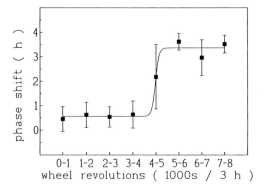

FIG. 3. Phase shifts (mean ± SD) in three–five-month-old male hamsters as a function of the number of revolutions made in a novel wheel during a three-hour pulse starting at zeitgeber time (ZT) 4. Data are replotted from Janik & Mrosovsky (1993), with a sigmoid curve fitted by GraphPad InPlot, with the lower and upper asymptotes fixed at 0.56 and 3.37 h respectively ($R^2 = 0.99$). An Aschoff (1965) type II procedure was used; the maximal phase shifts with this procedure were similar to those obtained with the Aschoff type I procedure (see Mrosovsky et al 1992). It should not be assumed that the threshold value, lying between 4000 and 5000, will be exactly the same in other strains or ages or sexes of hamsters, or with different test conditions, wheel sizes, etc. The construction of similar curves for particular situations is advocated to assist in the interpretation of data on non-photic effects on circadian rhythms.

whether other non-photic inputs to the clock depend on the same or different mechanisms.

Activity-response curves and their implications

Before the results of physiological manipulations can be interpreted with confidence, it is essential to appreciate that brief bouts of activity are relatively ineffective non-photic phase shifters (Honrado & Mrosovsky 1989, Reebs & Mrosovsky 1989b). Although it is not known what the critical feature of confinement to a novel wheel is, passing a threshold number of wheel revolutions *predicts*, but does not necessarily *cause*, the occurrence of a large phase shift (Fig. 3). In the particular conditions of our tests, hamsters making fewer than 4000 revolutions in the middle of the subjective day are unlikely to show a phase shift, whereas those making more than 5000 have a high probability of shifting (Fig. 3).

It has been realized that when phase shifting is produced by a drug, it is necessary to know whether the drug is simply a fancy way of producing activity or is having a more direct effect on the circadian system. As mentioned above, tests with hamsters whose movements are restricted indicate that phase shifts produced by triazolam depend on activity, or a correlate. In contrast, those produced by chlordiazepoxide (Biello & Mrosovsky 1993) and melatonin (Redman & Roberts 1991) do not seem to be mediated in this way. In other cases direct tests have not been done, but at least the possibility of behavioural mediation of phase-shifting effects has been recognized as a possibility (Tsujimaru et al 1992, Edgar et al 1993).

The converse, that loss of non-photic phase shifting following a manipulation may depend on decreased activity or arousal in response to non-photic stimuli, does not seem to have been generally appreciated. For example, after chemical lesions of the intergeniculate leaflet (IGL), triazolam no longer produces phase shifting (Johnson et al 1988), but no evidence has been provided to show that this benzodiazepine induces the same amount of activity in the lesioned and non-lesioned hamsters. After electrolytic lesions of this area, hamsters become less active and do not run as much when confined to a novel wheel (Janik & Mrosovsky 1994, Wickland & Turek 1992). Johnson et al (1988) wrote: 'the requirement of an intact LGN [lateral geniculate nucleus] for triazolam to shift circadian phase suggests that the LGN may be a site through which stimuli gain access to the circadian clock'. An alternative is that stimuli reach the circadian clock by some other route and the IGL lesions simply prevent hamsters from becoming sufficiently active or aroused for adequate stimulation of those pathways.

Although experiments with lesions have not yet established that triazolam-induced phase shifts depend on information to the SCN being routed through the IGL, ironically the conclusions that have been drawn from such experiments are probably correct. Supportive evidence comes from several other types of experiment.

The involvement of c-*fos*

Non-photic phase shifting can induce expression of *fos*-related genes in the IGL (Janik & Mrosovsky 1992; Fig. 4). This experiment was guided by a knowledge of the behavioural predictors of large phase shifts (Fig. 3). Animals that ran more than 5000 revolutions in a novel wheel were selected and compared with those left in their home cages. No expression of *fos* was found in the SCN, and this accords with the results of others working with other non-photic stimuli (Mead et al 1992), though in some of those experiments one cannot be sure that the manipulations would have resulted in phase shifts in the conditions used for the study of c-*fos* (Cutrera et al 1993). Induction of immediate-early genes in the IGL, but not in the SCN itself, suggests that information about non-photic events is conveyed from the IGL to the SCN in some way that does not involve induction of *fos*. One obvious candidate is the NPY projection from the IGL to the SCN (Mikkelsen 1990, Card & Moore 1991, Mikkelsen & O'Hare 1991, Moore 1992), especially since the PRC for NPY injections into the SCN (Albers & Ferris 1984) has some rough similarities to those for novelty-induced activity and triazolam. However, before getting carried away by this line of reasoning, it is necessary to consider the possible involvement of increased activity in phase shifts produced by NPY.

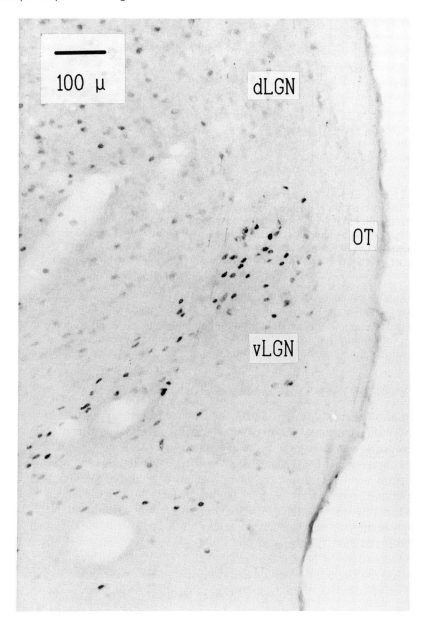

FIG. 4. Fos-immunoreactivity in the IGL of a hamster following a three-hour pulse of novelty-induced wheel running. Antibody provided by Drs D. Hancock and G. Evan; photomicrograph by Dr J. D. Mikkelsen.

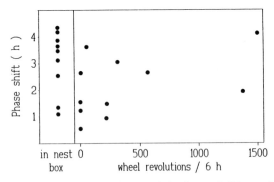

FIG. 5. Phase shifts of hamsters given SCN injections of 400 ng of NPY at ZT 4 in an Aschoff type II test procedure. Data from Biello et al (1994) are plotted as a function of the number of wheel revolutions made in the six hours following the injections. The left panel shows shifts from hamsters confined to their nest boxes for this six-hour period. Phase shifts of confined hamsters were not significantly different from those in hamsters with access (but not confined) to wheels. The correlation between the number of wheel revolutions and the extent of phase shifts was not significant ($r = 0.46$, $P = 0.15$ two-tailed test, slope not significantly different from 0). Note also that the number of wheel revolutions are below the values associated with phase shifts in Fig. 3. However, in the graph and all calculations, one point is excluded, that from a test in which an animal ran 10 983 revolutions after receiving NPY; the associated shift of 3.05 h may well have been mediated by activity or a correlate. Inclusion of this animal would have reduced the correlation between activity and phase shifts to even less significant values.

NPY-induced phase shifts and motor activity

In Albers & Ferris's paper (1984), the actograms illustrating shifts in response to NPY show prominent bouts of wheel running immediately after the injections. Their work was done before the chronobiological effects of activity/arousal were discovered (Mrosovsky 1988, Mrosovsky & Salmon 1987) and they did not study activity systematically. If phase shifts produced by NPY depend on the animal becoming active, NPY is not a good candidate for *mediating* shifts produced by novelty-induced wheel running; administration of NPY would simply be another way, a biochemical way, of initiating or *mimicking* such shifts.

To investigate this we measured wheel running in hamsters given NPY injections to the SCN at about zeitgeber time (ZT) 4. There was no correlation between the amount of running and the extent of the phase shift. Also, in every case except one, the animals made fewer than 4000 revolutions in the six hours following the injections (cf., Fig. 3). Other hamsters were confined to small nest boxes for six hours after receiving NPY, but this did not prevent NPY from producing significant phase advances (Fig. 5). These results show that NPY can produce phase advances in the absence of appreciable amounts of wheel running. These necessary preliminaries fuelled our enthusiasm for NPY as a candidate for conveying non-photic input to the SCN.

Co-localization of Fos and NPY

In our studies of immediate-early gene expression in the IGL, D. Janik noticed that cells with Fos-like immunoreactivity seemed to occupy a field roughly similar in extent to the NPY field described by Morin et al (1992). To look into this more closely, we double-labelled sections with antibodies to Fos and antibodies to NPY. Brains of hamsters that had run more than 5000 revolutions during three hours of confinement to a novel wheel were sent with brains of control hamsters to J. D. Mikkelsen in Copenhagen for blind assessment. Using a different antibody and greater dilutions than in Janik & Mrosovsky's (1992) initial work, he confirmed that immediate-early genes are turned on in the IGL by this manipulation. It also became apparent that about 70% of the NPY-immunopositive cells also expressed c-*fos*.

Antibodies to NPY

If activation of NPY-containing cells in the IGL and subsequent release of NPY at terminals in the SCN mediates non-photic phase shifting, it should be possible to block such shifts by interfering with the ability of NPY to bind to receptors. Unfortunately, there are no readily available well-studied NPY antagonists, so we tried another approach, injecting antiserum to NPY into the SCN before placing animals in a novel wheel. These experiments were done by S. M. Biello using a well-characterized antiserum generously provided by J. M. Polak (Allen et al 1983). In view of the dose–response requirements for non-photic shifts (Fig. 3), it was essential that injection of the antiserum should not interfere with the hamsters' responses to the novel wheel. Fortunately, the treated animals generally ran as much as those given normal rabbit serum. Of course, some hamsters, even untreated ones, decline the opportunity to run when placed in a novel wheel; such sluggards are not expected to show phase shifts (Fig. 3). However, from data on animals that made more than 5000 revolutions both on their tests with NPY antiserum and on those with normal serum, it was clear that the phase-shifting effect of confinement to a novel wheel can be greatly attenuated without there being reductions of wheel running during the pulse (e.g., Fig. 6; Biello et al 1994).

Does exercise increase NPY release?

So far, we know that immediate-early genes are expressed in NPY-containing cells in the IGL after a well-characterized non-photic phase-shifting stimulus; we know that these NPY neurons send axons to the SCN; and we know that interfering with the action of NPY at the terminals of these axons by introducing antiserum to NPY greatly attenuates the phase shifts. To complete this story, it would be nice to demonstrate an increased release of NPY in the SCN in

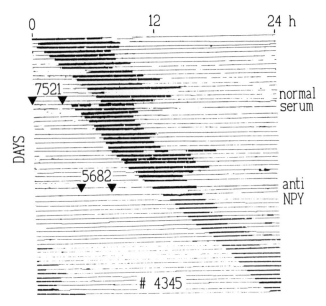

FIG. 6. Actogram from Esterline Angus records of a hamster in continuous darkness given injections to the SCN shortly before being placed in a novel wheel for three hours. First injection, normal rabbit serum; second injection, antiserum to NPY. Numbers by triangles show the number of wheel revolutions made in the three-hour pulses. Pen deflections show wheel running at other times. (Data from Beillo et al 1994.)

response to a non-photic stimulus. Initial attempts to detect this immunocytochemically have not been promising, and we have turned to other methods. It is interesting to note that NPY increases in a number of hypothalamic areas in rats in a state of chronic negative energy balance produced by a combination of intense wheel running and limited rations (Lewis et al 1993). Whether a single bout of running increases NPY concentration or NPY turnover in the SCN has not been investigated.

Limitations and problems

Gaps

A major gap in our understanding of the physiology of non-photic effects is how information reaches the IGL. Another obvious limitation is that the experiments cited on NPY have been done with only one species, and mostly in the advance portion of the PRC. A topic almost untouched at the physiological level is the interaction between photic and non-photic phase-shifting events. Some striking interactions have been demonstrated at the behavioural level (Mrosovsky 1991, Ralph & Mrosovsky 1992).

Direct input or coupling?

Another question is whether the NPY pathway to the SCN mediates simple phase-shifting effects or represents a coupling pathway to the light-entrainable oscillator in the SCN from a non-photically entrainable *oscillator* elsewhere. Because all the wheel-running activity, not just a component, can be entrained non-photically in hamsters (Mrosovsky 1988, Rusak et al 1988, Reebs & Mrosovsky 1989a, Mrosovsky et al 1989), one must infer that the light-entrainable oscillator is affected by non-photic stimuli. There could be direct influences on the light-entrainable oscillator(s) in the SCN, or a coupling influence from another oscillator (Mrosovsky 1988, Rusak et al 1988).

If circadian rhythms could be entrained by non-photic means in animals with SCN lesions, this would support the idea of there being some non-photic oscillator outside the SCN. It has been reported (Mistlberger 1992) that hamsters with SCN lesions do entrain to a two-hour stay each day in a different cage, but because of the low levels of activity and the lack of stability in period and in phase angle during the days when the pulses were given, the data are not compelling. If an animal is activated by being in a new cage, it is possible that it will rest more in the few hours after it gets back to its home cage. Non-random distribution of activity over a 24 h period might result in this way from masking effects. Without clear free runs after withdrawal of the putative zeitgeber, rhythms of activity during its presence become harder to interpret. Finally, persistence of non-photically entrained rhythms was weaker or absent in the hamsters with the larger lesions; this suggested that there are some non-photically entrainable oscillators in or near the SCN (Mistlberger 1992).

Canbeyli et al (1991) found that SCN-lesioned hamsters whose rhythmicity had been restored by transplants of fetal SCN failed to show a phase shift in response to triazolam. They suggested that afferents outside the SCN mediate the shifting effects of triazolam. Although the results are consistent with this interpretation, the loss of phase shifting in response to triazolam is not instructive because the amounts of activity induced by the drug were not adequately quantified. Even if the amounts of activity after the triazolam injections had reached levels associated with shifting in intact animals, a failure to find shifting in the transplanted animals may reveal more about the ability of transplanted tissue to fulfil normal functions than about the normal site of action of triazolam.

Dark pulses

A potential problem with the view that non-photic phase shifts associated with activity require the NPY projections to the SCN comes from experiments with dark pulses. Because preventing wheel running greatly attenuates the phase-shifting effect of dark pulses, it has been concluded that phase shifts produced by dark pulses are mediated by activity or some correlated variable (Reebs et al 1989, Van Reeth & Turek 1989). If this is so, lesions of the LGN should attenuate phase shifts after dark pulses. However, Harrington & Rusak (1986) reported that this occurs only when six-hour dark pulses start at circadian time (CT) 5–7, and not at other times. This half answer is difficult to interpret. The advance portion of Harrington & Rusak's dark pulse PRC for control hamsters peaked at CT 11–13, whereas previously Boulos & Rusak (1982) found the peak for six-hour dark pulses some six hours earlier in the same species. It is possible that some dark pulse PRCs are mediated by activity, but that others are not, or not to the same extent. Some experimenters use procedures likely to promote activity, such as 'jostling the cages to awaken all animals' at the start of the pulse (Ellis et al 1982). Another complication is the differences in PRCs for different durations of dark pulses (see Boulos & Rusak 1982, Ellis et al 1982). After a long dark pulse starting towards the end of the subjective day, the lights will tend to come on in an advance portion of the photic PRC. The background intensity of light is another poorly researched variable. Mark Twain's comments about the ability of researchers to cast much darkness on a subject might be doubly apt for this case.

Summary

This paper makes two main points. The first is that work on the physiological basis of non-photic effects on circadian rhythms must be informed by an understanding of the behaviour that is associated with phase shifting (e.g., Fig. 3). This methodological point is probably more important than the second and more substantive point: that the NPY projection from the IGL to the SCN is the mechanism through which certain non-photic events affect the circadian clock in hamsters.

With respect to the latter point, the suggestion is that NPY is a non-photic *gateway* to the SCN, not that it is the *pathway*. By what route behavioural and neural information arrives at that gateway remains a blank. A major difficulty in mapping the complete path from the non-photic inputs to the SCN is that there is no clear starting point. With photic entrainment in mammals, the eye is the obvious starting point. With non-photic shifting events, such as confinement to a novel wheel, it is not known whether it is the induced activity, the associated arousal, or some other variable that is critical for phase shifting. Perhaps with non-photic entrainment it will be possible to start at the other end of the system, at the clock itself, and follow the thread back to the periphery, and so discover more about pathways and inputs. The recent findings with NPY and c-*fos* put the end of that thread into our hands.

Acknowledgements

I thank Daniel Janik, Stephany Biello and Jens Mikkelsen for collaboration, ideas and comments. Peggy Salmon and Kelly Foster assisted with the experiments. J. M. Polak gave us NPY antiserum. The Medical Research Council of Canada provided support.

References

Albers HE, Ferris CF 1984 Neuropeptide Y: role in light–dark cycle entrainment of hamster circadian rhythms. Neurosci Lett 50:163–168

Allen YS, Adrian TE, Allen JM et al 1983 Neuropeptide Y distribution in the rat brain. Science 221:877–879

Aschoff J 1965 Response curves in circadian periodicity. In: Aschoff J (ed) Circadian clocks. North-Holland Publishing Company, Amsterdam, p 95–111

Biello SM, Mrosovsky N 1993 Circadian phase-shifts induced by chlordiazepoxide without increased locomotor activity. Brain Res 622:58–62

Biello SM, Janik D, Mrosovsky N 1994 Neuropeptide Y and behaviourally induced phase shifts. Neuroscience 62:273–279

Boulos Z, Rusak B 1982 Circadian phase response curves for dark pulses in the hamster. J Comp Physiol A Sens Neural Behav Physiol 146:411–417

Canbeyli RS, Romero M-T, Silver R 1991 Neither triazolam nor activity phase advance circadian locomotor activity in SCN-lesioned hamsters bearing fetal SCN transplants. Brain Res 566:40–45

Card PJ, Moore RY 1991 The organization of visual circuits influencing the circadian activity of the suprachiasmatic nucleus. In: Klein DC, Moore RY, Reppert SM (eds) Suprachiasmatic nucleus. Oxford University Press, New York, p 51–76

Cutrera RA, Kalsbeek A, Pévet P 1993 No triazolam-induced expression of *Fos* protein in raphe nuclei of the male Syrian hamster. Brain Res 602:14–20

Edgar DM, Miller JD, Prosser RA, Dean RR, Dement WC 1993 Serotonin and the mammalian circadian system. II. Phase-shifting rat behavioural rhythms with serotonergic agonists. J Biol Rhythms 8:17–31

Ellis GB, McKlveen RE, Turek FW 1982 Dark pulses affect the circadian rhythm of activity in hamsters kept in constant light. Am J Physiol 242:R44–R50

Gillette MU 1991 SCN electrophysiology *in vitro*: rhythmic activity and endogenous clock properties. In: Klein DC, Moore RY, Reppert SM (eds) Suprachiasmatic nucleus. Oxford University Press, New York, p 125–143

Harrington ME, Rusak B 1986 Lesions of the thalamic intergeniculate leaflet alter hamster circadian rhythms. J Biol Rhythms 1:309–325

Honrado GI, Mrosovsky N 1989 Arousal by sexual stimuli accelerates the re-entrainment of hamsters to phase advanced light–dark cycles. Behav Ecol Sociobiol 25:57–63

Janik D, Mrosovsky N 1992 Gene expression in the geniculate induced by a nonphotic circadian phase shifting stimulus. NeuroReport 3:575–578

Janik D, Mrosovsky N 1993 Nonphotically induced phase shifts of circadian rhythms in the golden hamster: activity-response curves at different ambient temperatures. Physiol & Behav 53:431–436

Janik D, Mrosovsky N 1994 Intergeniculate leaflet lesions and behaviorally induced phase shifts of circadian rhythms. Brain Res 631:174–182

Janik D, Mikkelsen JD, Mrosovsky N 1994 Cellular colocalization of Fos and NPY in the intergeniculate leaflet after nonphotic phase-shifting events. In preparation

Johnson RF, Smale I, Moore RY, Morin LP 1988 Lateral geniculate lesions block circadian phase-shift responses to a benzodiazepine. Proc Natl Acad Sci USA 85:5301–5304

Lewis DE, Shellard L, Koeslag DG et al 1993 Intense exercise and food restriction cause similar hypothalamic neuropeptide Y increases in rats. Am J Physiol 264:E279–E284

Mead S, Ebling FJP, Maywood ES, Humby T, Herbert J, Hastings MH 1992 A nonphotic stimulus causes instantaneous phase advances of the light-entrainable circadian oscillator of the Syrian hamster but does not induce the expression of *c-fos* in the suprachiasmatic nuclei. J Neurosci 12:2516–2522

Mikkelsen JD 1990 Projections from the lateral geniculate nucleus to the hypothalamus of the Mongolian gerbil (*Meriones unguiculatus*): an anterograde and retrograde tracing study. J Comp Neurol 299:493-508

Mikkelsen JD, O'Hare MMT 1991 An immunohistochemical and chromatographic analysis of the distribution and processing of proneuropeptide Y in the rat suprachiasmatic nucleus. Peptides 12:177-185

Mistlberger RE 1992 Nonphotic entrainment of circadian activity rhythms in suprachiasmatic nuclei-ablated hamsters. Behav Neurosci 106:192-202

Moore RY 1992 The enigma of the geniculohypothalamic tract: why two visual entraining pathways? J Interdiscip Cycle Res 23:144-152

Morin LP 1991 Neural control of circadian rhythms as revealed through the use of benzodiazepines. In: Klein DC, Moore RY, Reppert SM (eds) Suprachiasmatic nucleus. Oxford University Press, New York, p 324-338

Morin LP, Blanchard J, Moore RY 1992 Intergeniculate leaflet and suprachiasmatic nucleus organization and connections in the golden hamster. Visual Neurosci 8:219-230

Mrosovsky N 1988 Phase response curves for social entrainment. J Comp Physiol A Sens Neural Behav Physiol 162:35-46

Mrosovsky N 1991 Double-pulse experiments with nonphotic and photic phase-shifting stimuli. J Biol Rhythms 6:167-179

Mrosovsky N, Salmon PA 1987 A behavioural method for accelerating re-entrainment of rhythms to new light-dark cycles. Nature 330:372-373

Mrosovsky N, Salmon PA 1990 Triazolam and phase-shifting acceleration re-evaluated. Chronobiol Int 7:35-41

Mrosovsky N, Reebs SG, Honrado GI, Salmon PA 1989 Behavioural entrainment of circadian rhythms. Experientia 45:696-702

Mrosovsky N, Salmon PA, Menaker M, Ralph MR 1992 Nonphotic phase shifting in hamster clock mutants. J Biol Rhythms 7:41-49

Prosser RA, Dean RR, Edgar DM, Heller HC, Miller JD 1993 Serotonin and the mammalian circadian system. I. *In vitro* phase shifts by serotonergic agonists and antagonists. J Biol Rhythms 8:1-16

Ralph MR, Mrosovsky N 1992 Behavioral inhibition of circadian responses to light. J Biol Rhythms 7:353-359

Redman JR, Roberts CM 1991 Entrainment of rat activity rhythms by melatonin does not depend on wheel-running activity. Soc Neurosci Abstr 17:673

Reebs SG, Mrosovsky N 1989a Effects of induced wheel running on the circadian activity rhythms of Syrian hamsters: entrainment and phase response curve. J Biol Rhythms 4:39-48

Reebs SG, Mrosovsky N 1989b Large phase-shifts of circadian rhythms caused by induced running in a re-entrainment paradigm: the role of pulse duration and light. J Comp Physiol A Sens Neural Behav Physiol 165:819-825

Reebs SG, Lavery RJ, Mrosovsky N 1989 Running activity mediates the phase-advancing effects of dark pulses on hamster circadian rhythms. J Comp Physiol A Sens Neural Behav Physiol 165:811-818

Rusak B, Mistlberger RE, Losier B, Jones CH 1988 Daily hoarding opportunity entrains the pacemaker for hamster activity rhythms. J Comp Physiol A Sens Neural Behav Physiol 164:165-171

Smith RD, Turek FW, Takahashi JS 1992 Two families of phase-response curves characterize the resetting of the hamster circadian clock. Am J Physiol 262: R1149-R1153

Takahashi JS, DeCoursey PJ, Bauman L, Menaker M 1984 Spectral sensitivity of a novel photoreceptive system mediating entrainment of mammalian circadian rhythms. Nature 308:186-188

Tsujimaru S, Ida Y, Satoh H et al 1992 Vitamin B_{12} accelerates re-entrainment of activity rhythms in rats. Life Sci 50:1843–1850

Turek FW 1988 Manipulation of central circadian clock regulating behavioral and endocrine rhythms with a short-acting benzodiazepine used in the treatment of insomnia. Psychoneuroendocrinology 13:217–232

Van Reeth O, Turek FW 1989 Stimulated activity mediates phase shifts in the hamster circadian clock induced by dark pulses or benzodiazepines. Nature 339:49–51

Wickland CR, Turek FW 1992 Lesions of the intergeniculate leaflet block activity-induced phase shifts in the circadian rhythm of activity in golden hamsters. In: Program and abstracts for the 3rd meeting of the Society for Research on Biological Rhythms, May 1992, p 65

DISCUSSION

Miller: It is implicit in your model that the gene encoding NPY must be a target gene for Fos, which in turn suggests that there's an AP1 binding site on the NPY gene.

Mrosovsky: I'm not sure there has to be an AP1 binding site on the NPY gene. There could be several intermediate steps.

Mikkelsen: It's likely that Fos turns on the NPY gene, even though a direct contact between c-*fos* expression and NPY mRNA expression has not been demonstrated. In other systems, such as the polymorph layer of the dentate gyrus, NPY mRNA expression is increased after seizures (Gall et al 1990). We have recently shown that *fos* is induced in NPY-immunoreactive cells (M. Woldbye & J. D. Mikkelsen, unpublished observations) after seizures and take this as evidence for the presence of an AP1 site(s) on the NPY gene.

Moore: Could I respond to your comments about our paper on IGL lesions (Johnson et al 1988). At the time we did that work it was not recognized that activity was a component of the response and we didn't look at activity; but, after all, the fact that we didn't look at it doesn't mean that the activity was not induced in the lesioned animals. Another piece of information that we published in that paper indicated that the IGL is a component of the non-photic pathway: R. F. Johnson made one set of lesions with an excitatory amino acid, and that agent initially stimulated the IGL prior to destroying it. This was done with the animals under anaesthesia and, in that situation, obviously without any locomotor activity, there were subsequent changes in the phase of the animals' wheel running that were dependent on the time that the lesion was made. This clearly implicates the IGL–geniculohypothalamic tract pathway in non-photic entrainment.

The IGL, like the SCN, surprisingly has relatively few inputs from other brain areas. The predominant input, other than that from the retina, is from the retrochiasmatic area, which is a crossroads for all kinds of inputs coming through the supraoptic commisures, including ones from the brainstem and from the forebrain. The retrochiasmatic area input is likely to be the most important input other than the visual input.

Mrosovsky: I should have emphasized that Johnson et al's paper (1988) was an early paper on IGL lesions; and the point you made about the phase shifts induced under anaesthesia is compelling. I just wanted to stress that one has to take activity into account in interpreting the results of such manipulations.

Turek: The type of activity may not be particularly important, though. We see major phase shifts in animals that do not have a running wheel, who have never seen a wheel. We give them triazolam and they are active afterwards, and though they are not running for 4000–5000 wheel revolutions they still show phase shifts. If you give a running wheel to animals that have never had one before, they show even larger phase shifts than an animal that has had a running wheel all the time and is given a novel one. We certainly agree with Dr Mrosovsky that the amount of activity is important but it does not have to be in a wheel. If you stimulate only a little activity you will not stimulate a phase shift.

Mrosovsky: We confine the hamsters to the running wheel because we then get a good measure of their activity. Basically, they can either sleep or run, but not much else. But we also have evidence that activity other than wheel running can have some effects (Reebs et al 1989, Honrado & Mrosovsky 1989).

Menaker: Have you looked at the interaction of non-photic stimulation with light?

Mrosovsky: The interaction between non-photic and photic effects will, I predict, become an important research area. Such interactions may be relevant to some of the anatomical complexity in the circadian system. Why isn't the retinohypothalamic tract sufficient to get light information into the SCN and to induce a phase shift? Why do we need all this other stuff? The answer could be that the circadian system, in addition to handling photic phase shifting, is doing several other things, one of which is arranging for interactions between photic and non-photic inputs. Hamsters put into a novel wheel just before being given a sub-saturating light pulse at CT 18 show attenuated phase shifts. That is an example of an interaction between photic and non-photic inputs—one which, incidentally, has possible implications for interpreting the blocking of photic effects with MK801, a substance which has been reported to activate rats, at least. Another type of interaction occurs when a non-photic pulse given at about CT 5 is followed by a photic pulse; the advance normally produced by the non-photic stimulus is virtually abolished. This is not a simple subtractive effect, because in hamsters the phase delay in response to a photic stimulus is about 1.5 h maximum, whereas the non-photically induced advance is around three hours, so a photic stimulus coming after a non-photic stimulus can eliminate at least 1.5 h of the advance (Mrosovsky 1991). There are complicated synergistic actions that may allow us to give 'double whammies' (photic and non-photic) to the system to achieve greater phase shifts.

Arendt: Is there any scope for considering changes in body temperature as a common pathway for a whole variety of agents?

Mrosovsky: There's certainly scope for that, but there is scope for a whole lot of other things that happen when the animal exercises—interleukins, corticosteroids and many other things change. I don't view a fishing expedition as a pleasant way to proceed unless there is hope of catching a big fish. We have done some work on temperature. We can't mimic the effects of running in a novel wheel by passive warming of the animal, but that may not be critical, because there is an argument about whether temperature goes up passively with exercise or rises to meet an elevated set point. Wickland & Turek's (1991) work also suggests that temperature is not a factor.

Turek: We can induce a rise in temperature which is similar to that observed in response to wheel running, but we don't see phase shifts in response to the temperature rise. We can restrain animals after treatment with triazolam and still see a rise in temperature, but no phase shifts are observed (Wickland & Turek 1991).

Menaker: Is it correct that enforced running doesn't produce phase shifts?

Mrosovsky: This is a highly relevant point, relevant also to Dr Block's attempts to induce phase shifts in *Bulla* by putting them in a jacuzzi (p 64). Mistlberger (1991) got only marginal entraining effects with rats forced to run on a treadmill.

Meijer: Forced wheel running does not induce phase shifts in hamsters either (M. J. De Vries & J. H. Meijer, unpublished results).

Menaker: That's an important result, because it eliminates a whole class of possible inputs.

Mrosovsky: If we put the sluggardly hamsters, the ones that don't spontaneously run in our novel wheels, into a cold environment (5–10 °C) they will run, but we do not correspondingly increase the percentage of animals that show a phase shift. In other words, some of these sluggardly animals (as defined by their disinterest in running in our normal warm test conditions) don't shift even after considerable running in the cold wheels. That, we think, is instructive. It's not necessarily the running *per se* which stimulates the shift. The reason for running seems to be important.

Miller: This discussion reminds me of conversations I have had with my colleague Dr Dale Edgar, who has speculated for some time about that bugbear for psychologists, the notion of volition in locomotion. Although this is anathema to many of us, perhaps there is an element here that you would not see in forced locomotion or in cold-induced locomotion. Are there other techniques by which you could increase this volitional running and convert your sluggards into runners? Operant-conditioned locomotion would be one possibility and another would be to reduce the body weight by 20%, which should cause a general increase in locomotion.

Mrosovsky: There are many ways in which one can get a hamster to run, but to isolate the volitional aspect one would need a situation in which the hamster wants to run but does not.

Roenneberg: I'm confused about the need for running. Could it be that it is not easy to make the sluggards nervous, even if you make them run? Is running essential or not?

Mrosovsky: The methodological point is that in normal warm conditions the running is predictive of a phase shift; this allows us to pick animals that we know will respond with shifts for our work with NPY antiserum. But is it the running *per se* which is responsible for the shift? Dr Meijer's work with enforced running and our experiment with running at reduced temperatures tells us that running is not *sufficient*. Once we can say that, we can go on to ask whether it is *necessary* at all, but we don't have the answer to that question yet. The motivational context in which the running occurs is important.

Daan: What happens when you block the running wheel?

Mrosovsky: We have done some experiments with blocked wheels (Honrado & Mrosovsky 1989, Reebs et al 1989). Phase shifting is reduced but by how much depends on whether or not the hamsters have access to their home cages. We don't really need to block the wheel, because we have the sluggards who won't run.

Roenneberg: But they might be a different type of animal. There may be sluggards who wouldn't shift even if you forced them to run, but also runners who would indeed shift if they are nervous in spite of not being allowed to run.

Czeisler: Or are perhaps just thinking about running!

Mrosovsky: Agreed; we should take individuals that we know from previous work will run and put them in a blocked wheel.

Armstrong: I wonder how far these findings will extend to other species. We have all seen articles in newspapers recommending jogging for jet lag on the basis of work in hamsters.

Curt Richter (1967) gave just about every physiological, endocrinological and neurological insult to wild and laboratory rats. For example, he induced prolonged sleep deprivation by forcing rats to swim in a tank for between 19 and 42 hours (see Fig. 22 in Richter 1967). Although these types of treatment often destroyed or interfered with the manifestation of behaviour, the effect was temporary, lasting for only a few days, and the onset of running always reappeared at the expected time, indicating that although locomotor behaviour had been interfered with by the stressor, the biological clock was still keeping time. Nevertheless, if you look carefully at Richter's published data, you can in fact find phase shifts of about an hour. Now, you might interpret this to support what Dr Mrosovsky has been saying, but I would argue that the phase shifts are small given the magnitude of the stress applied to the animal.

In Dr Menaker's laboratory in Oregon, I investigated immobilization stress in C57BL mice and saw exactly the same phenomenon as that reported by Richter for rats. Our mice ($n=6$) were immobilized for 30 min at the same time every day. In one mouse, although wheel-running behaviour was interfered with for a few days at the time when the onset of the activity rhythm coincided with the 30 min immobilization, the rhythm reappeared at exactly the time predicted from the pre-immobilization running pattern. In other cases, there was no effect on running at activity onset, but some disruptions at the end of activity (see Fig. 5 in Armstrong & Redman 1985). With Long-Evans hooded rats, we've tried 30 min of immobilization, novelty stress and daily handling (Barrington et al 1993). In some rats you do get effects (entrainment, period changes and phase shifts) but I would underline the fact that the most sensitive time is at subjective dawn, the end of the active period, and not at subjective dusk.

Turek: Was 30 min the maximum time of immobilization?

Armstrong: Yes.

Turek: You need to immobilize hamsters for three hours to get a phase shift.

Armstrong: In terms of corticosteroid release, 20 min is fine in rats. Indeed, after several days of daily stress at the same time, corticosterone will rise in anticipation of the stressor.

Turek: Adrenalectomized hamsters will show phase shifts in response to an activity-inducing stimulus, so there is no reason to invoke the involvement of corticosteroids in this species.

Armstrong: I'm not saying they are necessarily involved in a causal way, but simply a marker of stress. Our rats were certainly stressed daily as indicated by plasma corticosterone concentrations.

Basically, I'm just wondering if there are marked differences between rodents, and the extent to which we can generalize from hamsters to other mammals, including humans. When I started thinking about this I went back to the origin of the golden hamster. As you may know, all the hamsters, *Mesocricetus auratus,* we have in Cambridge, England, in Toronto and in Virginia, USA came from one male and three females many years ago, as the following quotation from Harrison & Bates (1991, p 265–266) shows:

'According to Reynolds (1954), this species was first brought back alive to Britain by J. H. Skene, the former Consul General at Aleppo, in 1880. It apparently bred in captivity for 30 years before dying out. In 1930, a female and 12 young were found by I. Aharoni near Aleppo; they were taken to the Hebrew University at Jerusalem. Arahoni (1932) stated that from one male and three females of this stock up to 150 specimens were raised in one year. Apparently, the vast numbers of hamsters now found as captive stock around the world are all derived from the pregnant female found by Arahoni.'

What this says about the genetics I won't go into here. I happened to be talking to an Israeli ecophysiologist, Professor Abraham Haim, and asked what had happened to all the hamsters, whether they had all died out or had been eaten! He told me that as far as he was aware they exist still, but are subterranean and hardly ever come up. A male will come to the surface about once a day, will sniff around and will go back down again unless there's a female in oestrus, when it will start running around. These are animals which would actually see very little light, in which case non-photic stimuli would be important, especially interaction with conspecifics underground. What I can't make sense of is the elegant work on photoperiodism in this species.

Mrosovsky: Non-photic effects have been demonstrated in scorpions (Hohmann et al 1990), sparrows (Reebs 1989) and mice (Edgar & Dement 1991). It would be convenient if they could be demonstrated in the rat because we know so much about rat physiology. Forcing rats to swim is on the aversive end of the volition–compulsion spectrum. Also, there may be important differences between species in what constitutes an appropriately arousing stimulus for phase shifting. Rats and hamsters differ in terms of what they like to do in a wheel. When you first give a rat the chance to run in a wheel, it runs rather little. It takes two or three weeks before it reaches a reasonable level of running (Cornish & Mrosovsky 1965). When you give a hamster the chance to run in a wheel, the first time it hits the wheel it engages in a prolonged marathonian bout of running that can last for hours. The rat is simply a less 'wheely' animal and we may have to look for other ways of producing non-photic shifts in rats.

Block: The 23 h enforced swimming is problematic because you are integrating over the advance and delay region of the PRC. A shorter bout of swimming might give a good phase shift.

Waterhouse: We've spoken at considerable length about the input to this system. The output, as I understand it, is activity. Is there evidence that other outputs from the clock are changed? Dr Arendt and Dr Lewy have found circumstances under which the activity seems to be adjusted in humans but other rhythms are not.

Mrosovsky: The hamster's phase response to light is shifted by our non-photic manipulation (Mrosovsky 1989). We tried, in collaboration with Dr Lewy, to check whether the melatonin rhythm was affected but we didn't have enough animals to be sure. If non-photic manipulation of circadian rhythms is to be valuable to people, it is important to learn whether or not many different rhythms are phase shifted.

Turek: With triazolam, we have shifted the circadian pre-ovulatory luteinizing hormone surge, a hormonal rhythm which is not dependent on activity, and we also shifted the body temperature rhythm, though I'll admit that the temperature rhythm may be masked or linked to the activity rhythm. So, I think it is clear that an activity-inducing stimulus can affect at least one hormonal and one metabolic rhythm, but I agree, it would be nice to have data on a few more metabolic and/or hormonal rhythms.

Mikkelsen: If we consider the non-photic and photic inputs to the SCN to be anatomically distinct, we also need to discuss whether and where those two inputs meet. So far it looks as if there are two parallel projections. Photic and non-photic stimuli induce *fos* expression in different cells in the IGL. Also, in the SCN the NPY input and the glutamatergic input may innervate two different populations of cells. It has not yet been shown that they go to the same population of neurons, and there's actually some indirect evidence that that's not the case. The distribution of *fos* expression in the SCN induced by light is different from the distribution of a protein involved in phosphorylation (Mikkelsen & Gustafson 1993), even though both are present in the ventrolateral part of the SCN. This may indicate that there are two populations of tightly coupled neurons in the SCN, one regulated by glutamate (photic) and coupled to c-*fos* expression, the other regulated by neuropeptide Y and/or serotonin (non-photic) and coupled to cyclic AMP.

Miller: There is evidence for co-innervation of vasoactive intestinal peptide (VIP) neurons by the geniculohypothalamic tract, the retinohypothalamic tract and the 5-HT projection from the dorsal and medial raphe. The same recipient cell could have two different signal transduction pathways for different zeitgebers. There's no reason why there necessarily have to be different populations of cells.

Mikkelsen: We have double-stained for Fos and VIP in hamsters and rats exposed to a light pulse at CT 14 and CT 18, and have not found more than about 10% of the VIP neurons containing Fos.

Lewy: Dr Mrosovsky, are responses to dark pulses mediated by increases in activity, and, if so, in what species?

Mrosovsky: I know of two relevant experiments on hamsters, both showing that restricting movement greatly attenuates the phase-shifting effects of dark pulses (Reebs et al 1989, Van Reeth & Turek 1989). Nevertheless, there may be different types of PRC produced by dark pulses. Some investigators (Ellis et al 1982) note in their methods section that they jostle the hamsters' cages in order to wake them up before giving them the dark pulse, to make sure the animals are aware of the dark pulse, whereas other investigators may not do that. Also, the PRCs produced by dark pulses are very variable. Harrington & Rusak (1986) obtained a peak advance at close to CT 12 whereas previously Boulos & Rusak (1982) got a peak 5–6 h earlier. The other work that I should mention is that by Dr Takahashi on the effects of dark pulses on cells in culture; unfortunately, those cells were not running around in their culture dish! There are too many discrepancies in the data from experiments with dark pulses to answer your question with confidence.

Roenneberg: Giving a dark pulse to *Gonyaulax* during the day produces a nice PRC. We discovered only recently that if you put the cells into darkness they cease to aggregate, but when you release them into light, they begin to aggregate again with an enormous density. This raises the question, what does a dark pulse really do? Perhaps in animals active at night activity is induced during a dark pulse, but in organisms active during the day activity might be reduced during the pulse and may come back with a huge surge afterwards. This brings us to the question about the importance of the onset and the end of pulses; which of them is providing the information, or are both doing something additively? In any case, *Gonyaulax*, which is far away from the hamster, seems to react with increased activity following dark pulses but not following light pulses in background dim light.

References

Armstrong SM, Redman J 1985 Melatonin administration: effects on rodent circadian rhythms. In: Photoperiodism, melatonin and the pineal. Pitman, London (Ciba Found Symp 117) p 188–207

Barrington J, Jarvis H, Redman JR, Armstrong SM 1993 Limited effect of three types of daily stress on rat free-running locomotor rhythms. Chronobiol Int 10:410–419

Boulos Z, Rusak B 1982 Circadian phase response curves for dark pulses in the hamster. J Comp Physiol A Sens Neural Behav Physiol 146:411–417

Cornish ER, Mrosovsky N 1965 Activity during food deprivation and satiation of six species of rodent. Anim Behav 13:242–248

Edgar DM, Dement WC 1991 Regularly scheduled voluntary exercise synchronizes the mouse circadian clock. Am J Physiol 261:R928–R933

Ellis GB, McKlveen RE, Turek FW 1982 Dark pulses affect the circadian rhythm of activity in hamsters kept in constant light. Am J Physiol 242:R44–R50

Gall C, Lauterborn J, Isackson P, White J 1990 Seizures, neuropeptide regulation and mRNA expression in the hippocampus. Prog Brain Res 83:371–390

Harrington ME, Rusak B 1986 Lesions of the thalamic intergeniculate leaflet alter hamster circadian rhythms. J Biol Rhythms 1:309–325

Harrison DL, Bates PJJ 1991 The mammals of Arabia, 2nd edn. Harrison Zoological Museum, Sevenoaks, Kent

Hohmann W, Michel S, Fleissner G 1990 Locomotor activity and light pulses as competing zeitgeber stimuli in the scorpion circadian system. Program and abstracts for the 3rd meeting of the Society for Research on Biological Rhythms, May 1992, p 52

Honrado GI, Mrosovsky N 1989 Arousal by sexual stimuli accelerates the re-entrainment of hamsters to phase advanced light–dark cycles. Behav Ecol Sociobiol 25:57–63

Johnson RF, Smale I, Moore RY, Morin LP 1988 Lateral geniculate lesions block circadian phase-shift responses to a benzodiazepine. Proc Natl Acad Sci USA 85:5301–5304

Mistlberger RE 1991 Effects of daily schedules of forced activity on free-running rhythms in the rat. J Biol Rhythms 6:71–80

Mikkelsen JD, Gustafson EL 1993 Distribution of phosphatase inhibitor-1-immunoreactive neurons in the suprachiasmatic nucleus of the Syrian hamster. Brain Res 623:147–154

Mrosovsky N 1989 Nonphotic enhancement of adjustment to new light–dark cycles: masking interpretation discounted. J Biol Rhythms 4:365–370

Mrosovsky N 1991 Double-pulse experiments with nonphotic and photic phase-shifting stimuli. J Biol Rhythms 6:167–179

Reebs SG 1989 Acoustical entrainment of circadian rhythms in house sparrows: constant light is not necessary. Ethology 80:172–181

Reebs SG, Lavery RJ, Mrosovsky N 1989 Running activity mediates the phase-advancing effects of dark pulses on hamster circadian rhythms. J Comp Physiol A Sens Neural Behav Physiol 165:811–818

Richter C 1967 Sleep and activity: their relation to the 24-hour clock. Proc Assoc Res New Ment Dis 45:8–29

Van Reeth O, Turek FW 1989 Stimulated activity mediates phase shifts in the hamster circadian clock induced by dark pulses or benzodiazepines. Nature 339:49–51

Wickland C, Turek FW 1991 Phase-shifting effect of triazolam on the hamster's circadian rhythm of activity is not mediated by a change in body temperature. Brain Res 560:12–16

Immediate-early genes and the neural bases of photic and non-photic entrainment

M. H. Hastings, F. J. P. Ebling, J. Grosse, J. Herbert, E. S. Maywood, J. D. Mikkelsen* and A. Sumova†

Department of Anatomy, University of Cambridge, Downing Street, Cambridge CB2 3DY, UK, *Institute of Medical Anatomy, Department B, University of Copenhagen, The Panum Institute, 3 Blegdamsvej, DK-2200 Copenhagen N, Denmark and †Institute of Physiology, Academy of Sciences of the Czech Republic, Prague, Czech Republic

Abstract. The expression of immediate-early genes (IEGs) within the mammalian suprachiasmatic nucleus (SCN) identifies individual light-responsive cells of the circadian system. Cells immunoreactive for products of IEGs form a neurochemically heterogeneous population, of which a few are VIP (vasoactive intestinal peptide)-immunoreactive or GRP (gastrin-releasing peptide)-immunoreactive, although the phenotypes of most of the others have yet to be determined. Dual-labelling experiments with anatomical tracers reveal that only a minority of efferent projection neurons of the SCN are immunoreactive for IEG products, and it is likely that the majority of the immunoreactive cells are interneurons or glia. Photic induction of IEGs is mediated via NMDA (*N*-methyl-D-aspartate) and non-NMDA glutamatergic receptors, the SCN expressing a topographically specific complement of subtypes of the NMDA receptor. Non-photic cues (arousal) can shift the clock but this is not associated with expression of IEGs, demonstrating that the proteins encoded by IEGs are probably involved in transducing photic cues, rather than shifting the clock *per se*. Their induction provides an anatomically explicit marker for circadian phase and photic sensitivity and so is useful in analyses of circadian function, for example, in the *tau* mutant hamster. Non-photic phase shifts are accompanied by adrenocortical activation, confirming the importance of arousal in shifting of the clock. The phase-shifting effect of arousal can be blocked by treatment with the serotonin receptor antagonist ketanserin, suggesting that ascending serotonergic input to the forebrain, possibly directly to the SCN, is an important mediator of entrainment by arousal.

1995 Circadian clocks and their adjustment. Wiley, Chichester (Ciba Foundation Symposium 183) p 175–197

In mammals, the circadian clock located within the suprachiasmatic nuclei can be synchronized both by light and by non-photic cues which arouse the animal during its inactive phase (Ralph 1995, Mrosovsky 1995, this volume). Comparison of the phase response curves (PRCs) generated by light and by arousal reveals important qualitative differences (Smith et al 1992). Light pulses are ineffective during subjective day, and produce phase delays or advances when presented during early or late subjective night, respectively. In contrast, non-photic cues characteristically induce phase advances when presented during subjective day and a variable pattern of delays or no shifts during subjective night. This implies that light and non-photic cues reset the clock through different neural/cellular mechanisms. If we are to understand the cellular basis of entrainment, we must exploit methods which resolve the activity of the suprachiasmatic pacemaker at the level of individual neurons. Electrophysiological studies of the suprachiasmatic nucleus (SCN) *in vivo* or in slice preparations *in vitro* have proved to be very useful in this respect (Gillette 1995, this volume). A second cellular function, the induced expression of immediate-early genes (IEGs), has also greatly advanced our understanding of the cellular organization of the SCN and its entrainment. Our aim here is to review the neural basis of circadian entrainment in mammals, paying particular attention to information obtained from studies of IEG expression in the SCN.

Expression of immediate-early genes in the suprachiasmatic nuclei

Cells express immediate-early genes transiently in response to discrete stimuli. Many of these genes code for transcription factors which are thought to coordinate the expression of a cascade of late-response genes and thereby direct longer-term adaptation to the stimulus. Within the nervous system, expression of IEGs is often associated with synaptic plasticity (Sheng & Greenberg 1990). On the basis of hybridization to mRNA and immunoreactivity against protein products, several laboratories have shown that light pulses are able to induce expression within the SCN of a number of IEGs, including c-*fos*, *egr-1* (*NGF-1A*), and *jun-B* (see reviews by Aronin & Schwarz 1991, Kornhauser et al 1992a). The transcriptional response is of interest for two reasons. It is spatially specific, being absent from the retinorecipient zones of the tectum and thalamus but very pronounced in the SCN. It is also temporally specific: light pulses are able to induce IEGs only when presented during subjective night, i.e., at times when they would shift the clock. This temporal restriction is very precise: light delivered 40 min before the onset of activity (circadian time [CT] 12) is ineffective, whereas light delivered 20 min after onset causes a full induction of IEGs (Mead et al 1992). Moreover, in groups of hamsters which have been exposed to single light pulses spanning a broad range of intensities, there are positive correlations between light intensity, the abundance of mRNA for c-*fos* and the magnitude of the induced phase shift (Kornhauser et al 1990). Of the products of IEGs, Egr-1 acts independently to direct transcription through a 'zinc-finger' motif, whereas members of the Fos and Jun families form 'leucine zipper' dimers which are then able to bind to DNA and direct transcription through the AP-1 site. Light pulses delivered during subjective night induce the expression of

AP1-binding activity, confirming that phase shifts are associated with changes in the transcriptional programme of light-responsive cells (Kornhauser et al 1992b). The expression of IEGs is therefore a cellular correlate of photic entrainment, and identifies a population of light-responsive cells in the SCN. This raises many questions, including: which types of cell express IEGs, which neurotransmitter or transmitters mediate this effect, and how is the expression of IEGs related to shifting of the clock?

Neuroanatomical identity of light-responsive cells expressing immediate-early genes

Although the pattern of retinal inputs to the SCN varies appreciably between species, in the rat, the Syrian hamster (*Mesocricetus auratus*) and the Siberian hamster (*Phodopus sungorus*) the distribution of Fos-immunoreactive cells maps to the region of retinal innervation of the SCN, as revealed by intraocular injection of cholera toxin B-subunit (Fig. 1), suggesting that the Fos-immunoreactive cells are on the input pathway to the circadian clock. In rats and hamsters arginine vasopressin (AVP)-immunoreactive perikarya are restricted to the dorsomedial zone of the SCN (Moore 1994, this volume), which does not receive a major retinal input, so they are unlikely to be Fos-immunoreactive. In contrast, the distribution of cells immunoreactive for both vasoactive intestinal peptide (VIP) and peptide histidine-isoleucine (PHI) overlaps the Fos immunoreactivity, and ultrastructural studies have shown that these cells receive direct retinal afferents (Ibata et al 1989, Tanaka et al 1993). It might therefore be expected that Fos-immunoreactive nuclei would be located within VIP/PHI-immunoreactive perikarya. Dual immunocytochemistry in the hamster shows that only a very small proportion of the Fos-immunoreactive nuclei are within PHI-immunoreactive perikarya; this is not unexpected because there are appreciably more Fos-immunoreactive than PHI-immunoreactive cells. However, only a small proportion (about 10%) of PHI-immunoreactive cells contain Fos-immunoreactive nuclei (Fig. 2a). This suggests that the VIP/PHI-immunoreactive cells are a heterogeneous population only some of which are directly responsive to light during subjective night. A small proportion ($<10\%$) of gastrin-releasing peptide (GRP)-immunoreactive cells, which are localized to the retinorecipient zone, are also Fos-immunoreactive, but again, in both hamster and rat, the majority of GRP-immunoreactive perikarya do not contain Fos-immunoreactive nuclei (Earnest et al 1993), and the neurochemical phenotype of the majority of Fos-immunoreactive cells awaits clarification. It is apparent from these studies that light pulses act upon a mixed population of neurons within the SCN.

Retrograde tract tracing of fluorescent microspheres (Bittman et al 1992) and cholera toxin B-subunit (J. D. Mikkelsen & N. Vrang, unpublished work) have been combined with visualization immunoreactivity to Fos and Egr in the SCN

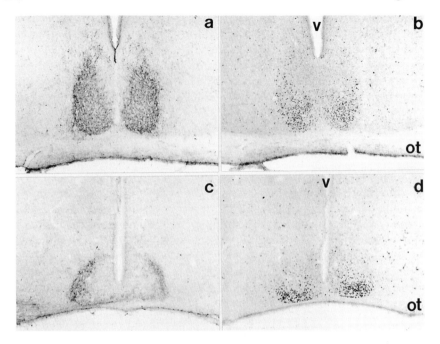

FIG. 1. Coronal sections through the caudal suprachiasmatic nuclei of (a, b) Syrian hamster (*Mesocricetus auratus*) and (c, d) Siberian hamster (*Phodopus sungorus*) following intraocular injection of cholera toxin B-subunit 48 h previously, and exposure to light for 15 min during subjective night, one hour before death. Serial sections were processed to reveal cholera toxin within the retinal terminal field (a, c), and Fos-immunoreactive nuclei (b, d). Note the overlapping distribution of Fos-immunoreactive cells and retinal input in the ventral, but not more dorsal, SCN. V, ventricle; ot, optic tract.

of rats and hamsters in an effort to determine whether cells expressing IEGs contribute to the efferent innervation of the SCN or are intrinsic interneurons. Despite extensive injections into known targets of SCN efferents (Kalsbeek et al 1993, Watts 1991), including the septum and preoptic hypothalamus and the paraventricular nuclei of the thalamus (PVT) and of the hypothalamus (PVN), few double-labelled cells have been detected (Fig. 2b). The frequency of double-labelled cells was highest after injection into the PVN, which is consistent with the massive projection of the SCN to the sub-PVN region. Nevertheless, the majority ($>75\%$) of cells identified as projecting neurons did not contain Fos- or Egr-immunoreactive nuclei. This finding is consistent with the results for PHI/VIP neurons, which represent one of the important efferent projections of the nucleus. If they had expressed IEGs, one would expect a significant number of projecting neurons to be Fos- and Egr-immunoreactive. It therefore

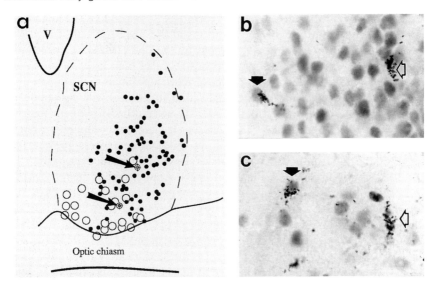

FIG. 2. (a) Camera lucida tracing demonstrating the relative distribution of PHI (peptide histidine-isoleucine)-immunoreactive perikarya and Fos-immunoreactive nuclei within a coronal section of the caudal SCN of a Syrian hamster following exposure to light for 15 min during subjective night. ●, Fos-immunoreactive nuclei ($n=84$); ○, PHI-immunoreactive perikarya ($n=25$); ⊙, dual-stained Fos-immunoreactive nuclei within PHI-immunoreactive perikarya. Note only two cells are dual stained (arrows). V, ventricle. (b, c) Photomicrographs of the SCN of a Syrian hamster which received a unilateral injection of cholera toxin B-subunit into the ipsilateral paraventricular nucleus of the hypothalamus to reveal projecting neurons by retrograde transport. The tissue was processed to reveal Fos-immunoreactive nuclei (dark spheres) and cytoplasmic accumulation of cholera toxin (granular reaction product). Open arrow, single-labelled; filled arrow, dual-labelled cells (J. D. Mikkelsen & N. Vrang, unpublished results).

appears that the light-responsive, IEG-expressing cells are, on the whole, components of the intrinsic circuitry of the SCN. Whether or not some are glia is unknown.

Glutamate, immediate-early genes and photic entrainment

An important question is which neurotransmitter or transmitters mediate photic input to the SCN. Behavioural studies in hamsters suggest a role for the excitatory amino acid glutamate, acting via NMDA (N-methyl-D-aspartate) and non-NMDA receptors (Colwell et al 1990, Vindlacheruvu et al 1992). This conclusion is supported by electrophysiological studies in which glutamatergic antagonists blocked the response of SCN neurons to stimulation of the optic nerve (Cahill & Menaker 1989, Kim & Dudek 1991). Interestingly, the relative contributions of the different types of receptor may depend on the membrane potential of the cells of the SCN, activation via the NMDA receptor requiring partial depolarization which may be gated by the non-NMDA receptor (Kim & Dudek 1991). Ultrastructural analysis has revealed high levels of glutamate immunoreactivity in the secretory vesicles of retinal terminals in the mouse SCN (Castel et al 1993), and intracerebroventricular infusion of the glutamatergic agonist NMDA induces Fos immunoreactivity within the SCN of the hamster (Ebling et al 1991). Furthermore, systemic or intracerebral administration of MK801 (a non-competitive antagonist at NMDA receptors) to hamsters attenuates the light-induced expression of Fos immunoreactivity within the ventral SCN (Abe et al 1991, Ebling et al 1991). Antagonists of the non-NMDA receptor also have a comparable effect upon Fos immunoreactivity (Abe et al 1992, Vindlacheruvu et al 1992), as might be predicted from the electrophysiological data. However, one important point is that even after treatment with antagonists specific for either NMDA or non-NMDA receptors, light pulses can still induce the expression of Fos in the dorsolateral retinorecipient SCN (Abe et al 1991, 1992, Ebling et al 1991, Vindlacheruvu et al 1992), even though the behavioural (phase-shifting) response to light is blocked completely. This indicates a pronounced heterogeneity in the functional neurochemistry of the retinoreceptive field. In addition, there may be appreciable differences between species; induction of Fos expression in the rat is much less sensitive to blockade by MK801 than it is in the hamster (J. D. Mikkelsen, unpublished results).

Glutamate is a widespread and abundant transmitter within the CNS. Its anatomical and chemical spectrum of actions depends upon the existence of many subtypes of receptor. NMDA receptors are heteromeric receptor–ion channel complexes formed by tissue-specific expression of genes encoding various subunits (NMDAR-1, 2a–d) (Kutsuwada et al 1992, Monyer et al 1992). All NMDA receptors contain the NMDAR-1 subunit and *in situ* hybridization reveals that it is expressed in the SCN of the rat and hamster (Fig. 3) (Mikkelsen et al 1993). The SCN does not express NMDAR-2a but does contain mRNA for NMDAR-2c in both species. Whereas mRNA for the NMDAR-1 subunit is widespread throughout the SCN, mRNA for the NMDAR-2c subunit is located predominantly in the dorsal region of the nucleus. This may be significant, because homologues of NMDAR-1 and NMDAR-2c subunits from mouse co-expressed in *Xenopus* oocytes are appreciably less sensitive to blockade by MK801 and D-amino-5-phosphopentanoate (a competitive inhibitor of NMDA) than receptors consisting of NMDAR-1 and either NMDAR-2a or NMDAR-2b (Kutsuwada et al 1992). The partial insensitivity of the SCN to NMDA antagonists may therefore be related to the limited distribution of the NMDAR-2c subunit: retinal activation of neurons (and induction of IEGs) in the dorsal part of the nucleus may be insensitive to the antagonists because those cells express the NMDAR-1/R-2c complex. It is not yet clear which subunit

Immediate-early genes and entrainment

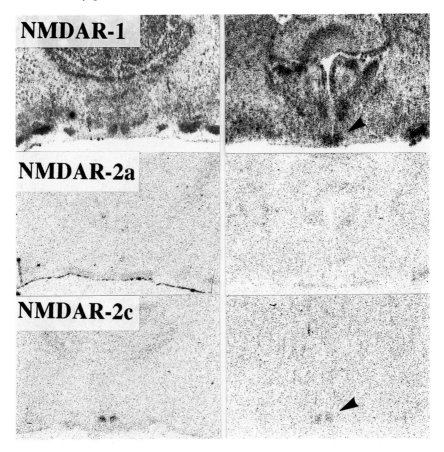

FIG. 3. Film autoradiograms showing *in situ* hybridization of oligonucleotide probes for subunits of the NMDA receptor in coronal sections of the SCN of the rat (*left*) and Syrian hamster (*right*). Arrows indicate position of SCN in hamster sections; note hybridization for NMDAR-1 and NMDAR-2c subunits and absence of the NMDAR-2a subunit in both species.

in addition to NMDAR-1 is expressed in the ventral SCN of the hamster where the induction of IEGs can be blocked by NMDA receptor antagonists. Given that prevention of Fos expression in this region correlates with loss of the phase-shifting response, it is likely that the receptor subtype in this area is involved in transduction of photic signals to the clock. Conversely, the role of the NMDAR-2c subunit in entrainment has yet to be confirmed. It may be the case that antagonism of either heteromeric receptor (R-1 + R-? and R-1 + R-2c) is sufficient to prevent entrainment of the clock. Duality of photic inputs is suggested by the recent observation that intracerebroventricular treatment with the cholinergic antagonist mecamylamine (which may have actions at NMDA receptors) prevents light-induced phase shifts in the hamster, and also blocks photic induction of Fos immunoreactivity in the dorsal retinorecipient zone of the SCN, without affecting Fos immunoreactivity in the ventral region (Zhang et al 1993)—the converse of the effect of MK801. Together, these results indicate that there are at least two pharmacologically and anatomically distinct pathways for photic entrainment in the SCN.

Induction of immediate-early genes and non-photic entrainment

Do the IEGs induced by light mediate phase shifts, or are they a correlate of photic activation of the retinorecipient SCN? One way to address this question is to determine whether or not all shifts of the circadian clock are accompanied by the expression of these genes. In Syrian hamsters, non-photic stimuli which arouse the animal produce phase advances in its free-running circadian activity rhythm. At CT 10, two hours before the projected onset of activity, simply handling the animal and injecting it subcutaneously with saline is sufficient to advance the clock by around one hour. However, this shift is not associated with expression of c-*fos*, c-*jun* or *egr-1* (Mead et al 1992). Phase advances can also be stimulated in hamsters by enforced running activity or by injection of the benzodiazepine triazolam, but neither treatment induces the expression of c-*fos* in the SCN (Janik & Mrosovsky 1992, Cutrera et al 1993), suggesting that either the clock does not really advance in parallel with the onset of activity, or the expression of these IEGs is not essential for shifting. In a test of whether handling and injection of saline at CT 10.20 advances the clock, hamsters were presented with a light pulse one hour after the arousing cue. If they were at CT 11.20 when the light pulse fell, i.e., no shift of the clock had happened, Fos immunoreactivity would not be induced in the SCN. However, if the central oscillator had advanced in parallel with overt activity, its phase would then be approximately CT 12.20 (assuming an advance of one hour, as observed in the activity traces), which is in the sensitive phase for the induction of Fos immunoreactivity. The combined arousal and light pulse induced Fos immunoreactivity within the SCN, confirming that arousal did indeed advance the clock, even though it was not associated with the expression of IEGs (Mead et al 1992). It is therefore likely that the expression of IEGs is not *necessary* for shifts to occur, and that when IEGs are induced, the response is related to transduction of the photic stimulus within the SCN, rather than shifting of the clock. This conclusion is supported further by the observation that intracerebroventricular administration of the cholinergic agonist carbachol produces phase shifts which follow a light-type phase response curve, but does not induce the expression of c-*fos* in the SCN (Colwell et al 1993). Nevertheless, these advances are blocked by glutamatergic antagonists, suggesting that glutamate is involved both in the transmission of photic cues into the SCN and in the execution of phase shifts.

Immediate-early gene expression in *tau* mutants

In the *tau* mutant strain of the Syrian hamster, the period of the circadian oscillator within the SCN is shortened to around 20 h in homozygotes (Ralph 1995, this volume). In addition, the magnitude of phase shifts induced by light and arousal are much greater than in wild-type animals (Mrosovsky et al 1992, M. R. Ralph, personal communication). For example, presentation of a 15 min light pulse to free-running homozygous mutants between CT 16 and CT 18 advances activity onset by about five hours, whereas the advance in wild-type hamsters is 40 min (J. Grosse, A. S. Loudon & M. H. Hastings, unpublished data). However, the abundance and distribution of Fos-immunoreactive nuclei induced by light pulses at CT 13, CT 16 or CT 18 are the same in *tau* mutants as in wild-type animals, indicating that the retinal inputs to the SCN in the two genotypes are comparable, and that their observed differential responsiveness to light is a property of the oscillator and not its entrainment pathway.

The cause of the shortened period in the *tau* mutants is unknown. Each circadian hour may be equivalent to 50 solar minutes, or the temporal programme may have been deleted by four circadian hours such that it consists of only 20 circadian hours, each of which is one solar hour in duration. It is possible that these alternatives could be discriminated by using the photic induction of Fos immunoreactivity to reveal the extent of subjective night. In both wild-type and *tau* mutant animals, Fos immunoreactivity is induced by light pulses delivered immediately after the onset of activity (J. Grosse, A. S. Loudon & M. H. Hastings, unpublished results). Preliminary studies show that in *tau* mutants it can also be induced by light at CT 2, a phase at which it is not induced in wild-types. This is consistent with the circadian period shortening through a deletion of part of the subjective day, subjective night in the mutant remaining of the same duration, in solar hours, as it is in the wild-type (about 12 solar hours). However, light at CT 2 does not shift the activity rhythm in mutants. Although this is consistent with results from wild-type hamsters, in which CT 2 is part of the dead zone of the PRC, it is paradoxical given the induction of c-*fos* expression in mutants given light at this phase. One possibility is that the temporal correlation between photic shifts and Fos immunoreactivity in the mutants may be disordered as a result of the mutation. Alternatively, the accuracy of the behavioural PRC as a marker of subjective night might be questioned. It is known from studies in wild-type hamsters that both the amplitude of the PRC and the observed duration of subjective night as defined by the behavioural shifts are plastic, depending upon prior photoperiodic exposures (Daan & Pittendrigh 1976, J. A. Elliott, personal communication). Consequently, the measurable duration of subjective night is a function of the amplitude of phase shifts: if a phase shift occurring at a phase late in the subjective night is too small to be measurable, the phase would be excluded from subjective night. It could be argued that photic induction of IEGs, as a phase-dependent event that is a function of circadian gating of the photic sensitivity of the SCN, provides a better marker for the duration of subjective night than behavioural responsiveness does. This proposition could be tested by mapping the relative extent of the phases at which light pulses can induce behavioural shifts and the expression of IEGs. By exposing wild-type hamsters to different lighting schedules, one would expect that the duration of subjective night as defined by behavioural shifts could be varied systematically. The important question is whether or not the duration of the zone of induction for IEGs shows an equivalent, parallel change. One possible outcome would be for the absolute duration of the zone of induction of IEGs to remain the same, independently of the duration of the zone of measurable behavioural responsiveness. Furthermore, the absolute duration of the IEG photo-inductive zone would be the same in *tau* mutants and in wild-type hamsters, an outcome consistent with the 'deletion' hypothesis. It is clear that the IEG-inducible zone has great potential value as a marker of subjective night that is independent of behavioural phase shifting.

The neural basis of non-photic entrainment

Non-photic entrainment is described in detail elsewhere in this volume, so we shall consider only its neural and neuroendocrine bases here. Non-photic shifts occur without expression of c-*fos*, *egr-1* or c-*jun*, although the possibility that they are accompanied by the expression of genes yet to be identified cannot be ruled out. In addition, arousal shifts the clock without associated phosphorylation of the transcriptional regulator CREB (cyclic AMP/calcium response-element-binding protein) (Sumova et al 1994a). Phosphorylation of CREB within the retinorecipient SCN is a correlate of light-induced phase shifts in the hamster (Ginty et al 1993), so it is clear that the cellular transduction of non-photic entrainment is different from that of entrainment by light. The rapid phosphorylation of CREB and induction of IEGs may be brought about by calcium-dependent kinases, activated after opening of the calcium channels integral to the NMDA receptor complex, whereas non-photic shifts probably occur independently of this cascade. They may be mediated by cAMP-dependent signal transduction, because treatment of the SCN *in vitro* with cyclic AMP (cAMP) or its analogues produces phase shifts similar to those caused by non-photic cues *in vivo* (Gillette 1995, this volume). However, to identify the cellular basis of non-photic shifts, it is first necessary to identify the afferent pathways to the SCN which are responsible for non-photic entrainment.

The predominant feature of non-photic zeitgebers is that they arouse the animal (Sumova et al 1994b). As noted earlier, handling and injection of the

Immediate-early genes and entrainment

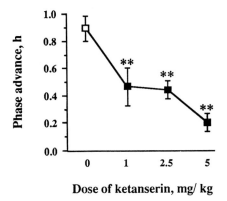

FIG. 4. Mean phase advances (± SEM) of wheel-running activity rhythm in free-running Syrian hamsters following arousal (handling and subcutaneous injection of saline) at CT 10, combined with prior injection of vehicle or ketanserin at CT 6. Animals were held in dim red light. Injection of vehicle at CT 6 did not affect the magnitude of the advance following arousal at CT 10 on the same circadian cycle (□). Injection of ketanserin (5 mg/kg) at CT 6 did not affect the phase of the free-running rhythm (data not shown). However, injection of ketanserin at CT 6 did impair the phase-shifting response to arousal at CT 10 (■). All animals were tested with vehicle and also with one (2.5 mg/kg, $n = 17$) or two doses (1, 5 mg/kg, $n = 11$) of ketanserin. **Significantly different ($P < 0.01$) from vehicle-injected hamsters (ANOVA and post hoc Dunnett's t-test). A. Sumova, unpublished results.

Syrian hamster with saline is sufficient to advance the light-entrainable oscillator of the SCN. This provides a useful protocol with which to investigate non-photic shifts because it is not confounded by pharmacological manipulation (Mead et al 1992, Hastings et al 1992). However, to investigate the involvement of arousal in entrainment, it is necessary to estimate the degree of arousal exhibited by an animal under experimental conditions. This has been achieved in the Syrian hamster by measurement of serum cortisol concentration, a peripheral index of arousal. Handling and injection of saline at CT 6 does not shift the clock and is not associated with adrenocortical activation, but at CT 10 the procedure both advances the clock and leads to a pronounced rise in serum cortisol within 15 min (Sumova et al 1994a). The correlation between phase shifts and arousal is strengthened further by comparison of the majority of hamsters which do show a phase shift in response to injection at CT 10 (around 80%) and the minority of animals which do not and also do not show an adrenocortical response. It remains to be determined whether the rise in cortisol is causal in phase shifting or simply a correlate of the aroused state. Nevertheless, the results confirm that the clock is reset under conditions of arousal, and it is likely that central pathways activated upon arousal mediate the shift. Activation of ascending serotonergic pathways is a characteristic feature of arousal (Jacobs & Azmitia 1992), and these may affect the SCN in two ways (Moore 1995, this volume). First, the SCN receive direct inputs from the mesencephalic raphe nuclei. Second, they receive NPY (neuropeptide Y)-immunoreactive fibres from the intergeniculate leaflet (IGL) of the thalamus, which in turn is innervated by the serotonergic raphe (the role of the IGL as a mediator of phase shifts has been reviewed recently by Moore [1992] and Mrosovsky 1995, this volume). Mediation of arousal-dependent shifts by direct or indirect serotonergic activation of the SCN neurons would be consistent with the adrenocortical activation coincident with arousal-induced shifts, which is also thought to depend upon heightened serotonergic neurotransmission within the hypothalamus.

In support of the hypothesis that serotonergic activation is necessary for arousal-induced phase shifting, both NPY and serotonergic agonists (particularly those acting at $5-HT_{1A}$ receptors) can shift the circadian clock *in vivo* and *in vitro*, producing PRCs with typically non-photic shapes (Gillette 1995, this volume). To demonstrate that *endogenous* serotonergic and NPY systems are involved in entrainment it is necessary to show that shifts cannot occur in the absence of normal functioning within those pathways. To test this, one can take advantage of the Syrian hamster's phase-dependent response to handling and injection. Injection of vehicle or the serotonergic antagonist ketanserin at CT 6 has no effect on the expression or phase of the hamster's free-running rhythm of activity, and injection of vehicle at CT 6 does not affect the advance following arousal at CT 10 (A. Sumova, unpublished results). However, treatment with ketanserin at CT 6 does block the advancing effect of a subsequent injection of saline at CT 10 (Fig. 4), indicating that unimpaired serotonergic neurotransmission is necessary for arousal to shift the circadian clock of the hamster.

The anatomical locus of this effect, which could include the SCN, IGL, raphe or forebrain, is not yet clear. Ketanserin has greatest affinity for the $5-HT_2$ (now $5-HT_{2A}$) receptor, but the SCN of rats express very little mRNA coding for this receptor (Roca et al 1993), the predominant area of expression of the $5-HT_{2A}$ receptor being the cerebral cortex (Julius et al 1990). However, ketanserin also has an affinity for the $5-HT_{1C}$ (now $5-HT_{2C}$) receptor, which is the predominant form of serotonergic receptor expressed by intrinsic neurons of the rat SCN (Hoffman & Mezey 1989, Roca et al 1993); the situation in hamsters is not known. It may be that systemically applied ketanserin acts directly upon the SCN to inhibit activation of $5-HT_{2C}$ receptors and thereby block arousal-induced shifts, but this interpretation apparently conflicts with the available literature suggesting that it is $5-HT_{1A}$ receptors which are involved in serotonin-mediated shifts. However, very little mRNA for the $5-HT_{1A}$ receptor is expressed in the rat SCN (Roca et al 1993). One way to reconcile these findings is to suggest that the $5-HT_{1A}$ receptors are synthesized in perikarya outwith the SCN, such that agents which act upon them may have a predominantly presynaptic action. In contrast, a putative $5-HT_{2C}$-dependent pathway could act directly on neurons of the SCN. This proposition could be tested by examining the effects on arousal-induced shifts of local, intracerebral application of serotonergic antagonists, for example in the SCN and raphe. Given that the 5-HT_{1A} receptor acts through inhibition of cAMP synthesis, whereas the predominant effect of 5-HT_{2C} receptor activation is to stimulate the turnover of inositol phosphate and diacylglycerol metabolism, a comparison of the importance of these pathways in shifting the clock may also throw further light onto the cellular mechanisms of entrainment by arousal.

Conclusion

It is clear that although the neurochemical phenotype and connectivity of cells of the SCN expressing IEGs are not yet established in full, the expression of these genes provides anatomically explicit markers of circadian phase and photosensitivity within the SCN. Further examination of the neurochemistry and connectivity of cells immunoreactive for products of IEGs has great potential for unravelling the circuitry of the SCN. The expression of IEGs is probably involved in the transduction of photic input within the SCN rather than shifting of the clock *per se*, but examination of the effects of various antagonists on photic induction of IEGs has nevertheless revealed the importance of glutamatergic neurotransmission as the mediator of photic input to the clock. It has also identified heterogeneity of light-responsive cells, which may be related to the topographically selective expression of specific NMDA receptor subtypes in the SCN. Arousal-induced shifts are mediated by a second cellular transduction pathway, as yet unidentified and distinct from that addressed by retinal afferents. It is likely that serotonergic activation of the SCN, directly from the raphe and/or indirectly via the IGL, is central to entrainment by arousal. Further comparison of the cellular pathways conveying photic cues and arousal through the SCN should ultimately converge at the oscillator itself.

Acknowledgements

Our studies cited in this paper were supported by awards from the Wellcome Trust, the Biotechnology and Biological Sciences Research Council, the Royal Society, the Science and Engineering Research Council and the European Science Foundation. The authors are very grateful to Mr J. Bashford and colleagues of the Audio Visual Media Unit, Department of Anatomy, University of Cambridge for photographic assistance.

References

Abe H, Rusak B, Robertson HA 1991 Photic induction of Fos protein in the suprachiasmatic nucleus is inhibited by the NMDA receptor antagonist MK-801. Neurosci Lett 127:9–12

Abe H, Rusak B, Robertson HA 1992 NMDA and non-NMDA receptor antagonists inhibit photic induction of Fos protein in the hamster suprachiasmatic nucleus. Brain Res Bull 28:831–835

Aronin N, Schwartz WJ 1991 A new strategy to explore molecular mechanisms of suprachiasmatic nucleus function. In: Klein DC, Moore RY, Reppert SM (eds) Suprachiasmatic nucleus: the mind's clock. Oxford University Press, New York, p 445–456

Bittman EL, Hastings MH, Ebling FJP 1992 Light activation of immediate-early gene expression occurs mainly in intrinsic neurons of the suprachiasmatic nucleus. Soc Neurosci Abstr 22:1229

Cahill GM, Menaker M 1989 Effects of excitatory amino acid receptor antagonists and agonists on suprachiasmatic nucleus responses to retinohypothalamic tract volleys. Brain Res 479:76–82

Castel M, Belenky M, Cohen S, Ottersen OP, Storm-Mathisen J 1993 Glutamate-like immunoreactivity in retinal terminals of the mouse suprachiasmatic nucleus. Eur J Neurosci 5:368–381

Colwell CS, Ralph MR, Menaker M 1990 Do NMDA receptors mediate the effects of light on circadian behaviour? Brain Res 523:117–120

Colwell CS, Kaufman CM, Menaker M 1993 Phase-shifting mechanisms in the mammalian circadian system: new light on the carbachol paradox. J Neurosci 13:1454–1459

Cutrera RA, Kalsbeek A, Pevet P 1993 No triazolam-induced expression of Fos protein in raphe nuclei of the male Syrian hamster. Brain Res 602:14–20

Daan S, Pittendrigh CS 1976 A functional analysis of circadian pacemakers in nocturnal rodents. J Comp Physiol 106:223–355

Earnest D, DiGiorgio SM, Olschowska JA 1993 Light induces expression of Fos-related proteins within gastrin-releasing peptide neurons in the rat suprachiasmatic nucleus. Brain Res 627:205–209

Ebling FJP, Maywood ES, Staley K et al 1991 The role of NMDA-type glutamatergic neurotransmission in the suprachiasmatic nuclei of the Syrian hamster. J Neuroendocrinol 3:641–652

Gillette MU, Medanic M, McArthur AJ et al 1995 Intrinsic neuronal rhythms in the suprachiasmatic nuclei and their adjustment. In: Circadian clocks and their adjustment. Wiley, Chichester (Ciba Found Symp 183) p 134–153

Ginty DD, Kornhauser JM, Thompson MA et al 1993 Regulation of CREB phosphorylation in the suprachiasmatic nucleus by light and a circadian clock. Science 260:238–241

Hastings MH, Mead SM, Vindlacheruvu RR, Ebling FJP, Maywood ES, Grosse J 1992 Non-photic phase shifting of the circadian activity rhythm of Syrian hamsters: the relative potency of arousal and melatonin. Brain Res 591:20–26

Hoffman BJ, Mezey E 1989 Distribution of serotonin $5HT_{1C}$ receptor mRNA in adult rat brain. FEBS (Fed Eur Biochem Soc) Lett 247:453–462

Ibata Y, Takahashi Y, Okamura H et al 1989 Vasoactive intestinal peptide (VIP)-like immunoreactive neurons located in the rat suprachiasmatic nucleus receive a direct retinal projection. Neurosci Lett 98:1–5

Jacobs BL, Azmitia EC 1992 Structure and function of the brain serotonergic system. Physiol Rev 72:167–229

Janik D, Mrosovsky N 1992 Gene expression in the geniculate induced by a nonphotic circadian phase shifting stimulus. NeuroReport 3:575–578

Julius D, Haung KN, Livelli TJ, Axel R, Jessell TM 1990 The $5HT_2$ receptor defines a family of structurally distinct but functionally conserved serotonin receptors. Proc Natl Acad Sci USA 87:928–932

Kalsbeek A, Teclemariam-Mesbah R, Pevet P 1993 Efferent projections of the suprachiasmatic nucleus in the golden hamster. J Comp Neurol 332:293–314

Kim YI, Dudek FE 1991 Intracellular electrophysiological study of suprachiasmatic nucleus in rodents: excitatory synaptic mechanisms. J Physiol 444:269–287

Kornhauser JM, Nelson DE, Mayo KE, Takahashi JS 1990 Photic and circadian regulation of c-*fos* gene expression in the hamster suprachiasmatic nucleus. Neuron 5:127–134

Kornhauser JM, Mayo KE, Takahashi JS 1992a Immediate-early gene expression in a mammalian circadian pacemaker: the suprachiasmatic nucleus. In: Young MW (ed) Molecular genetics of biological rhythms. Marcell Decker, New York, p 271–307

Kornhauser JM, Nelson DE, Mayo KE, Takahashi JS 1992b Regulation of *jun*-B messenger RNA and AP-1 activity by light and a circadian clock. Science 255:1581–1584

Kutsuwada T, Kashiwabuchi N, Mori H et al 1992 Molecular diversity of the NMDA receptor channel. Nature 358:36–41

Mead S, Ebling FJP, Maywood ES, Humby T, Herbert J, Hastings MH 1992 A nonphotic stimulus causes instantaneous phase advances of the light-entrainable circadian oscillator of the Syrian hamster but does not induce the expression of c-*fos* in the suprachiasmatic nuclei. J Neurosci 12:2516–2522

Mikkelsen JD, Larsen PJ, Ebling FJP 1993 Distribution of *N*-methyl-D-aspartate (NMDA) receptor messenger RNAs in the rat suprachiasmatic nucleus. Brain Res 632:329–333

Monyer H, Sprengl R, Schoepfer R et al 1992 Heteromeric NMDA receptors: molecular and functional distinction of subtypes. Science 256:1217–1221

Moore RY 1992 The enigma of the geniculo-hypothalamic tract: why two visual entraining pathways? J Interdiscip Cycle Res 23:144–152

Moore RY 1995 Organization of the mammalian circadian system. In: Circadian clocks and their adjustment. Wiley, Chichester (Ciba Found Symp 183) p 88–106

Mrosovsky N 1995 A non-photic gateway to the circadian clock of hamsters. In: Circadian clocks and their adjustment. Wiley, Chichester (Ciba Found Symp 183) p 154–174

Mrosovsky N, Salmon PA, Menaker M, Ralph MR 1992 Non-photic phase shifting in hamster clock mutants. J Biol Rhythms 7:41–49

Ralph MR Circadian pacemakers in vertebrates. In: Circadian clocks and their adjustment. Wiley, Chichester (Ciba Found Symp 183) p 67–87

Roca AL, Weaver DR, Reppert SM 1993 Serotonin receptor gene expression in the rat suprachiasmatic nuclei. Brain Res 608:159–165

Sheng M, Greenberg ME 1990 The regulation and function of c-*fos* and other immediate early genes in the nervous system. Neuron 4:477–485

Smith RD, Turek FW, Takahashi JS 1992 Two families of phase-response curves characterise the resetting of the hamster circadian clock. Am J Physiol 31:R1149–R1153

Sumova A, Ebling FJP, Maywood ES, Herbert J, Hastings MH 1994a Non-photic circadian entrainment in the Syrian hamster is not associated with phosphorylation of the transcriptional regulator CREB within the suprachiasmatic nucleus, but is associated with adrenocortical activation. Neuroendocrinology 59:579–589

Sumova A, Ebling FJP, Herbert J, Maywood ES, Moore EM, Hastings MH 1994b Non-photic entrainment of circadian rhythms. Adv Pineal Res 7, in press

Tanaka M, Ichitani Y, Okamura H, Tanaka Y, Ibata Y 1993 The direct retinal projection to VIP neuronal elements in the rat SCN. Brain Res Bull 31:637–640

Vindlacheruvu RR, Ebling FJP, Maywood ES, Hastings MH 1992 Blockade of glutamatergic neurotransmission in the suprachiasmatic nucleus prevents cellular and behavioural responses of the circadian system to light. Eur J Neurosci 7:673–679

Watts AG 1991 The efferent projection of the suprachiasmatic nucleus: anatomical insights into the control of circadian rhythms. In: Klein DC, Moore RY, Reppert SM (eds) The suprachiasmatic nucleus: the mind's clock. Oxford University Press, New York, p 77–106

Zhang Y, Zee PC, Kirby JD, Takahashi JS, Turek FW 1993 A cholinergic antagonist, mecamylamine, blocks light-induced Fos immunoreactivity in specific regions of the hamster suprachiasmatic nucleus. Brain Res 615:107–112

DISCUSSION

Cassone: Because you can block light-induced phase shifting with an anti-glutaminergic agent and still get *fos* expression, and, conversely, because you can get non-photically-induced phase shifts without *fos* expression, would it be fair to say that *fos* at least is neither necessary nor sufficient for phase shifting of the clock in the hamster?

Hastings: Following discussions I've had with Bill Schwartz, my feeling is that expression of IEGs is not necessary for shifting of the clock. When we give a light pulse we are giving a metabolic insult to the SCN: studies with 2-deoxyglucose autoradiography have demonstrated a marked increase in metabolism following presentation of a light pulse during subjective night (Schwartz et al 1980). The coincident induction of expression of IEGs is probably a housekeeping response to that metabolic demand, rather than a transcriptional response that is necessary for the implementation of a phase shift.

If Fos works through an AP1 binding site to alter the transcription of other genes, those transcriptional events may occur several hours after the light pulse has been delivered. The temporal characteristics of the *fos* response are therefore probably too slow to be causal to a shift. We know that the clock can shift in response to a non-photic stimulus within about one hour (Mead et al 1992). If a light-induced phase shift can occur as rapidly, the Fos-mediated events would occur too late to be causal.

Takahashi: Dr Cassone mentioned blockage of phase shifts by antagonists. These experiments generally involve sub-saturating light pulses with which you can see a block at the behavioural level. If the light intensity is increased 10-fold, the antagonist can't block the phase shift. To see the pharmacological blockade you have to titrate the light intensity against the antagonist. That tells you that you are definitely not blocking everything in the first place to get this behavioural effect. You have to be careful about using the results of this kind of blocking experiment to prove or rule out a role for a particular agent such as Fos.

In some work that we did with Dr Turek (Zhang et al 1993), we used an antagonist called mecamylamine, which blocks light-induced phase shifting in hamsters in the same way as MK801. Mecamylamine blocked *fos* induction in the dorsal SCN but the ventral SCN signal was intact. This is the complement of what Dr Hastings showed us here. Again, this is a peculiar type of correlation. If you block part of a *fos* signal and you are working in a situation where you are titrating the antagonist against light, you can see a reduction in the phase-shifting behaviour. Mecamylamine, like MK801, won't block the phase shift if you raise light intensity 10-fold.

I would disagree with what Dr Hastings said about the role of IEGs, that their expresssion is probably just a housekeeping response to a light pulse. We can't come to a logical conclusion either way on the basis of the existing information. There is a nice correlation between induction of IEGs and behavioural phase shifting, but it has not yet been possible to do a clean experiment in which *fos* is selectively removed rather than a pharmacological antagonist, for example, being used. We don't have results proving that Fos is not part of the pathway.

Hastings: As I said, if the light-induced shifts occur as rapidly as the arousal-induced shifts, within one hour, we may be able to discount *fos* as causal, but I don't know if that is the case. Even if the protein is maximal after one hour, it still has to act on another transcriptional step before the effect is complete.

Takahashi: Yes, but you can easily see AP1-binding activity in the SCN within an hour of light treatment. There's not sufficient time resolution to make a distinction. With a two-pulse paradigm you might see resetting (a delay) within two hours in a hamster, but an advance may not occur that quickly, for example.

Hastings: Would you say that the area of *fos* induction that is lost with mecamylamine is comparable to the area that's retained after treatment with MK801? Is it solely in the dorsal retinorecipient zone?

Turek: Yes.

Hastings: Then it would be interesting to know the effect of mecamylamine upon distinct forms of NMDA receptors, made up of the different subtypes.

Mrosovsky: Dr Hastings suggested that a chunk may be cut out of the cycle, that there's a deletion of a particular phase of the cycle. I understand that the PRCs produced by light in the *tau* mutant and the wild-type hamster are generally similar in shape despite some differences in amplitude. From that one might argue there is a uniform compression of the cycle in the mutant, not a deletion of a particular phase.

Menaker: I was about to say that. If you do the experiment in a certain way, the amplitudes are not different. The PRC to saturating light pulses that Takahashi et al (1984) got years ago from wild-type hamsters and the one that K. Shimomura in my lab (unpublished results) has just got from *tau* mutants lie right on top of each other if you plot them in circadian time. That is really the opposite result of the one suggested by Dr Hastings. It has to be said that we are keeping our animals differently and that we know that these differences can have effects on phase shifting responses to light pulses.

K. Shimomura has found that although the holding conditions affect the amplitude of the phase response, they also affect *fos* induction in parallel. *fos* isn't somehow protected from the effects of previous conditions. Indeed, the larger the amplitude of phase shift that you get, the more cells in the SCN show *fos* induction. I don't buy the argument that *fos* is somehow more fundamental than the behaviour.

Hastings: I agree with you: it's an alternative, and it's important that we have alternatives. If we simply base everything on behaviour, where the amplitude of shifts can be reduced as a function of previous treatments, we may not be able to measure them and so will underestimate the extent of the light-responsive zone of the circadian cycle. The alternative of course is that we can use the induction of c-*fos* to map this zone. What you say about *fos* exhibiting an amplitude effect in parallel to that shown by behavioural shifts is interesting but leaves open the question of the duration of the *fos*-inductive zone. It may be that the *fos*-inductive zone of the oscillator is a single timed interval, initiated at around CT 12 or whatever and finishing 12 solar hours later, running independently of the circadian cycle.

Menaker: That would be absolutely fascinating if it were the case.

Hastings: We need to go back to those paradigms where you can use photoschedules with non-24 h period lengths (T cycles) or different 24 h photoperiods to alter the apparent extent of the interval over which light induces phase shifts, and look to see how the duration of the *fos*-inductive zone changes, if at all. This should be done just in wild-types to determine whether as we change alpha and/or the amplitude of the PRC we see corresponding changes in the duration of subjective night as defined by the induction of c-*fos*.

Block: When c-*fos* was first discussed there was at least an assumption that it would play a role in the circadian system. It was exciting not because it was a useful marker but because it might be a series element in an input pathway to the pacemaker. We have heard two different views on this. There's emerging evidence that inhibitors of transcription and protein synthesis don't block light-induced phase delays (Bogart 1992). Phase advances are more complicated because of the time of the cycle at which they occur. Is there an experiment that will allow us to move past this issue, either to put c-*fos* aside or to embrace it as a series element in the pathway?

Miller: There is a mouse that lacks *fos* (M. Ralph, personal communication 1992).

Takahashi: The *fos*-less mouse is not a good example. A whole set of genes, we see at least five, are induced by light in addition to *fos*.

Miller: It could be used to address the question of whether *fos* in particular is essential.

Takahashi: But AP1 activity, which is the form Fos becomes as an active agent, is composed of gene products from many combinations of IEGs. It is difficult to take out AP1-binding activity effectively when there is a redundant signal—a single knock-out is not enough. Perhaps you could take out the AP1 signal by introducing competing oligonucleotides complementary to the AP1-binding site.

Gillette: I believe the photic signal isn't received by the SCN in the *fos*-less mouse. You can't ask this question about the role of *fos* in the SCN, because you have already compromised the photoreception in the retina. Is this correct?

Ralph: The *fos*-less mouse seems to respond to a short light pulse with a phase delay or advance. We haven't constructed a complete PRC, but the mice can certainly respond to light. In earlier experiments other *fos*-less mice showed abnormal entrainment. If there are five or so inducible transcription factors, and you think that some of those might be involved in the phase-shifting response, you would predict that if you knocked out one you would not necessarily knock out the whole response to light, but you might expect to see a *change* in the response to light.

Redfern: Dr Hastings, ketanserin, at the dose you gave, would be expected to reduce blood pressure. This might explain the temporary decrease in locomotor activity that you saw.

Hastings: It may well do. Clearly, the systemic treatment potentially has ambiguous outcomes: local intracerebral administration will be necessary to identify sites and mechanisms of action. Nevertheless, we feel that the general approach in which we have the animal doing something and then we try to prevent it is the most direct way to determine what is happening during non-photic shifts.

Miller: Your work on ketanserin is interesting, but there are some paradoxical phenomena to consider here. We showed some time ago that there is a tremendous amount of 8-OH-DPAT (8-hydroxy-2-[di-*n*-propylamino]tetralin) binding in the rat SCN. Dr Reppert has found that the amount of 5-HT$_{1A}$ message is minimal in the SCN. So, what on earth is 8-OH-DPAT binding to?

Hastings: The 5-HT$_{1A}$ receptors could be presynaptic, the message encoding them being synthesized outside the SCN, possibly in the raphe nuclei.

Miller: I don't think that's it. The terminal autoreceptor on 5-HT axon endings is the 5-HT$_{1B}$ receptor, which does not bind 8-OH-DPAT with high affinity. 8-OH-DPAT does not bind to the 5-HT transporter with high affinity either. Thus, it is unlikely that the high affinity (2 nM) 8-OH-DPAT binding was presynaptic.

Also, mRNA for the 5-HT$_{1C}$ (now termed 5-HT$_{2C}$) receptor is abundant in the SCN. However, again some years ago, we found that DOB (dimethoxy-bromoamphetamine), which is a high affinity agonist for both the 5-HT$_{2A}$ and the 5-HT$_{2C(1C)}$ receptor, had absolutely no phase-shifting effect at all in the SCN. It's difficult for me to believe that the 5-HT$_{2C(1C)}$ receptor could be involved in mediating serotonin-induced phase shifts.

Hastings: So you would say that the 5-HT$_{2C}$ receptor for which mRNA is present may have an entirely different function or is redundant?

Miller: I think it has a different function; we have been unable to find a function for it. We have been looking at a recently cloned 5-HT receptor (Lovenberg et al 1993) which is thought (from polymerase chain reaction [PCR] studies and some debatable *in situ* hybridization) to exist in the SCN. 5-Carboxamidotryptamine (5-CT), which is a wonderful phase-shifting agent in the SCN, binds excellently to this receptor expressed in transfected cells, as does 5-HT and also 8-OH-DPAT, but at lower affinity. The effects of 5-CT can be antagonized by mesulergine, ritanserin and NAN-190 (1-[2-methoxyphenyl]-4-[4-(2-phthalimido)butyl]piperazine). We already know that NAN-190 antagonizes the effects of 8-OH-DPAT in the SCN. The important thing here is ritanserin. Ritanserin is quite similar to ketanserin—they are both considered to be 5-HT$_{2A}$ antagonists. In the SCN slice, 8-OH-DPAT given at CT 6 causes a 2–3 h phase shift. Pindolol, the classical 5-HT$_{1A}$ antagonist, does not affect the phase shift. Ritanserin, however, blocks it completely. Ritanserin is the most effective antagonist we have found for this new receptor, which we have dubbed 5-HT$_7$. Thus, I would bet that you would find ritanserin an even better antagonist than ketanserin. The 5-HT$_7$ receptor has peculiar pharmacology; as far as agonists go, it's 5-HT$_{1A}$-like, but with respect to antagonists it is 5-HT$_{2A}$-like. We believe that this pharmacology explains the effectiveness of 8-OH-DPAT as a phase-shifting agent, even though 5-HT$_{1A}$ receptors are scarce in the SCN. Similarly, we can now explain why ritanserin (and perhaps ketanserin) is a good antagonist, even though the 5-HT$_{2A}$/5-HT$_{2C(1C)}$ agonist DOB is ineffective as an agonist in the SCN. We believe that 5-HT$_7$ (or another receptor with essentially identical pharmacology) is the 5-HT receptor responsible for 5-HT-mediated phase advances.

Hastings: That's interesting, but I would point out that in our *in vivo* model we are interested in what endogenous 5-HT is addressing. The spatial and temporal pattern of action of endogenous 5-HT will be different from that resulting from bath application of drugs.

Reppert: Dr Miller, do you feel that this new 5-HT receptor mediates the phase-shifting effects of serotonin on SCN in culture?

Miller: If it's not 5-HT$_7$ that's involved, it's something with a pharmacology similar to that of 5-HT$_7$.

Menaker: What does *in situ* hybridization tell you?

Miller: The PCR work indicated that the 5-HT$_7$ receptor message was present in SCN punches. With *in situ* hybridization Erlander found a capsular distribution of 5-HT$_7$ mRNA surrounding the SCN (M. G. Erlander, personal communication 1993). We are trying to resolve this issue with Northern blots, with which, so far, we have found weak expression of 5-HT$_7$ message in the SCN.

Armstrong: Dr Hastings, in rats we do not see such large phase shifts as you do with saline at CT 9–12, though we do with melatonin. I also wonder whether the Cambridge hamsters are queer. Throughout the literature on hamsters, saline controls in experiments in which drugs such as triazolam or bicuculline are administered don't have that phase-shifting effect. Is there something particularly strange about the Cambridge hamsters?

Hastings: In a recent paper by Meyer et al (1993) controls for chlordiazepoxide injected with saline at CT 8 and CT 10 show phase shifts. These hamsters are, I assume, a North American strain. Also, this arousal works over a very narrow range, and the stimulus is weak in comparison with triazolam or voluntary running. If there is a tendency to group data from saline controls or to measure at infrequent intervals, it would be quite easy to miss this if one were not looking for it.

Turek: We have certainly never seen phase shifts of this magnitude in hamsters injected with saline.

When I did some work in Belgium with Belgian hamsters, I noticed how aggressive they were and how much more they were aroused by a stimulus.

In fact, they were so aggressive I decided to have my hamsters sent over from America so I would be bitten less often. Perhaps your hamsters are more arousable. I would expect to see a phase shift to an activity-inducing stimulus given at CT 8 or CT 10 because presentation of a running wheel or an injection of triazolam near these CTs will induce a phase advance in the circadian activity rhythm.

Takahashi: Dr Hastings' records show that his hamsters do not run as intensely as Lakeview hamsters. It appears that a Wright strain hamster, on average, runs a lot less than a Lakeview hamster. Thus, an activity-inducing agent may be more effective in a Wright strain hamster.

Hastings: I don't see this as a problem. We happen to have these animals, and, in our hands, they are a good, non-pharmacological model for arousal-induced shifts.

Armstrong: The problem comes in your generalizing to all rodent species and all the experiments in other laboratories around the world, and saying that effects of pharmacological agents applied at that time are due purely to arousal.

Hastings: With our weak stimulus it is clear that the late subjective day part of the PRC (CT 8–12) is similar to Dr Turek's triazolam-induced PRC; the differences are at CT 4 and CT 6. Although the stimulus is applied at CT4 or CT6, we don't actually know when the stimulus causes the clock to shift. This may be a difficulty in comparing PRCs. If a particular stimulus causes certain after-effects in the animal, the shift may occur because these after-effects impinge upon a sensitive phase several hours after the initial stimulus was presented.

Turek: Do you get arousal with saline injection at CT 4 and CT 6?

Hastings: No, at least not in terms of corticosteroid secretion or activity.

Turek: Then why do you resort to that hypothesis? You are not getting the activity, so you are not getting the phase shift.

Hastings: Yes, but what I'm saying is that a stronger stimulus, which may be the one you use, does produce the phase shift simply because it's a stronger stimulus and you do get arousal. I would argue that with our weak stimulus at CT 8 or CT 10 we are tapping into the same mechanism that you activate earlier on in the subjective day.

Czeisler: Isn't the mechanism active even in adrenalectomized animals?

Turek: That's correct, so the response is unlikely to be mediated by adrenal corticosteroids.

Hastings: I have never said that it was; corticosteroid increases are an independent marker for arousal.

Czeisler: An increase in corticosteroids may be more a marker of pain, anxiety and stress than of activation, particularly with reference to potential differences between volitional activity and forced activity. It's important to use the appropriate terminology.

Hastings: I agree. If one's definition of stress is any agent which increases the secretion of corticosteroids, then this handling and injection is stressful.

However, I prefer to use the term arousal because observation of the animals' behaviour suggests that they are not under stress: they neither freeze nor become hyperactive. They are however responsive to their external environment and one might assume that brainstem systems governing arousal have been activated. To consider whether arousal can occur in the absence of stress is an interesting argument.

References

Bogart B 1992 Cellular mechanisms of circadian pacemaker entrainment in the mollusc *Bulla*. PhD thesis, University of Virginia, Charlottesville, VA, USA

Lovenberg TW, Baron BM, de Lecea L et al 1993 A novel adenylate cyclase-activating serotonin receptor (5-HT$_7$) implicated in the regulation of mammalian circadian rhythms. Neuron 11:449–458

Mead S, Ebling FJP, Maywood ES, Humby T, Herbert J, Hastings MH 1992 A nonphotic stimulus causes instantaneous phase advances of the light-entrainable oscillator of the Syrian hamster but does not induce the expression of c-*fos* in the suprachiasmatic nuclei. J Neurosci 12:2516–2522

Meyer EL, Harrington ME, Rahmani T 1993 A phase response curve to the benzodiazepine chlordiazepoxide and the effect of geniculo-hypothalamic tract ablation. Physiol Behav 53:237–243

Schwartz WJ, Davidsen LC, Smith CB 1980 *In vivo* metabolic activity of a putative circadian oscillator, the rat suprachiasmatic nucleus. J Comp Neurol 189:157–167

Takahashi JS, DeCoursey PJ, Bauman L, Menaker M 1984 Spectral sensitivity of a novel photoreceptive system mediating entrainment of mammalian circadian rhythms. Nature 308:186–188

Zhang Y, Zee PC, Kirby JD, Takahashi JS, Turek FW 1993 A cholinergic antagonist, mecamylamine, blocks light-induced Fos immunoreactivity in specific regions of the hamster suprachiasmatic nucleus. Brain Res 615:107–112

End note added in proof by M. H. Hastings: Our preliminary studies gave a clear difference between wild-type and *tau* mutant hamsters in the ability of a light pulse at CT 2 to induce the expression of *fos* in the SCN. This was consistent with a deletion model for the short period mutation. Further investigation of this effect has since yielded several unexpected results (Grosse J, Loudon ASI, Hastings MH 1994 Behavioural and cellular responses to light of the circadian system of *tau* mutant and wild-type Syrian hamsters. Neuroscience, in press). Light pulses at CT 0, CT 1 or CT 3 given to free-running wild-type and *tau* mutant hamsters produced no significant phase shifts in free-running activity rhythms, demonstrating that subjective night defined by behavioural responses to light has a precise termination at the same phase in the two strains. However, in both strains, light presented at the same phase seven cycles later did induce the expression of *fos* in the retinorecipient SCN of some animals. Overall, there was a general decline in the mean number of light-induced Fos-immunoreactive nuclei from CT 22 onwards, but the response was very variable between animals. In contrast to the onset at CT 12, it was not possible to identify a sharp termination of the phase of *fos* inducibility in either wild-type or mutant hamsters. The activity traces revealed that those animals in which light induced the strongest *fos* response had been active shortly before presentation of the pulse, regardless of the circadian phase of the light pulse. In contrast, animals which had been inactive for some time, and in which extrapolation of activity traces for previous cycles suggested that their activity phase (alpha) was over when the light was presented, showed a weak *fos* response. For both wild-type and mutant animals there was a significant ($P<0.01$) negative correlation between the amount of *fos* expression induced by light and the time elapsed between the end of the activity phase and presentation of the light pulse. Given that Fos immunoreactivity is present after light pulses delivered in late subjective night/early subjective day which do not induce an overt phase shift, we conclude that photic induction of *fos* in the SCN is not sufficient for a behavioural shift to occur. Furthermore, it appears that in addition to circadian phase, the animal's activity state influences the *fos* response to light. On inspection, it was apparent that the mutant animals ($n=7$) originally exposed to light at CT 2 had alphas of long duration whereas the wild-type controls ($n=4$) had short alphas. These results offer an alternative interpretation of the original finding at CT 2 which does not support the 'deletion' hypothesis for the short period mutation.

Recent studies have shown that ritanserin given at CT 6 is as effective as ketanserin in blocking advances to arousal at CT 10. Ritanserin did not affect the phase of the free-running rhythm when given alone, neither did it affect light-induced advances when given at CT 14, with a 15 min light pulse at CT 18 (A. Sumova, unpublished results).

Interaction between the circadian clocks of mother and fetus

Steven M. Reppert

Laboratory of Developmental Chronobiology, Children's Service, Massachusetts General Hospital and Harvard Medical School, Boston, MA 02114, USA

Abstract. In mammals, a unidirectional communication exists between the biological clocks of the mother and fetus. As a biological clock begins oscillating in the suprachiasmatic nuclei of the fetus, redundant circadian signals entrain the fetal clock to the prevailing light–dark cycle. Recent studies have revealed an activatable dopamine system within the fetal hypothalamus which may serve as a final common pathway by which maternal signals entrain the fetus. An entrained biological clock during fetal life makes the developing mammal better prepared for life in the outside world.

1995 Circadian clocks and their adjustment. Wiley, Chichester (Ciba Foundation Symposium 183) p 198–211

Studies in several mammalian species show that the biological clock in the suprachiasmatic nuclei is oscillating in the fetus and that its oscillation is in phase with the environmental light–dark cycle. Entrainment of the fetal suprachiasmatic nuclei involves a unidirectional communication of circadian signals from mother to fetus. Here, I review these studies and detail recent results which suggest a cellular mechanism through which maternal circadian signals entrain the fetal suprachiasmatic nucleus (SCN).

In mammals, circadian rhythms in physiological and behavioural processes are not overtly expressed until postnatal life (Reppert 1987). Deguchi (1975) first suggested that a biological clock might be oscillating in the mammalian fetus before rhythms are expressed. He determined the phase of a rhythm monitored postnatally under constant conditions in rats and used this to estimate the phase of the biological clock at earlier developmental stages. Deguchi's findings suggested that a circadian clock oscillates in the fetal rat at or before birth and that its phase is coordinated with the dam.

Postnatal paradigms alone cannot prove that a biological clock actually functions *in utero*. It is possible that some rhythmic aspect of the birth process itself could start or set the timing of the developing clock. Demonstrating prenatal circadian function requires a method that can measure an intrinsic, functionally relevant property of the clock itself. A method proven useful for monitoring the oscillatory activity of the SCN in adult rats is ^{14}C-labelled 2-deoxyglucose (2-DG) autoradiography, which provides a biochemical means of visualizing the metabolic (functional) activity of discrete structures in the central nervous system *in vivo* (Schwartz 1991).

Reppert & Schwartz (1983) used the 2-DG method to demonstrate a circadian rhythm of metabolic activity in the SCN of fetal rats. This study showed a clear circadian variation of metabolic activity in the fetal SCN that is 'in time' (coordinated with the rhythm in the dam and with the external lighting cycle). This fetal rhythm of 2-DG metabolism can be detected as early as Day 19 of gestation (two–three days before birth; all gestational ages have been standardized to Day 0 = day of sperm positivity) (Reppert & Schwartz 1984). A rhythm of 2-DG metabolism in the fetal SCN has also been demonstrated *in vitro* (Shibata & Moore 1988), suggesting that the metabolic activity rhythm in the SCN is generated endogenously in the fetus and not passively driven by the mother.

Neurogenesis of the rat SCN occurs between Days 13 and 16 of gestation (Altman & Bayer 1978, Ifft 1972). A few synapses first appear in the SCN on Day 19 of gestation, but the vast majority appear postnatally (Lenn et al 1977, Moore & Bernstein 1989). A clear day–night rhythm of action potentials in the SCN is first apparent in fetal hypothalamic slices on Day 21 of gestation (Shibata & Moore 1987) at a time when the metabolic activity rhythm is quite prominent. Thus, the SCN begins displaying circadian oscillations in metabolic and electrical activity when the neurons are virtually devoid of neural connections. This early circadian activity in the absence of neural connections has helped fuel the notion that individual neurons in the SCN function as individual circadian pacemaking units (Moore & Bernstein 1989).

The mother entrains the fetus by communicating circadian signal(s) to the fetus (Fig. 1). Blind pregnant rats were used to show that environmental lighting acts through the maternal circadian system to entrain the rhythm of metabolic activity in the fetal SCN (Reppert & Schwartz 1983); the fetal rhythm was synchronous with the circadian time of the blind mothers and not affected directly by ambient lighting. A variety of prenatal and postnatal rhythms have been used by workers in several laboratories to confirm the prenatal coordination of maternal and fetal rhythms in several species (Davis 1989, Reppert et al 1989).

Experiments in which the maternal suprachiasmatic nuclei were destroyed early in gestation in rats and hamsters have shown that entraining signals for the fetus originate in the maternal SCN (Davis & Gorski 1988, Honma et al 1984a,b, Reppert & Schwartz 1986a, Weaver & Reppert 1989). Maternal SCN lesions do not destroy the timekeeping capability of the fetal clock; instead, they cause desynchronization of the individual fetal clocks within a litter, while circadian rhythms persist in each of the offspring (Fig. 2).

We have shown that during the postnatal period, rat pups born to SCN-lesioned dams respond appropriately (entrain) to light–dark cycles despite the

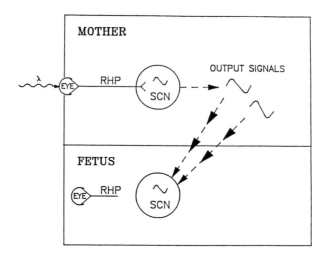

FIG. 1. A conceptual model of maternal–fetal communication of circadian phase. Light-induced neural signals are conveyed to the mother's suprachiasmatic nucleus (SCN) by her retinohypothalamic pathway (RHP), entraining her circadian rhythms. Maternal output signals then entrain the fetal clock at a time when the innervation of the fetal SCN by the RHP is incomplete.

absence of prenatal maternal entrainment (D. R. Weaver & S. M. Reppert, unpublished work). Furthermore, pups from sham-operated and SCN-lesioned dams express free-running rhythms with similar cycle lengths (Reppert & Schwartz 1986a). Thus, normal circadian functioning—light–dark entrainment and free-running cycle length—in adulthood does not require prenatal maternal entrainment.

Because the *in utero* environment provides a rich source of rhythmic hormones from the mother, such as prolactin, corticosterone and melatonin, it seemed possible that the maternal signal entraining the fetus was a hormone. Melatonin, the principal hormone of the pineal gland, was a prime candidate for the maternal signal communicating phase information to the fetus (Reppert 1982). There is a robust rhythm of melatonin in the maternal circulation and, because melatonin can cross the placenta, this melatonin rhythm is reflected in the fetal circulation (Klein 1972, McMillen & Nowak 1989, Reppert et al 1979, Yellon & Longo 1987, 1988, Zemdegs et al 1988).

However, maternal pinealectomy (which reduces circulating melatonin to at least below measurable levels [Lewy et al 1980]) does not abolish maternal coordination of fetal circadian phase in rats (Reppert & Schwartz 1986a), hamsters (F.C. Davis, personal communication) or spiny mice (D. R. Weaver & S. M. Reppert, unpublished results). Furthermore, removal of the maternal adrenals, thyroid–parathyroids, pituitary or ovaries (in separate experiments) on or before Day 7 of gestation (the time when maternal SCN lesions are disruptive) does not abolish the clear day–night rhythm of metabolic activity in the fetal rat SCN (Reppert & Schwartz 1986b). The maternal eyes, a potential source of both neural and endocrine signals, are also not necessary for the communication, since dams enucleated on Day 2 of gestation synchronize the circadian clocks of their fetuses (Reppert & Schwartz 1983).

Another approach by which to determine which aspect of maternal rhythmicity entrains the fetus is to restore rhythmicity artificially in an SCN-lesioned dam (who has no endogenous rhythmicity). The normal circadian rhythm in food consumption is disrupted by SCN lesions (Nagai et al 1978). Restricted access to food in SCN-lesioned rats artificially produces a rhythm in food consumption and can also induce rhythms in locomotor activity and temperature (Kreiger et al 1977, Stephan 1981, Weaver & Reppert 1989). Experiments involving a food access restriction paradigm (food cue) in SCN-lesioned pregnant rats have shown that the rhythmic ingestion of food can entrain the fetuses of SCN-lesioned dams (Weaver & Reppert 1989).

Rhythmic food ingestion would cause rhythmic fluctuations in nutrient levels in the blood, which could in turn induce physiological responses to the presence of nutrients. In SCN-lesioned dams, food restriction may generate a nutrient-related signal which entrains the fetuses in a manner like that occurring in intact dams. Alternatively, other rhythms (e.g., in temperature or activity) generated by restricted feeding of SCN-lesioned dams could be involved in setting the phase of the fetus.

Because the fetus is exposed to a multitude of maternal rhythms (behavioural rhythms, as well as hormonal), several maternal rhythms may act in concert to entrain the fetal biological clock (Fig. 1). Thus, eliminating any one of these rhythms would not be sufficient to disrupt maternal entrainment. Redundancy of maternal entraining signals may explain some recent, seemingly contradictory, results.

Although the maternal pineal gland is not necessary for maternal entrainment of the fetus, Davis & Mannion (1988) have convincingly shown that timed injections of pharmacological doses of melatonin into SCN-lesioned Syrian hamsters restore fetal synchrony. This finding supports a redundancy hypothesis: although melatonin is not necessary for fetal synchronization, it is one of several circadian signals capable of synchronizing the fetal biological clock.

Another line of evidence implicating melatonin as an important mediator of information about time of day from mother to fetus is the discovery of putative melatonin receptors in the fetal SCN of several species (Reppert et al 1988, Weaver et al 1988, Williams et al 1991). Thus, there is an anatomical substrate for a direct action of maternal melatonin on the fetal biological clock.

Recent studies have begun to address the issue of how diverse maternal signals might act at a cellular level to set the time of the fetal SCN. As a first approach to address this issue, neurotransmitter receptor gene expression has been mapped in the fetal SCN. Surprisingly, a high level of D_1 dopamine receptor mRNA expression was discovered in the fetal rat SCN (Weaver et al 1992). Furthermore, activation of these D_1 dopamine receptors induced a high level of expression of the immediate-early gene c-*fos* (Weaver et al 1992).

Cocaine has been administered to pregnant rats to show that dopamine receptors in the fetal SCN can activate c-*fos* expression. Cocaine was used because it acts on monoamine release and reuptake at presynaptic terminals (Koe 1976, Langer & Arbilla 1984) and so should affect only receptors receiving an endogenous monoamine input. Administration of cocaine to the mother (10-30 mg/kg) activated c-*fos* gene expression in the fetal SCN in a dose-dependent manner. The induction of c-*fos* mRNA in the fetal SCN appeared to be the result of a direct action of cocaine within the fetus, because cocaine administered directly to rat fetuses delivered by Caesarean section on Day 20 of gestation also induced c-*fos* expression in the SCN. Interestingly, the ability of cocaine to induce c-*fos* expression in the fetal biological clock was not dependent on the time of day of administration, unlike the effects of light on c-*fos* expression in the SCN of adult rodents (Aronin et al 1990, Earnest et al 1990, Kornhauser et al 1990, 1992, Rae 1989, Rusak et al 1990).

Induction of c-*fos* expression by cocaine in the fetal SCN was shown to occur through D_1 dopamine receptor activation (Weaver et al 1992): pretreatment with the D_1 receptor antagonist SCH-23390 (0.2 mg/kg) blocked the cocaine-induced expression of c-*fos* in the fetal SCN, and the D_1-selective agonist SKF-38393 (10 mg/kg) induced high levels of c-*fos* expression in the fetal SCN. An SKF-38393-induced increase in c-*fos* expression has also been shown in the SCN of fetal mice and Syrian hamsters, as well as in fetal rats. Thus, D_1 dopamine receptor-mediated c-*fos* induction occurs in three rodent species in which maternal-fetal entrainment has been demonstrated.

Recent studies in Syrian hamsters have shown that prenatal activation of D_1 receptors can actually entrain the fetus. Timed injections of SKF-38393 to pregnant, SCN-lesioned hamsters from Days 11 to 15 of gestation clearly set the phases of the pups' circadian rhythms (Viswanathan et al 1994). Thus, dopamine activation can affect the timing of the fetal circadian pacemaker.

The c-*fos* response to D_1 dopamine receptor stimulation in the fetal rat SCN is apparent on Day 18 of gestation, when the nuclei are immature and devoid of synapses, as discussed above. None the less, neurotransmission could occur in the absence of conventional synapses, because monoamines can be released from growth cones (Lauder 1988). Tyrosine hydroxylase-positive neurons and fibres which have been reported in the region of the SCN in fetal rat brain could provide a source of dopamine for D_1 receptor activation within the fetal SCN (Reisert et al 1990, Umgrumov et al 1989a,b).

In the adult, exposure to light at night induces expression of c-*fos* mRNA and protein (Fos) in the SCN (Aronin et al 1990, Earnest et al 1990, Kornhauser et al 1990, 1992, Rae 1989, Rusak et al 1990) and causes phase shifts in circadian rhythms (Klein et al 1991). Furthermore, the quantitative features of light-induced c-*fos* expression and light-induced phase shifts are very similar, suggesting that Fos is a molecular component of the photic entrainment pathway for mammals (Kornhauser et al 1990, 1992). A parallel situation may occur in the fetus. In the fetal SCN, activation of D_1 dopamine receptors induces c-*fos* expression and entrains developing circadian rhythms. Thus, an activatable fetal dopamine system may be an important mechanism by which maternal signals entrain the fetal SCN (Fig. 2), representing a final common pathway by which diverse entraining stimuli converge upon the fetal SCN.

As mentioned above, the two identified entraining signals are rhythmic food ingestion and timed injections of melatonin (Davis & Mannion 1988, Weaver & Reppert 1989). The rhythmic ingestion of food by the mother could provide a rhythmic precursor pool of tyrosine, driving rhythmic fetal brain dopamine production (Arevalo et al 1987, Garabal et al 1988), while maternally derived melatonin could directly modulate the release of dopamine in the fetal SCN, as demonstrated in other systems (Dubocovich 1988, Zisapel & Laudon 1982).

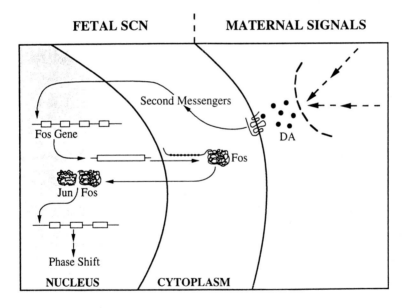

FIG. 2. A hypothetical model of a fetal dopamine system serving as a final common pathway for the action of maternal entraining signals on the fetal SCN. Maternal signals act upon a dopaminergic system within the fetal SCN to regulate local dopamine release. Dopamine (DA) binds to its D_1 receptors in the SCN and, via second messenger systems, activates expression of c-*fos* and other immediate-early genes. Fos and Jun proteins are translocated to the nucleus, dimerize and interact with AP1 binding sites on target genes, ultimately leading to a phase shift.

During the postnatal period, the maternal circadian system of altricial rodents maintains or reinforces the coordination of phase that has been established during the prenatal period (Takahashi et al 1989). In rats, the postnatal maternal influence persists until the pup develops the potential for direct light–dark entrainment through its own eyes (via the retinohypothalamic tract) at the end of the first week of life (Duncan et al 1986, Stanfield & Cowan 1976, Terubayashi et al 1985).

The existence of prenatal communication of circadian phase in mammals suggests that this phenomenon is of adaptive value. Maternal entrainment would presumably coordinate the pups to the dam and to the environment, so that when physiological and behavioural rhythmicity later develops, the rhythms are expressed in proper relationship to one another and to the 24 h day. This allows the young animal to assume its temporal niche more easily and may promote survival (Hudson & Distel 1989).

References

Altman J, Bayer SA 1978 Development of the diencephalon in the rat. I. Autoradiographic study of the time of origin and settling patterns of neurons of the hypothalamus. J Comp Neurol 182:945-947

Arevalo R, Castro R, Palarea MD, Rodriguez M 1987 Tyrosine administration to pregnant rats induces persistent behavioral modifications in the male offspring. Physiol & Behav 39:477-481

Aronin N, Sagar SM, Sharp FR, Schwartz WJ 1990 Light regulates expression of a Fos-related protein in rat suprachiasmatic nuclei. Proc Natl Acad Sci USA 87:5959-5962

Davis FC 1989 Use of post-natal behavioral rhythms to monitor pre-natal circadian function. In: Reppert SM (ed) Development of circadian rhythmicity and photoperiodism in mammals. Perinatology Press, Ithaca, NY, p 45-66

Davis FC, Gorski RA 1988 Development of hamster circadian rhythms. III. Role of the maternal suprachiasmatic nucleus. J Comp Physiol 162:601-610

Davis FC, Mannion J 1988 Entrainment of hamster pup circadian rhythms by prenatal melatonin injections. Am J Physiol 255:R439-R448

Deguchi T 1975 Ontogenesis of a biological clock for serotonin: acetyl coenzyme A N-acetyltransferase in the pineal gland of rat. Proc Natl Acad Sci USA 72:2914-2920

Dubocovich ML 1988 Pharmacology and function of melatonin receptors. FASEB (Fed Am Soc Exp Biol) J 2:2765-2773

Duncan MJ, Banister MJ, Reppert SM 1986 Developmental appearance of light–dark entrainment in the rat. Brain Res 369:326-330

Earnest DJ, Iadarola M, Yeh HH, Olschowka JA 1990 Photic regulation of c-*fos* expression in neural components governing the entrainment of circadian rhythms. Exp Neurol 109:353-361

Garabal MV, Arevalo RM, Diaz-Palarea MD, Castro R, Rodriguez M 1988 Tyrosine availability and brain noradrenaline synthesis in the fetus: control by maternal tyrosine ingestion. Brain Res 457:330-337

Honma S, Honma K, Shirakawa T, Hiroshige T 1984a Effects of elimination of maternal circadian rhythms during pregnancy on the postnatal development of circadian corticosterone rhythm in blinded infantile rats. Endocrinology 114:44-50

Honma S, Honma K, Shirakawa T, Hiroshige T 1984b Maternal phase setting of fetal circadian oscillation underlying the plasma corticosterone rhythm in rats. Endocrinology 114:1791–1796

Hudson R, Distel H 1989 Temporal pattern of suckling in rabbit pups: a model of circadian synchrony between mother and young. In: Reppert SM (ed) Development of circadian rhythmicity and photoperiodism in mammals. Perinatology Press, Ithaca, NY, p 83–103

Ifft JD 1972 An autoradiographic study of the time of final division of neurons in rat hypothalamic nuclei. J Comp Neurol 144:193–204

Klein DC 1972 Evidence for the placental transfer of ^{3}H-acetyl melatonin. Nature 237:117–118

Klein DC, Moore RY, Reppert SM (eds) 1991 The suprachiasmatic nucleus: the mind's clock. Oxford University Press, New York

Koe BK 1976 Molecular geometry of inhibitors of the uptake of catecholamines and serotonin in synaptosomal preparations of rat brain. J Pharmacol Exp Ther 199:649–661

Kornhauser JM, Nelson DE, Mayo KE, Takahashi JS 1990 Photic and circadian regulation of c-*fos* gene expression in the hamster suprachiasmatic nucleus. Neuron 5:127–134

Kornhauser JM, Nelson DE, Mayo KE, Takahashi JS 1992 Regulation of *jun*-B messenger RNA and AP-1 activity by light and a circadian clock. Science 255:1581–1583

Kreiger DT, Hause L, Krey LC 1977 Suprachiasmatic nuclear lesions do not abolish food-shifted circadian adrenal and temperature rhythmicity. Science 197:398–399

Langer SZ, Arbilla S 1984 The amphetamine paradox in dopaminergic neurotransmission. Trends Pharmacol Sci 4:387–390

Lauder JM 1988 Neurotransmitters as morphogens. In: Boer GJ, Feenstra MGP, Mirmiran M, Swaab DF, Van Haaren F (eds) Biochemical basis of functional neuroteratology. Elsevier Science Publishers, Amsterdam (Prog Brain Res 73) p 365–387

Lenn NJ, Beebe B, Moore RY 1977 Post-natal development of the suprachiasmatic nucleus of the rat. Cell Tissue Res 178:463–475

Lewy AJ, Tetsuo M, Markey SP, Goodwin FK, Kopin IJ 1980 Pinealectomy abolishes plasma melatonin in the rat. J Clin Endocrinol & Metab 50:204–207

McMillen IC, Nowak R 1989 Maternal pinealectomy abolishes the diurnal rhythm in plasma melatonin concentrations in the fetal sheep and pregnant ewe during late gestation. J Endocrinol 120:459–464

Moore RY, Bernstein ME 1989 Synaptogenesis in the rat suprachiasmatic nucleus demonstrated by electron microscopy and synapsin I immunoreactivity. J Neurosci 9:2151–2162

Nagai K, Nishino T, Nakagawa S, Fukuda Y 1978 Effects of bilateral lesions of the suprachiasmatic nuclei on the circadian rhythm of food intake. Brain Res 142:384–389

Rae MA 1989 Light increases Fos-related protein immunoreactivity in the rat suprachiasmatic nuclei. Brain Res Bull 23:577–581

Reisert I, Schuster R, Zienecker R, Pilgrim C 1990 Prenatal development of mesencephalic and diencephalic dopaminergic systems in the male and female rat. Dev Brain Res 53:222–229

Reppert SM 1982 Maternal melatonin: a source of melatonin for the immature mammal. In: Klein DC (ed) Melatonin rhythm generating system. Karger, Basel, p 182–191

Reppert SM 1987 Circadian rhythms: basic aspects and pediatric implications. In: Styne DM (ed) Current concepts in pediatric endocrinology. Elsevier Science Publishers, Amsterdam, p 91–125

Reppert SM, Schwartz WJ 1983 Maternal coordination of the fetal biological clock *in utero*. Science 220:969–971

Reppert SM, Schwartz WJ 1984 The suprachiasmatic nuclei of the fetal rat: characterization of a functional circadian clock using ^{14}C-labeled deoxyglucose. J Neurosci 4:1677–1682

Reppert SM, Schwartz WJ 1986a The maternal suprachiasmatic nuclei are necessary for maternal coordination of the developing circadian system. J Neurosci 6:2724–2729

Reppert SM, Schwartz WJ 1986b Maternal endocrine extirpations do not abolish maternal coordination of the fetal circadian clock. Endocrinology 119:1763–1767

Reppert SM, Chez RA, Anderson A, Klein DC 1979 Maternal–fetal transfer of melatonin in a non-human primate. Pediatr Res 13:788–791

Reppert SM, Weaver DR, Rivkees SA, Stopa EG 1988 Putative melatonin receptors in a human biological clock. Science 242:78–81

Reppert SM, Weaver DR, Rivkees SA 1989 Perinatal function and entrainment of a circadian clock. In: Reppert SM (ed) Development of circadian rhythmicity and photoperiodism in mammals. Perinatology Press, Ithaca, NY, p 25–44

Rusak B, Robertson HA, Wisden W, Hunt SP 1990 Light pulses that shift rhythms induce gene expression in the suprachiasmatic nucleus. Science 248:1237–1240

Schwartz WJ 1991 SCN metabolic activity *in vivo*. In: Klein DC, Reppert SM, Moore RY (eds) The suprachiasmatic nucleus: the mind's clock. Oxford University Press, New York, p 144–156

Shibata S, Moore RY 1987 Development of neuronal activity in the rat suprachiasmatic nucleus. Dev Brain Res 34:311–315

Shibata S, Moore RY 1988 Development of fetal circadian rhythm after disruption of the maternal circadian system. Dev Brain Res 41:313–317

Stanfield B, Cowan WN 1976 Evidence for a change in the retinohypothalamic projection in the rat following early removal of one eye. Brain Res 104:129–133

Stephan FK 1981 Limits of entrainment of periodic feeding in rats with suprachiasmatic lesions. J Comp Physiol 143:401–410

Takahashi K, Ohi K, Shimoda K, Yamada N, Hayashi S 1989 Postnatal maternal entrainment of circadian rhythms. In: Reppert SM (ed) Development of circadian rhythmicity and photoperiodism in mammals. Perinatology Press, Ithaca, NY, p 67–82

Terubayashi H, Fujisawa H, Itoi M, Ibata Y 1985 HRP-histochemical detection of retinal projections to the hypothalamus in neonatal rats. Acta Histochem Cytochem 18:433–438

Umgrumov MV, Taxi J, Tixier-Vidal A, Thibault J, Mitskevich MS 1989a Ontogenesis of tyrosine hydroxylase-immunopositive structures in the rat hypothalamus. An atlas of neuronal cell bodies. Neuroscience 29:135–156

Umgrumov MV, Tixier-Vidal A, Taxi J, Thibault J, Mitskevich MS 1989b Ontogenesis of tyrosine hydroxylase-immunopositive structures in the rat hypothalamus. Fiber pathways and terminal fields. Neuroscience 29:157–166

Viswanathan N, Weaver DR, Reppert SM, Davis FC 1994 Entrainment of the fetal hamster circadian pacemaker by prenatal injections of the dopamine agonist SKF38393. J Neurosci 14:5393–5398

Weaver DR, Reppert SM 1989 Periodic feeding of SCN-lesioned pregnant rats entrains the fetal biological clock. Dev Brain Res 46:291–296

Weaver DR, Namboodiri MAA, Reppert SM 1988 Iodinated melatonin mimics melatonin action and reveals discrete binding sites in fetal brain. FEBS (Fed Eur Biochem Soc) Lett 228:123–127

Weaver DR, Rivkees SA, Reppert SR 1992 D_1-dopamine receptors activate c-*fos* expression in the fetal suprachiasmatic nuclei. Proc Natl Acad Sci USA 89:9201–9204

Williams LM, Martinoli MG, Titchener LT, Pelletier G 1991 The ontogeny of central melatonin binding sites in the rat. Endocrinology 128:2083–2090

Yellon SM, Longo LD 1987 Melatonin rhythms in fetal and maternal circulation during pregnancy in sheep. Am J Physiol 252:E799–E802

Yellon SM, Longo LD 1988 Effect of maternal pinealectomy and reverse photoperiod on the circadian melatonin rhythm in the sheep and fetus during the last trimester of pregnancy. Biol Reprod 39:1093–1099

Zemdegs IZ, McMillen IC, Walker DW, Thorburn GD, Nowak R 1988 Diurnal rhythms in plasma melatonin concentrations in the fetal sheep and pregnant ewe during late gestation. Endocrinology 123:284–289

Zisapel N, Laudon M 1982 Dopamine release induced by electrical field stimulation of rat hypothalamus *in vitro*: inhibition by melatonin. Biochem Biophys Res Commun 104:1610–1616

DISCUSSION

Mikkelsen: I wonder whether those cells in the SCN that express c-*fos* in response to dopaminergic stimulation are the same cells that are activated by light. Have you done any experiments at the time when the retinohypothalamic tract is present and when dopamine still stimulates c-*fos* expression to see whether it's an additive response?

Reppert: We are looking at that right now. Preliminary evidence suggests that the *fos* response stops soon after birth. We are trying to pin down when that stops and when light induction of *fos* begins.

Mikkelsen: Another way to address this question would be to localize c-*fos*-containing neurons in more detail. We have done immunostaining for D_1 receptor-related phosphoprotein, DARRP-32. Surprisingly, this protein is present in the rat SCN, but the DARRP-32-positive cells appear to be located medially to the light-induced Fos-positive cells and the retinal afferents.

Reppert: In the fetus, resolution is too poor to say for sure, but, on the basis of film and some emulsion autoradiographs, it appears that we are activating *fos* throughout the SCN.

Armstrong: If it's dopamine in the maternal circulation that's the signal to the fetus, where do you think it's coming from?

Reppert: We do not feel that it is dopamine in the maternal circulation that activates the fetal clock. We feel that dopamine is released locally from an endogenous pool in or near the fetal SCN. There are tyrosine hydroxylase-positive fibres in the region of the fetal SCN, which some authors claim disappear early in postnatal life. Food cues, for example, could affect this fetal dopamine system by increasing circulating levels of tyrosine which, in turn, could increase dopamine production within fetal brain.

Lemmer: My question was almost the same. What is the compound which activates the fetal SCN physiologically? The results with cocaine are convincing, because it crosses the placental barrier, but dopamine is unlikely to cross.

Reppert: That's what I'm saying. It doesn't cross the placenta; other signals affect a local dopamine pool within or near the fetal SCN, altering the amount of transmitter available to act locally within the fetal SCN.

Lemmer: Do you have any idea about the nature of this transmitter?

Reppert: The two identified entraining signals for the fetus are rhythmic food ingestion and timed injections of melatonin. These two signals could act by the same or different mechanisms. In some systems, melatonin can influence local dopamine concentrations; there are melatonin receptors in the fetal SCN, so the potential for that interaction exists. Melatonin could also act on the fetal SCN through its own receptor, independent of the dopamine system. As already mentioned, rhythmic ingestion of food by the mother would provide a rhythmic precursor pool of tyrosine which could drive fetal brain dopamine production.

Gillette: Can you discriminate the times at which the fetal SCN is sensitive to these signals? Do you know that dopamine is acting at night, for example? Also, with melatonin you have shown sensitivity to morning and afternoon entrainment; is that restricted pattern characteristic of the communication between the mother and the fetus?

Reppert: There is no circadian 'gate' in terms of when dopamine will induce c-*fos* in the fetal SCN as there is for light activation of c-*fos* in the adult SCN. For setting the phase of the fetal SCN, injections of the D_1 dopamine agonist, SKF-38393, were given over a four-day period, so we cannot say much about phase response characteristics. However, it appears that the resultant phase of developing circadian rhythms lines up differently, relative to injection time, depending on whether a D_1 dopamine agonist or melatonin is injected.

Turek: If the response to dopamine is different from that to melatonin, how does it compare with the entrainment response to light?

Reppert: That is difficult to say. What aspect of light can you say the signal mimics?

Turek: You could look at the pups' phase relationship relative to the light–dark cycle that the mother saw.

Reppert: The pups 'see' only what their mother sees and communicates, but her SCN is lesioned in these experiments. The only thing the mother communicates is the injections.

Turek: With an intact mother, exposed to a particular light–dark cycle, does the phase relationship of the pups' rhythm look more like the phase relationship to melatonin injections or dopamine injections?

Reppert: It's more similar to dopamine than melatonin.

Menaker: The only similar situation that I know of is the isolated retina of *Xenopus*, which Cahill & Besharse (1991) have studied so beautifully. Could you comment on the similarities between your system and theirs, in which dopamine also acts as a resetting signal?

Reppert: I'm not sure that there are that many similarities, because signalling in their system is through the D_2 receptor and not the D_1 receptor.

Menaker: It's the only other system that I know of in which dopamine acts on the clock that way. Do you consider the fact that a different receptor is involved to be significant? After all it's a frog, not a rat!

Reppert: All the action is going to occur downstream of the receptor, so the receptor type is relevant.

Miller: The D_4 subtype is also fairly abundant in the rat SCN. The D_4 subtype supposedly has pharmacology similar to that of the D_2 type, and might be expected to suppress cAMP. So in the adult there is one receptor which increases cAMP, D_1, and one which may decrease cAMP, D_4. Therefore, you might not expect agents such as amphetamine or cocaine to be effective in adults, because of the reciprocal effect on cAMP. We tried for years to produce phase shifts with amphetamine and never succeeded. Perhaps your results could be explained in terms of the developmental time-course, with D_1 being expressed earlier than D_4. Have you tried giving a D_1 agonist to adults?

Reppert: We have not done that; somebody needs to.

Miller: I understood that D_1 agonists do produce phase shifts in fetuses (Viswanathan & Davis 1993), that you showed that the mother does not expresss c-*fos* in the SCN after treatment with a D_1 agonist, and that it is an open question whether or not a D_1 agonist (as opposed to amphetamine) will produce a phase shift in adult behavioural rhythms.

Arendt: People keep saying that SCN cells are born oscillating. Does anybody know at what point this happens?

Reppert: That's very difficult to say. The last division of SCN neurons occurs between embryonic Days 13 and 16 in rats. Deciding when to call a cell an SCN cell on the basis of the expression of various peptides or neurotransmitters is difficult. We have been trying to see whether we can use the D_1 dopamine receptor to identify progenitor SCN cells before they have undergone their final division.

Arendt: Melatonin gets into breast milk, in goats at least, and probably in humans also (Deveson 1990). That has all sorts of implications for human development, particularly in the light of Davis & Mannion's (1988) work. Do you think this might have a role in setting the human system?

Reppert: Not postnatally, though melatonin might well have a phase-setting role prenatally. We have mapped melatonin receptors in post mortem human material and have found high levels of melatonin receptor expression in the fetal SCN as early as 24 weeks of gestation. Melatonin probably has a phase-setting role *in utero*, but it's much less clear, even from animal work, what melatonin does postnatally.

Lemmer: Are there any implications from your results for the offspring of cocaine-dependent mothers? Also, dopamine receptor blockers are widely used as neuroleptics; are these drugs known to have any effects?

Reppert: It is too early to say. It's unclear what the long-term behavioural effects are on babies born to drug-abusing mothers. There are many acute, immediate effects that disappear within the first couple of weeks after birth. One might speculate that there could be longer term more subtle effects on the circadian system, such as sleep–wake abnormalities or learning disorders.

Waterhouse: Human babies are sometimes born a little late and sometimes quite early, so one can see the development of behavioural circadian rhythms. You showed a figure with two separate oscillators interacting to form the circadian pacemaker or oscillator. Bearing in mind the work of Kleitman & Engelman (1953) and ourselves (Tenreiro et al 1991) on the development of circadian rhythms—which seem to free run and to have ultradian components—is that telling us something about the processes involved?

Reppert: What one may be seeing during early development is the progressive coupling of the circadian pacemaker to ultradian oscillators that results in the eventual expression of an overt circadian rhythm.

Waterhouse: We have done a study of babies from 26 weeks of age for up to 17 weeks (Tenreiro et al 1991). The babies were in incubators, in approximately constant conditions, and were monitored every hour of the day and night. We found that circadian rhythms sometimes appeared but then disappeared or took on a new period. There was a suggestion that the oscillator(s) were beginning to 'learn' how to become circadian, but they didn't suddenly appear as stable circadian rhythmicities.

Reppert: This brings us back to the whole issue of whether individual cells have this capability as opposed to a network. I am not sure that developmental studies in humans have helped to answer that question.

Moore: We know almost nothing about the development of the human SCN. We know that the SCN is present early in fetal life, but different mammals show great differences in synaptogenesis. Synaptogenesis begins in the hamster much earlier than it does in the rat and is complete almost immediately after birth. It is not possible to generalize from one rodent species to another, so we can't begin to extrapolate to humans.

Lewy: I validated the melatonin levels in Dr Illnerová's milk samples, using our gas chromatography–mass spectrometry assay (Lewy & Markey 1978) and was impressed with the robust circadian rhythm. Many mothers come to us because their infants are keeping them awake all night. There are several current trends in breast feeding that may account for this, if melatonin has an entraining effect. One of these trends is that mothers are pumping breast milk during the day for night-time feedings. Hundreds of years ago mothers used to sleep with their infants and suckle them all night long. Under those conditions, babies slept fairly well at night, perhaps in part because they were fed with mother's night-time milk (containing melatonin) at night and daytime, melatonin-free milk during the day.

Hastings: Fred Davis tells me that a single injection of melatonin given prenatally is sufficient to synchronize the activity rhythms of hamster litters born to mothers bearing lesions of the SCN (F.C. Davis, personal communication). When that injection hits the litter, if the fetuses are desynchronized and are all brought to the same phase point, then a range of phase shifts, both advances and delays of up to 12 h, must occur. Alternatively, Dr Reppert, do you consider that the melatonin is able to stop the clock in all cases, and that all of the fetuses start up again from a unique, synchronized phase?

Reppert: That's a very interesting point. During early development there may not be a daily sensitivity of the circadian pacemaker to the phase-shifting effects of entraining agents like there is in the adult. Perhaps early light 'history' is important for the phase response characteristics so typical of the older animal.

References

Cahill GM, Besharse JC 1991 Resetting the circadian clock in cultured *Xenopus* eyecups: regulation of retinal melatonin rhythms by light and D_2 dopamine receptors. J Neurosci 11:2959–2971

Kleitman N, Engelman TG 1953 Sleep characteristics of infants. J Appl Physiol 6:269–282

Tenreiro S, Dowse HB, D'Souza S et al 1991 The development of ultradian and circadian rhythms in premature babies maintained in constant conditions. Early Hum Dev 27:33–52

Davis FC, Mannion J 1988 Entrainment of hamster pup circadian rhythms by prenatal melatonin injections. Am J Physiol 255:R439–R448

Deveson SL 1990 The effects of photoperiod and melatonin on seasonal breeding in goats. PhD thesis, University of Surrey, Guildford, UK

Lewy AJ, Markey SP 1978 Analysis of melatonin in human plasma by gas chromatography negative chemical ionization mass spectometry. Science 201:741–743

Viswanathan N, Davis FC 1993 Prenatal injections of D_1 dopamine receptor agonist SKF38393 entrain circadian ryhthms of hamster pups. Soc Neurosci Abstr 19:1488

Alterations in the circadian system in advanced age

Fred W. Turek*†, Plamen Penev*, Yan Zhang*, Olivier Van Reeth*, Joseph S. Takahashi* and Phyllis Zee*†

*NSF Science and Technology Center for Biological Timing, Department of Neurobiology and Physiology, Northwestern University, 2153 North Campus Drive, Evanston, IL 60208-3520 and †Department of Neurology, Northwestern University Medical School, 233 E. Erie St., Suite 500, Chicago, IL 60611, USA

Abstract. In addition to light, a variety of non-photic stimuli can induce phase shifts in the circadian clock of rodents. We have examined the effects of advanced age on the response of the circadian clock to both photic and non-photic stimuli in old hamsters (i.e., over 16 months of age). Among the age-related changes in the circadian rhythm of locomotor activity are: (1) alterations in the phase angle of entrainment to the light–dark cycle; (2) an altered response to the phase-shifting effects of light pulses; (3) changes in the time it takes to re-entrain to a new light–dark cycle; and (4) a loss of responsiveness to the phase-shifting or entraining effects of stimuli which induce an acute increase of activity. Many of the effects of ageing on the circadian clock system can be simulated in young animals by depleting brain monoamine levels, suggesting that ageing alters monoaminergic inputs to the clock. Some of the age-related changes in the response of the clock to an activity-inducing stimulus can be reversed by implanting old animals with fetal suprachiasmatic nuclear tissue. Determining the physiological basis of age-related changes in the responsiveness of the clock to both internal and external stimuli, and the mechanisms by which normal circadian functioning can be restored, should lead to new insight into the functioning of the circadian clock and may suggest new approaches to the normalization of disturbed circadian rhythms.

1995 Circadian clocks and their adjustment. Wiley, Chichester (Ciba Foundation Symposium 183) p 212–234

Age-related changes have been documented in endocrine, metabolic and behavioural circadian rhythms in a variety of animal species, humans included (Brock 1991, Czeisler et al 1991). Most studies have been limited largely to measurements of the effects of age on the amplitude of circadian rhythms (Ingram et al 1982, Van Gool et al 1987). For example, circadian rhythms in temperature (Sacher & Duffy 1978), corticosterone (Nicolau & Milcu 1977), serum testosterone (Tenover et al 1988) and melatonin (Reiter et al 1981) are all dampened in old rats, and reduced differences in the rhythm of locomotor activity between light and dark conditions have been reported in mice and rats

(Martin et al 1985, Peng et al 1980, Sacher & Duffy 1978). In comparison with young mice, old mice spend more time asleep during the normal period of activity and are awake more during the normal sleep period (Welsh et al 1986). Witting et al (1993) have made the interesting observation that during entrainment to a light–dark cycle, the amplitude of the activity rhythm of old rats could be increased by increasing the light intensity. Although at a given light intensity (ranging from 3.5–445 lux) the amplitude of the sleep–wake rhythm was lower in old than in young rats, the amplitude of the rhythm in old rats exposed to the brightest light was comparable to that of young rats under the dimmest light intensity. The authors concluded that reduction of the amplitude of circadian rhythms with ageing might, in principle, be compensated for partially by increasing the intensity of daytime ambient illumination (Witting et al 1993).

Alterations of light–dark fluctuations or the amplitude of circadian rhythms do not necessarily imply that a change is intrinsic to the circadian pacemaker system. These changes can also be explained by age-related changes that are either 'upstream' or 'downstream' from the circadian clock. For example, age-related changes in amplitude or entrainment of circadian rhythms could be due to a decrease in the sensory perception of light or an alteration in the mechanisms regulating various physiological processes that lie between the clock and the overt rhythmic output. Nevertheless, there is convincing evidence in rodents that the circadian clock itself is altered in senescence, because the free-running periods of the circadian rhythms of locomotor activity (Davis & Menaker 1980, Morin 1988, Pittendrigh & Daan 1974) and sleep–wakefulness (Van Gool et al 1987) shorten with age in hamsters and rats kept under constant lighting conditions. Support for age-related changes in the clock itself has recently been provided by Wise and her colleagues, who found that ageing alters the circadian rhythms of glucose utilization and α_1-adrenergic receptor levels in the suprachiasmatic nucleus (SCN) and that these changes are correlated with changes in the circadian rhythm of luteinizing hormone (LH) release (i.e., the preovulatory 'LH surge') that are observed with ageing in female rats (Weiland & Wise 1990, Wise et al 1987, 1988). In addition, changes in neuropeptides in the human SCN have been associated with senility and Alzheimer's disease (Swaab et al 1988), and patients suffering from Alzheimer's disease show marked disturbances in the extent and timing of daily locomotor activity (Satlin et al 1991) as well as fragmentation of the daily sleep–wake cycle (Prinz et al 1982).

The physiological mechanisms that underlie age-related changes in the circadian system are unknown. However, a dysfunction of hypothalamic activity with advanced age has been associated with a number of other age-related changes in physiology and metabolism, among which a major deficiency appears to be a decrease in the ability of the hypothalamus to secrete catecholamines, particularly dopamine and noradrenaline (Meites 1990). Indeed, drugs which reduce hypothalamic catecholamines produce many of the age-related decrements of body function, whereas drugs which increase hypothalamic catecholamine concentrations can inhibit or reverse physiological changes normally associated with advanced age in rats (Meites 1990).

Over the last few years we have examined the effects of ageing on the response of the circadian system of hamsters to both photic and non-photic stimuli which can influence the phase of the circadian clock. In addition to describing the results from those studies, we also review findings from studies in which we have used a pharmacological approach to mimic some of the effects of ageing on the circadian clock of young animals and to reverse at least some of the effects of ageing on the circadian clock by transplanting fetal SCN tissue into the region of the SCN of old animals.

Age-related changes in the response of the circadian system to light

Entrainment to a light cycle

To determine the effects of age on the phase relationship between the circadian rhythm of locomotor activity and the light–dark cycle during stable entrainment, we recorded locomotor activity continuously from 10 young (2–5 months old) and 12 middle-aged (13–16 months old) male golden hamsters (*Mesocricetus auratus*) exposed to a light–dark cycle of 14 hours of light and 10 hours of dark for about six weeks (Zee et al 1992). Older animals' activity began significantly earlier than young ones. Whereas 8/12 old hamsters began activity *before* or *when* the lights were turned off during entrainment to a 14 : 10 h light–dark cycle, all 10 of the young hamsters began activity 10–50 minutes *after* lights-out (Fig. 1). This clear alteration in the phase relationship between the activity rhythm and the light–dark cycle is consistent with the finding that the free-running period of the activity rhythm of hamsters is shortened with age. In a different group of hamsters, the period of the activity rhythm in old animals (Rosenberg et al 1991) was significantly less ($P<0.05$) than in 3–5-month-old animals (24.04 ± 0.05 vs. 24.20 ± 0.06 h). Similar findings on the effects of age on circadian period in this species have been reported by others (see above).

The phase-shifting effects of light

An important approach that has been used to study the response of the circadian system to light has been to expose animals kept in constant darkness to a single pulse of light at different circadian times and to plot any resultant phase shifts in a measurable circadian rhythm, such as the rhythm of locomotor activity. In hamsters, as in other species, exposure to a pulse of light during the late subjective night induces phase advances in the activity rhythm whereas light pulses during the early subjective night induce phase delays (Ellis et al 1982,

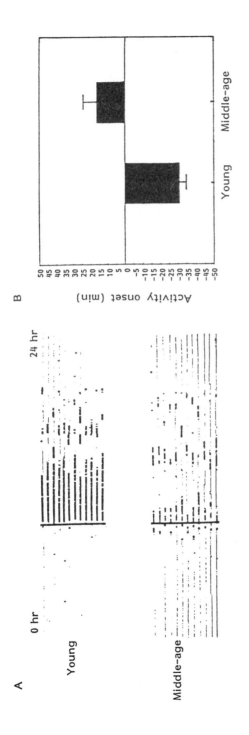

FIG. 1. (A) Representative activity records of young (2–5 month, *top*) and middle-aged (13–16 month, *bottom*) hamsters entrained to a dim 14:10 h light–dark cycle. Successive days are plotted from top to bottom. The time at which the lights were turned off is indicated by a solid vertical line. The onset of activity in the young hamster was precise, occurring about 20 min after the lights went out, whereas in the middle-aged animal the onset of activity was irregular and usually occurred before the lights went out. (B) Mean (±SE) onset of wheel-running activity in relation to the time at which the lights were turned off in 10 young and 12 middle-aged hamsters entrained to a dim 14:10 h light–dark cycle. A positive value indicates activity onset occurring before lights-out, and a negative value indicates activity onset after lights-out. From Zee et al (1992).

Takahashi et al 1984). Light pulses throughout most of the subjective day have no effect on the phase of the free-running activity rhythm. To determine if ageing alters the phase-shifting effect of a light pulse on the circadian rhythm of activity, we exposed young (2–5 month) and old (16–21 month) hamsters which had been kept in constant darkness to single one-hour pulses of light of an intensity (500 lux) sufficient to induce maximum phase shifts in young animals (Rosenberg et al 1991). The light pulses were presented at various circadian times so as to generate a complete phase response curve. Young and old animals showed phase shifts of similar magnitude in response to light pulses administered in the phase delay region and the times at which light did not induce phase shifts in the activity rhythm were the same in the two groups. However, in old animals, phase advances in response to light pulses given at circadian time (CT) 16 and 19 were significantly larger than in young animals. Unexpectedly, unusually large phase shifts were observed in old animals in response to light pulses administered at CT 16, with 73% showing shifts greater than 300 min whereas none of the young animals showed phase shifts of this magnitude (Fig. 2). Large phase shifts near the transition region from phase delays to advances indicates that this phase position may be unstable in old animals. As has been suggested earlier (Earnest & Turek 1982), such instability around CT 16 may be related to the degree of coupling between different oscillators that make up the circadian system in rodents.

Rate of re-entrainment

To determine if ageing alters the rate of re-entrainment to phase shifts in the light–dark cycle, we exposed young (3–4 month) and middle-aged (13–15 month) hamsters entrained to a 14 : 10 h light–dark cycle to both an eight-hour advance and an eight-hour delay in the light cycle (Zee et al 1992). The intensity of light during the 14 h period of light was dim (3–5 lux) to avoid any possible masking effects bright light might have on the expression of locomotor activity while the animals were re-entraining to the new light cycles. Middle-aged hamsters entrained significantly faster ($P < 0.01$) to a phase advance in the light cycle than did young hamsters (12.8 ± 2.1 vs. 18.2 ± 4.4 days). In contrast, young animals entrained significantly faster ($P < 0.01$) to a phase delay in the light cycle than older animals (16.3 ± 4.6 vs. 23.3 ± 3.5 days). These age-related changes in the rate of re-entrainment of the activity rhythm to a changed light–dark cycle suggest that ageing may alter not only the pattern of entrainment to a stable light–dark cycle but also the ability of the circadian clock to be phase shifted following a shift in the synchronizing light cycle. Such alterations in the rate of re-entrainment may be due to changes in the free-running period or a change in the phase-shifting effects of light that are associated with ageing in the hamster circadian clock.

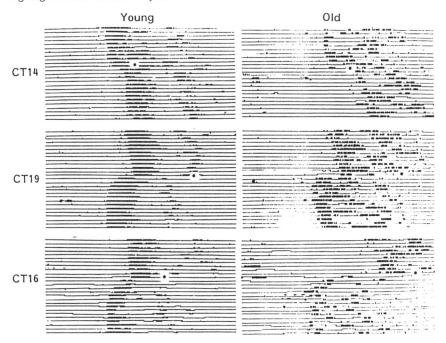

FIG. 2. Representative activity records from young (2–5 month) and old (16–21 month) hamsters in constant darkness exposed to a one-hour pulse of light (*) at circadian time (CT) 14, 16 or 19. Successive days are plotted from top to bottom. From Rosenberg et al (1991).

Age-related changes in the response of the circadian system to activity-inducing stimuli

Through a series of experiments, we and others have demonstrated that various pharmacological and non-pharmacological stimuli which induce an increase in locomotor activity can induce phase shifts in the circadian rhythm of locomotor activity in hamsters with a free-running circadian rhythm kept under constant lighting conditions. These stimuli include short-acting benzodiazepines (e.g., alprazolam, triazolam and midazolam), dark pulses and brief access to a running wheel (Mrosovsky 1988, Turek 1989, Van Reeth & Turek 1989a). The similarity of the PRCs generated in response to various activity-inducing stimuli suggested that similar mechanisms mediate the effects of these agents on the circadian system and that stimulated physical activity itself can have important entraining effects on the clock. The hypothesis that a stimulated increase in locomotor activity is responsible for mediating the effects on the circadian clock of various activity-inducing stimuli is further supported by the observation that immobilization or restriction of activity for a period after treatment totally blocks both benzodiazepine-induced and dark pulse-induced phase shifts in the circadian activity rhythm (Reebs & Mrosovsky 1989, Van Reeth & Turek 1989a). Although restraint during the normal period of inactivity does not induce phase shifts in the circadian clock (Van Reeth & Turek 1989a), we have recently demonstrated that a three-hour period of immobilization during the normal period of intense activity induces phase delays in the rhythm of locomotor activity (Van Reeth et al 1991). In addition, whereas restraint can block the phase shifts normally induced by activity-inducing stimuli, such as dark pulses or injections of triazolam, it does not block the effects of two stimuli (light and cycloheximide) that induce phase shifts in the clock in the absence of any acute increase in activity (Van Reeth et al 1991). Thus, the ability of immobilization to block stimulus-induced phase shifts in the circadian clock appears to be specific to stimuli that induce an acute increase in locomotor activity.

Taken together, these results indicate that stimulation of activity during the normal rest period, or the prevention of activity during the normal period of activity, can induce phase shifts in the circadian clock, and support the hypothesis that feedback signals during both the resting and active phases of the circadian rhythm of locomotor activity can influence the circadian pacemaker underlying this rhythm. The full extent of the effects of ageing on the response of the clock to these feedback signals remains to be determined, but, as detailed below, it is clear that ageing does have marked effects.

Phase-shifting effects of an activity-inducing stimulus

In view of our finding that the response of the circadian system to light pulses is altered in old age, we addressed the question of whether ageing also alters the response of the circadian system to activity-inducing stimuli. Young (2–4 month-old) and old (17–22-month-old) hamsters with free-running circadian systems, kept in constant light, were subjected to either injections of triazolam (5 mg/kg) or a six-hour pulse of darkness at times when these stimuli are known to induce phase advances or phase delays in the circadian rhythm of locomotor activity (Van Reeth & Turek 1992). As expected, the activity rhythm of young animals was advanced or delayed in response to injections of triazolam at CT 6 or 21, respectively (Fig. 3). Similarly, a six-hour period of darkness beginning at CT 6 or 18 also induced, respectively, a phase advance or phase delay in the activity rhythm. Surprisingly, however, phase shifts in the activity rhythm were reduced or not observed in old animals treated with triazolam or exposed to a six-hour dark pulse regardless of the time of treatment (Fig. 3).

Because an acute increase in activity appears to be necessary for triazolam or exposure to a dark pulse to induce a phase shift in the circadian clock (see above), one explanation for the surprising finding that these stimuli do not induce phase shifts in old animals is that they do not induce the acute increase in activity

Ageing and the circadian system 219

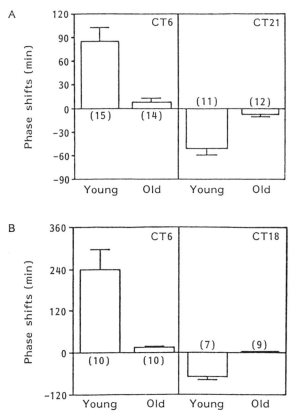

FIG. 3. (A) Mean (±SE) phase shifts in activity rhythms of young (2-4 month) and old (17-22 month) hamsters with free-running circadian systems kept in constant light that received, at either CT 6 or CT 21, a systemic injection of 5 mg/kg triazolam. A positive value, above the zero line, indicates an advance in the onset of locomotor activity; a negative value, below the line, indicates a delay. Values in parentheses are the number of trials for each group. Mean phase shifts in activity rhythm in response to an injection of triazolam were significantly greater ($P<0.001$) in young than in old hamsters at both circadian times tested. (B) Mean (±SE) phase shifts in activity rhythms of young and old hamsters maintained in constant light that were subjected to a six-hour dark pulse beginning at CT 6 or CT 18. Mean phase shifts in activity rhythm in response to a dark pulse were significantly greater ($P \leqslant 0.001$) in young than in old hamsters at both circadian times tested. From Van Reeth et al (1993a).

normally associated with their presentation. However, this does not appear to be the case; it is very clear from the activity records of old animals that an injection of triazolam does indeed induce an acute increase in locomotor activity which, from evaluation of the running wheel records, appears to be similar to that observed in young animals. Moreover, we have recently quantified the acute increase in total locomotor activity for the six-hour period after an injection of triazolam at CT 6 in both young and old hamsters and have found no differences (Van Reeth & Turek 1992). Thus, although triazolam induces comparable acute increases in locomotor activity in young and old hamsters, this activity-inducing stimulus induces phase shifts in the rhythm of locomotor activity only in young animals.

Phase response curves of the circadian rhythm in locomotor activity produced by two protein synthesis inhibitors, cycloheximide and anisomycin, in hamsters maintained in constant light or constant darkness (Inouye et al 1988, Takahashi & Turek 1987, Wollnik et al 1990) are similar in shape to those produced by benzodiazepines, dark pulses or running on a wheel. Interestingly, although advanced age is associated with a lack of a phase-shifting response to triazolam or a dark pulse, there is no age-related alteration to the phase-shifting effect of anisomycin. In contrast to the other stimuli, the protein synthesis inhibitors do not stimulate with an acute increase in locomotor activity. Thus, despite the similarities in the PRCs for these various stimuli in young animals, age-related changes in responsiveness to them appear to be specific to those stimuli that are associated with an acute increase in locomotor activity.

Entrainment to an activity-inducing stimulus

In young hamsters, repeated injections of benzodiazepines every 23.34, 23.72, 24.00 or 24.72 h entrain the circadian clock to the exact period of the injections (Van Reeth & Turek 1989b). In view of our finding that single injections of benzodiazepines induce small or no phase shifts in the activity rhythm of old hamsters, despite the fact that they induce similar acute changes in the activity state of young and old animals, we investigated whether the circadian clock of old hamsters is also less sensitive to the entraining effects of injections of benzodiazepine on a circadian basis. Young blind (2–5 month), and old blind (14–18 month) hamsters received repeated injections of triazolam (or vehicle) every 24.67 h for 19 days and the effects of this treatment on the period of the circadian rhythm of wheel running were determined during and after treatment (Van Reeth et al 1993). Whereas vehicle injections did not entrain the activity rhythm of young or old hamsters, six out of seven young hamsters were entrained by the triazolam treatment. In contrast, only one out of seven old animals became entrained to the triazolam injections. These results, together with findings on the response of old animals to the acute effects of activity-including stimuli, indicate that the circadian system of old hamsters becomes selectively unresponsive to synchronizing signals mediated by the activity–rest state, and suggest that ageing is associated with a weakened coupling between the activity–rest cycle and the circadian clock.

Effects of depleting brain monoamines on the circadian rhythm of activity

In view of the fact that catecholamine activity declines in the brain (including the hypothalamus) of old animals, we have initiated a series of studies to determine if depleting brain monoamines in young hamsters can mimic any of the age-related changes in circadian clock functioning. Because of our uncertainty as to which monoamine, or combination of monoamines, might be involved in age-related changes in the circadian system, our initial strategy was to use a non-selective monoamine depletor; we chose reserpine, a potent but non-selective monoamine depletor (Penev et al 1993, 1994). In young (2–5-month-old) hamsters, reserpine (4 mg/kg) caused: (1) a decrease in locomotor activity; (2) a phase advance in the onset of activity during entrainment to a 14:10 h light–dark cycle; (3) a loss of responsiveness to the phase-shifting effects of triazolam (Fig. 4); (4) an enhanced response to the phase-shifting effects of light pulses presented around CT 16 (Fig. 5); and (5) a reduction in the concentrations of serotonin, noradrenaline and dopamine in the hypothalamus, striatum and pons/medulla. The changes in the circadian rhythm of activity, and its response to phase-shifting stimuli, are all similar to those that occur spontaneously in old animals. These initial findings support the hypothesis that a depletion in monoaminergic activity in the brain may underlie some of the age-related changes in the circadian clock system of rodents.

Transplants of SCN tissue restore the response of old hamsters to the phase-shifting effects of triazolam

Our finding that in advanced age there is a loss of responsiveness of the circadian clock to the phase-shifting effects of activity-inducing stimuli, coupled with the

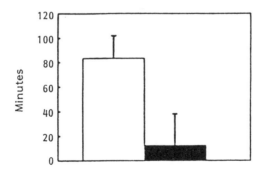

FIG. 4. Mean ($\pm$SEM, $n=8$ per group) phase shift in the activity rhythm of young (2–5 month) hamsters injected with triazolam at CT 6 while free-running in constant darkness. Eight to 10 days before treatment with triazolam the animals received a single subcutaneous injection of reserpine (2.5 mg/kg; solid bar) or vehicle (open bar). Significantly different, $P<0.05$. From Penev et al 1994.

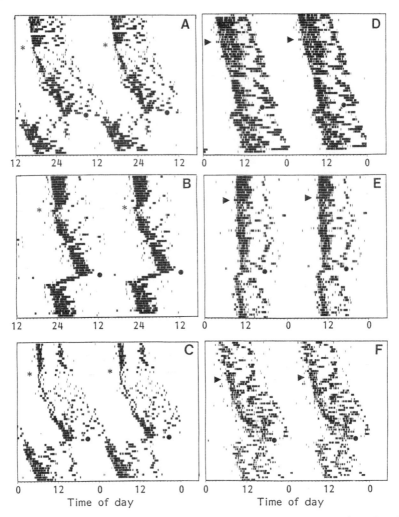

FIG. 5. Double-plotted wheel-running records from three reserpine-treated (A, B and C) and three vehicle-treated (D, E and F) young (2–5 month) hamsters kept in constant darkness. The asterisks and arrowheads on the records denote the day on which the animals received an injection of either reserpine (2.5 mg/kg,*) or vehicle (▶). The solid circles on the records indicate the exact times at which the animals were exposed to a 15 min pulse of light that was timed to occur near CT 16. The large phase shifts in the reserpine-treated animals were never observed in the vehicle-injected animals. From Penev et al 1993.

findings of others that fetal suprachiasmatic nuclei transplanted into the hypothalamus can develop anatomically and functionally in host brains (Lehman et al 1991, Ralph 1991), led us to ask whether age-associated changes in the circadian system could be reversed by fetal SCN transplants. Old hamsters (14–25 months of age) with free-running circadian systems kept in constant light were given a first intraperitoneal injection of triazolam (5 mg/kg) at CT 6 and were kept in constant light for at least 10 days before receiving transplants of cerebellar or fetal SCN tissue in the area of the SCN. Cerebellar grafts were used as a control because it has been demonstrated previously that they are unable to restore rhythmicity in SCN-lesioned animals (Lehman et al 1987). After surgery, the animals were allowed to recover before they were again given an injection of 5 mg/kg triazolam at CT 6. The experiment continued for at least 10 days after this second triazolam injection.

Before transplant, only 7/36 old hamsters showed phase shifts larger than 30 min in response to the injection of triazolam at CT 6, and the mean phase shift in all the old animals combined was 9 ± 4 min ($n = 36$). There was no significant difference between the mean phase shifts in control (15 ± 6 min) and experimental (6 ± 6 min) animals ($P = 0.3$). Only 29 animals remained for pair-wise comparisons after transplantation, the others having died or adopted an activity rhythm too erratic or fragmented for analysis. After transplantation with cerebellar tissue, 5/11 old hamsters showed phase shifts slightly larger than 30 min in response to triazolam; the mean phase shift in this group of 11 was 33 ± 11 min, a value not statistically different from the mean phase shift observed before transplantation ($P = 0.1$). In contrast, SCN grafts had a significant effect on the phase-shifting response to triazolam: 15/18 old hamsters that received fetal SCN tissue showed phase shifts larger than 30 min and the mean phase shift of this group averaged 72 ± 12 min, a value significantly larger than the mean phase shift before transplantation ($P < 0.001$). The phase shifts produced by triazolam after transplantation were significantly larger in SCN-grafted animals than in cerebellum-grafted animals ($P = 0.02$). Responses to triazolam were similar in SCN-grafted animals whether the graft was in the third ventricle or in the hypothalamus, and the response to triazolam did not seem to vary as a function of the presence of an intense cell immunoreactivity within the graft or the development of abundant interactions between the grafted SCN fibres and the host brain.

Our results indicate that SCN grafts can develop functionally in the brain of old hosts, and that they can at least partially restore some age-associated changes in the response of the circadian clock to activity-inducing stimuli. However, we do not yet know how the donated tissue is contributing to the circadian system to restore responsiveness; its role could range from one of simply facilitating the aged host SCN (by restoring some input, intra-SCN or output activity) so that it can again express a response to an acute increase in activity, to one of providing a new pacemaker in the circadian system which is itself able to regulate the hamster's circadian behaviour. Further studies are necessary to determine the physiological mechanism by which function is restored, and the extent to which SCN transplants restore circadian function in old animals.

Conclusions

Elucidation of the physiological basis of age-related changes in the responsiveness of the circadian clock, and the mechanisms which underlie the ability of various experimental manipulations to restore normal circadian functioning to old animals, should lead to new insight into the mechanisms by which environmental stimuli alter clock functioning. Because disturbances in circadian rhythmicity are a hallmark of ageing in humans, it is important to determine whether restoration of normal circadian functioning in the elderly will have a positive effect on the health and well-being of the individual. Determining how the response of the circadian clock to external and internal stimuli is altered with age may lead to new clinical approaches for the treatment of disorders of circadian rhythmicity in the elderly.

Acknowledgements

This work was supported by the National Science Foundation Science and Technology Center on Biological Timing and by National Institutes of Health grants R01-AG-09297 and R01-AG-10870. The excellent secretarial assistance of Ms Pauline Jasim is acknowledged.

References

Brock MA 1991 Chronobiology and aging. J Am Geriatr Soc 39:74–91

Czeisler CA, Chiasera AJ, Duffy JF 1991 Research on sleep, circadian rhythms and aging: applications to manned spaceflight. Exp Gerontol 26:217–232

Davis FC, Menaker M 1980 Hamsters through time's window: temporal structure of hamster locomotor rhythmicity. Am J Physiol 239:R149–R155

Earnest DJ, Turek FW 1982 Splitting of the circadian rhythm of activity: effects of exposure to constant darkness and subsequent re-exposure to constant light. J Comp Physiol 145:405–411

Ellis GB, McKlveen RE, Turek FW 1982 Dark pulses affect the circadian rhythm of activity in hamsters kept in constant light. Am J Physiol 242:R44–R50

Ingram DK, London ED, Reynolds MA 1982 Circadian rhythms and sleep: effects of aging in laboratory rats. Neurobiol Aging 3:287–297

Inouye ST, Takahashi JS, Wollnik F, Turek FW 1988 Inhibitor of protein synthesis phase shifts a circadian pacemaker in the mammalian SCN. Am J Physiol 255:R1055–R1058

Lehman MN, Silver R, Gladstone WR, Kahn RM, Gibson M, Bittman EL 1987 Circadian rhythmicity restored by neural transplant: immunocytochemical characterization of the graft and its integration with the host brain. J Neurosci 7:1626–1638

Lehman MN, Silver R, Bittman EL 1991 Anatomy of suprachiasmatic nucleus grafts. In: Klein DC, Moore RY, Reppert SM (eds) The suprachiasmatic nucleus: the mind's clock. Oxford University Press, New York, p 349–374

Martin JR, Fuchs A, Bender R, Harting J 1985 Altered light–dark activity difference with aging in two rat strains. J Gerontol 44:2–7

Meites J 1990 Aging: hypothalamic catecholamines, neuroendocrine–immune interactions and dietary restriction. Proc Soc Exp Biol Med 195:304–311

Morin LP 1988 Age-related changes in hamster circadian period, entrainment and rhythm splitting. J Biol Rhythms 3:237–248

Mrosovsky N 1988 Phase response curves for social entrainment. J Comp Physiol A Sens Neural Behav Physiol 162:35–46

Nicolau GY, Milcu S 1977 Circadian rhythm of corticosterone and nucleic acids in the rat adrenals in relation to age. Chronobiologia 4:136

Penev PD, Turek FW, Zee PC 1993 Monoamine depletion alters the entrainment and the response to light of the circadian activity rhythm in hamsters. Brain Res 612: 156–164

Penev PD, Zee PC, Turek FW 1994 Monoamine depletion blocks triazolam-induced phase advances of the circadian clock in hamsters. Brain Res 637:255–261

Peng MT, Jiang MJ, Hsu HK 1980 Changes in running-wheel activity, eating and drinking and their day/night distribution throughout the life span of the rat. J Gerontol 35:339–347

Pittendrigh CS, Daan S 1974 Circadian oscillations in rodents: a systematic increase of their frequency with age. Science 186:548–550

Prinz PN, Peskind ER, Vitaliano PP et al 1982 Changes in the sleep and waking EEGs of nondemented and demented elderly subjects. J Am Geriatr Soc 30:86–93

Ralph MR 1991 Suprachiasmatic nucleus transplant studies using the *tau* mutation in golden hamsters. In: Klein DC, Moore RY, Reppert SM (eds) The suprachiasmatic nucleus: the mind's clock. Oxford University Press, New York, p 341–348

Reebs S, Mrosovsky N 1989 Running activity mediates the phase-advancing effects of dark pulses on hamster circadian rhythms. J Comp Physiol A Sens Neural Behav Physiol 165:811–818

Reiter RJ, Craft CM, Johnson JE Jr et al 1981 Age associated reduction in nocturnal pineal melatonin levels in female rats. Endocrinology 109:1295–1297

Rosenberg R, Zee P, Turek FW 1991 Phase response curve to light in young and old hamsters. Am J Physiol 261:R491–R495

Sacher GA, Duffy PH 1978 Age changes in rhythms of energy metabolism, activity and body core temperature in *Mus musculus* and *Peromyscus*. In: Samis HV, Capobianio A (eds) Aging and biological rhythms. Plenum Press, New York, p 105–124

Satlin A, Teicher MH, Lieberman HR, Baldessarini RJ, Volicer L, Rheaume Y 1991 Circadian locomotor activity rhythms in Alzheimer's disease. Neuropsychopharmacology 5:115–126

Swaab DF, Fisser B, Kamphorst W, Troust D 1988 The human suprachiasmatic nucleus: neuropeptide changes in senium and Alzheimer's disease. Basic Appl Histochem 32:43–54

Takahashi JS, Turek FW 1987 Anisomycin, an inhibitor of protein synthesis, perturbs the phase of a mammalian circadian pacemaker. Brain Res 405:199–203

Takahashi JS, DeCoursey PJ, Bauman L, Menaker M 1984 Spectral sensitivity of a novel photoreceptive system mediating entrainment of mammalian circadian rhythms. Nature 308:186–188

Tenover JS, Matsumoto AM, Clifton DK, Bremner WJ 1988 Age-related alterations in the circadian rhythm of pulsatile luteinizing hormone and testosterone secretion in healthy men. J Gerontol 43:163–169

Turek FW 1989 Effects of stimulated activity on the circadian pacemaker of vertebrates. J Biol Rhythms 4:135–147

Van Gool WA, Witting W, Mirmirian M 1987 Age-related changes in circadian sleep–wakefulness rhythms in male rats isolated from time cues. Brain Res 413:384–387

Van Reeth O, Turek FW 1989a Stimulated activity mediates phase shifts in the hamster circadian clock induced by dark pulses or benzodiazepines. Nature 339:49–51

Van Reeth O, Turek FW 1989b Administering triazolam on a circadian basis entrains the activity rhythm of hamsters. Am J Physiol 256:R639–R645

Van Reeth O, Turek FW 1992 Aging alters feedback effects of the activity–rest cycle on the circadian clock. Am J Physiol 263:R981–R986

Van Reeth O, Hinch D, Tecco JM, Turek FW 1991 The effects of short periods of immobilization on the hamster circadian clock. Brain Res 545:208–214

Van Reeth O, Zhang Y, Reddy A, Zee P, Turek FW 1993 Aging alters the entraining effects of an activity-inducing stimulus on the circadian clock. Brain Res 607:286–292

Weiland NG, Wise PM 1990 Aging progressively decreases the densities and alters the diurnal rhythms of alpha-1 adrenergic receptors in selected hypothalamic regions. Endocrinology 126:2392–2397

Welsh DK, Richardson GS, Dement WC 1986 Effect of age on the circadian pattern of sleep and wakefulness in the mouse. J Gerontol 41:579–586

Wise PM, Walovitch RC, Cohen IR, Weiland NG, London DE 1987 Diurnal rhythmicity and hypothalamic deficits in glucose utilization in aged ovariectomized rats. J Neurosci 7:3469–3473

Wise PM, Cohen IR, Weiland NG, London DE 1988 Aging alters the circadian rhythm of glucose utilization in the suprachiasmatic nucleus. Proc Natl Acad Sci USA 85:5305–5309

Witting W, Mirmiran M, Bos NPA, Swaab DF 1993 Effect of light intensity on diurnal sleep–wake distribution in young and old rats. Brain Res Bull 30:157–162

Wollnik F, Turek FW, Majewski P, Takahashi JS 1990 Phase shifting the circadian clock with cycloheximide: response of hamsters with an intact or split rhythm of locomotor activity. Brain Res 496:82–88

Zee PC, Rosenberg RS, Turek FW 1992 Effects of aging on entrainment and rate of resynchronization of the circadian locomotor activity. Am J Physiol 263:R1099–R1103

DISCUSSION

Moore: One explanation for all of your results on ageing hamsters is that they are simply thiamine deficient. That would be an easy thing to test, obviously. We have found no abnormalities in the circadian systems of old rats, up to three years of age.

Turek: Have you tested responses to light pulses?

Moore: No; we have simply looked at free-running and entrained conditions, but we observed phase advances and no decrease in the amplitude of the rhythm. The animals seem perfectly all right until about three or four days before they die.

Turek: Witting et al (1993) have reported age-related changes in the response of the rat to light.

Mikkelsen: The most simple explanation for the differences between young and old animals is that some of the cells in the SCN are degenerating or even dying.

Turek: I'm not sure that's the most simple explanation—I thought you were going to ask if the hamsters had cataracts! There are indications from other laboratories that there are fewer vasopressin cells in the SCN of old animals. We have not done any systematic histology to address this question.

Takahashi: We have done one quantitative anatomical study of total CREB (cyclic AMP response element-binding protein)-positive neurons in the SCN in relation to phospho-CREB responsiveness in the SCN. There is only a modest and non-significant reduction, about 10%, in total CREB-positive SCN neurons in old (around 18-month-old) hamsters. However, there was a 35–40% reduction in the number of neurons staining for phospho-CREB in response to a saturating light pulse at CT 19. Thus, the changes upstream of *fos* can be seen at the level of phosphorylation of CREB in the SCN.

Roenneberg: You described an increase in the phase-shifting reponse of old hamsters given a light pulse at CT 16. I had understood that the old hamsters show a decreased response to light pulses associated with decreased c-*fos* expression.

Turek: The phase-shifting response to a five-minute, low intensity light pulse is decreased in old hamsters, as is the c-*fos* response. The huge phase shift seen in old animals but not young animals was in response to a one-hour pulse of bright light given at CT 16. These are two totally different phenomena.

Takahashi: The half-saturating light intensity in old animals is increased, shifted to a higher light intensity, but the maximum response is also increased in old animals.

Menaker: But is that only when the light pulse is given at CT 16?

Turek: One only sees large phase shifts in old animals in response to bright light at CT 16. This response, we believe, is totally separate from the age-related change in sensitivity to light.

Menaker: It's also quite different from the *tau* mutant. The *tau* mutant's PRC increases in amplitude with increasing time in constant darkness not just at CT 16 but in the whole delay region and to a lesser extent in the advance region as well.

Turek: That's a good point. In our original work we found larger phase shifts in the old animals than in the young animals in response to light at CT 19, but the shifts were not of this magnitude. The phase shifts at CT 19 were only about an hour greater in old animals than in young ones. Perhaps in even older animals the advance region might be larger than in young animals.

Czeisler: If the increased magnitude of the phase-shifting response is due to reduction in the pacemaker's amplitude of oscillation, you wouldn't necessarily detect the difference between the type 1 and the type 0 PRC at all initial phases. This would explain why large phase shifts are not always observed in older animals. Phases of initial stimulus application that are not in the most sensitive region of the PRC don't change all that much when you are just at the transition from type 1 to type 0 resetting.

Cassone: Do other aspects of the circadian system show increased amplitude with age? I was thinking in particular of your transplantation work, where transplants do not reinstate the melatonin rhythm from the pineal gland after lesion of the SCN.

Turek: This is a fantastic model system, and there is a lot we have to do. For example, if we can reverse the response to a light pulse at CT 16, getting a normal phase shift where we now see these large phase shifts in old animals, we won't attribute that to any reinnervation of the transplanted SCN by the retinohypothalamic tract, because there's no evidence that occurs. Instead, such a result would indicate that the transplanted SCN is affecting the host SCN. The fact that the response is augmented in old hamsters with a transplanted SCN while Rae Silver did not see phase shifts in response to triazolam in SCN-lesioned animals bearing a transplanted SCN, indicates that the transplanted suprachiasmatic nuclei are not themselves receiving the activity-inducing signal. Our old animals with a transplanted SCN are clearly receiving the activity-inducing signal and responding to it, which makes us think that it's the host SCN which is responding and not the transplanted tissue.

Ralph: If the restoration of responsiveness to triazolam in aged, triazolam-insensitive hosts by SCN grafts is due to the effect of the implant on the intact, elderly host SCN, the implication is that some sort of trophic factor is improving the 'health' of the host. Another possibility is that if the implanted SCN is being entrained by the host, and there's two-way communication between these, you effectively have a hybrid young–old SCN. If the acute response to triazolam, not the phase shift *per se*, can be communicated to the graft, and the graft can respond with a phase shift, it could communicate the shift back to the host. The whole system might respond in a younger fashion.

Turek: I agree; that is a possible interpretation.

Menaker: Do you think implantation of cerebellar tissue is an adequate control? Because the cerebellar implant had some effect, I would like to see you use extra-SCN hypothalamus as a control.

Turek: The SCN contains a lot of growth factors whereas cerebellar tissue is deficient in growth factors. At this stage, I wouldn't rule out the possibility that our results are due to rather non-specific effects, an effect of a thiamine or a growth factor, perhaps. In a sense, that does not matter to me; I want to know the mechanism of course, but I won't mind if it doesn't involve some magical circadian signal from the transplanted SCN.

Gillette: We have just completed a study on aged rats (Satinoff et al 1993) in which we followed the rhythms in activity, drinking and body temperature for at least three months, many of them for six months. All the rats were over two years of age. After following the rhythms we monitored electrical activity rhythms in SCN slices, and found they were dampened. In addition to a dampening of the overall rhythm in the slice we saw spurious peaks. The tissue is very different from that of young rats and is difficult to handle.

Turek: If we progress further in making our old animals appear young, it would be interesting to send you their SCN tissue.

Dunlap: Martha Gillette has suggested that SCN slices show large phase shifts because they lack inhibitory feedbacks from the rest of the brain. Could the same thing be happening in the ageing hamster? Could the SCN itself be normal but lack inhibitory feedback, in which case the ageing hamster is just a model for an SCN slice!

Turek: Our results, in conjunction with Martin Ralph's, suggest the ageing hamster could in some ways be viewed as a partially SCN-lesioned animal. A slice is certainly an example of a partial SCN lesion, albeit a big partial SCN lesion, so there may well be some relationship. We are looking at such fundamentally different systems that we shouldn't draw too many analogies. After Martha Gillette's presentation there was discussion of the differences between the effects she saw in culture and those seen in the whole animal. This was nit-picking. I'm amazed that we see any similarity at all. You take a chunk from the brain and stick it in a dish, and it behaves in some respects like the whole animal does—that's what we should be remarking on, not slight differences in the PRCs.

Miller: Were the effects of reserpine reversible?

Turek: I expect that they would be, but we didn't continue the studies for long enough to be sure. The effects on the activity rhythm were still evident a month afterwards. I understand that reserpine's effects do not last as long in rats.

Moore: The effects of reserpine on monoamines can last a very long time; days to weeks.

Miller: If you're concerned about catecholamines, you should test 6-OH-dopamine, to knock out noradrenaline and dopamine.

Turek: We shall be knocking out one at a time in our next experiments.

Lemmer: Reserpine is a very dirty drug. At the high dose you used, 2.5 mg/kg, it initially has an additional sympathomimetic effect, releasing an enormous amount of noradrenaline. If you want to decrease the catecholamine content completely you also have to inhibit tyrosine hydroxylase. Also, there is a feedback loop, involving induction of tyrosine hydroxylase when the endogenous concentration is lowered. Reserpine increases catecholamine turnover; thus, if reserpine is given without an inhibitor of the tyrosine hydroxylase the noradrenaline turnover goes up and partially compensates for amine loss. Moreover, reserpine reduces the concentrations of noradrenaline, dopamine and serotonin. These multiple effects make it impossible to draw definite conclusions from your results.

Turek: Would those effects last 10–15 days?

Lemmer: At a concentration of 3 mg/kg, the reduction in amine content will last about two weeks. Even in humans with the low doses of reserpine that are used for treatment of hypertension, you can clearly observe a decrease in platelet serotonin up to three weeks after treatment has finished.

Miller: That's what Dr Turek wants, though. The concentration increase occurs only immediately after drug administration.

Lemmer: Reserpine will increase turnover after two weeks. There will be partial compensation for the loss of amines.

Miller: Dr Turek, you gave the age of your 'old' hamsters, but you didn't say how old they were in relation to a hamster's total lifespan.

Turek: Most of our studies start when the animals are 15–16 months of age. Total lifespan is a little difficult to decide. By the age of 20–22 months, we are beginning to lose a fair number of them, but we have seen hamsters live as long as 30 months in our laboratory. This is one of our great difficulties—we do not all age in the same fashion and the same is true of hamsters. A sample of 15-month-old hamsters is not a homogeneous population; some look better or are healthier and have better activity records than others.

Miller: Did you see any shortening of period with age?

Turek: Yes. As others have, we have observed a slight shortening of the free-running period, but because the changes are small this is not a good experimental variable to try to manipulate.

Mrosovsky: Foster (1993) has done longitudinal studies on *rd/rd* mice to check whether phase shifting responses alter with age. He has studied mice for up to two years and has comparable data on the control (wild-type) animals. What is the longitudinal change in phase shifting in those control animals?

Menaker: As far as I recall, age has no effect on the amplitude of the phase response, at least in experiments done in this way, but phase shifts at CT 16 were not measured (Provencio et al 1994). They were checking the effects on clear advances and delays.

Mrosovsky: The effects of ageing may differ between hamsters and mice.

Edmunds: I was brought up to believe that constant light would be deleterious to circadian organization and, consequently, be reflected in an organism's longevity. As far as I can tell, in insects, in rodents, in almost every group that has been looked at, even within one species, results are contradictory (Natelson et al 1991; see Edmunds 1994). Is there a consensus that a breakdown of circadian organization, whatever that may mean, reduces lifespan?

Turek: That experiment has not really been done adequately. You cannot disrupt circadian organization and look at the effects on longevity in an animal that you keep in a laboratory situation for the rest of its life, in a totally germ-free environment with ample food, pampered to the greatest extent, without any of the dangers that the animal would have to face in the wild. I would like to know whether destruction of the SCN would affect the longevity of an animal given a challenging environment.

Edmunds: Even within one species, such as *Drosophila*, different experiments give entirely different results—neutrality, increased longevity, or decreased longevity. How can this kind of variation in experimental results be explained?

Turek: The conditions in which the experiments are done and the animals are kept will vary. We are in rather an embarrassing position. Although we say that the circadian clock is fundamental for the health and well-being of the organism, we don't have data indicating that removal or disruption of the circadian clock system causes major medical and health problems, because we haven't done the experiments.

Edmunds: Similarly, contradictory results are reported for plants.

Menaker: I don't agree. The early experiments on plants that Highkin and others did are actually quite consistent (Highkin 1960).

Edmunds: They have been challenged by more recent work from a number of laboratories. Our *Euglena*, for example, loves constant light!

Menaker: These experiments are old, but there were a lot of them and they form a consistent body of literature which has been largely ignored. That work ought to be repeated.

Redfern: What is known about disturbances of clock mechanisms or circadian abnormalities in patients with Alzheimer's disease? I am thinking particularly, for example, of patients who exhibit nocturnal wandering.

Turek: There are disruptions of the sleep–wake cycle in healthy elderly people, who are typically more awake during the night and often fall asleep during the day, and the discrepancies are even greater in demented individuals. In demented patients of the Alzheimer type there is further fragmentation of the sleep–wake cycle, which has important implications not only for the person suffering, but also for the care-givers. If we could improve temporal organization in such people, whether pharmacologically or by other means, we would also help the care-givers whose own sleep–wake cycles are disturbed. One of the main reasons people put their elderly relatives into nursing homes is that they can no longer cope with the sleep–wake problems and the impact they have on the sleep–wake cycle of other family members, including the children.

Moore: Although patients with Alzheimer's disease certainly have disrupted sleep–wake patterns, there is no substantial evidence that they have disrupted circadian functioning. They tend to be awakened early in the night by something and, because they don't know where they are, they don't know what's going on and they don't know it's night, they get up and wander around for a while. A much more substantial study has to be done before we can interpret these behaviour patterns as a disruption of circadian functioning.

Waterhouse: There is evidence from Okawa et al (1991) that regular social interaction can decrease these symptoms, but to what extent, if any, circadian rhythms have been entrained does not appear to have been established.

Turek: I would agree that there is not a lot of information demonstrating circadian disruptions in patients suffering from Alzheimer's disease, but there aren't many studies of them either. We are dealing with an absence of experiments rather than an absence of data, but I agree that we need to be cautious and not interpret all of the problems of Alzheimer's patients as a clock problem.

Lewy: Dr Robert Sack and I, as well as others, have shown that melatonin levels decrease with ageing (Sack et al 1986). There are some interesting but not definitive studies showing circadian abnormalities with ageing in humans.

Czeisler: We are studying the changes that occur in the circadian systems of older people (Czeisler et al 1992). Even in individuals who have no complaints about sleep, there are changes in the circadian system which are observable in constant routine conditions such as decreases in the amplitude of the body temperature and melatonin rhythms and advances in the entrained phase of the circadian system that are similar to those Dr Turek has described. There is increasing evidence for age-related changes and abnormalities in the output of the human circadian pacemaker. Our preliminary data suggest these changes are even more marked in older (>65 years) people with sleep disorders (Czeisler et al 1989, Dumont et al 1990). This relationship is not necessarily causal but the correlations that Dr Turek has reported are consistent with our results.

Moore: Tim Monk has shown that in healthy old humans circadian function is maintained quite well (Monk et al 1991).

Czeisler: Dr Turek made the point that different individuals respond differently to ageing—some people's hair goes grey, some lose their hair and some people's hair doesn't change at all. One of the hallmarks of ageing is increased variability in many physiological parameters, and this should be no exception. We see some older people who have just as robust an amplitude as young people, but on average, and we now have data from about 25 older people and 25 young ones, there is a significant reduction in endogenous circadian temperature amplitude and an advance of entrained phase (Czeisler et al 1992).

Mrosovsky: Changes in amplitude could be on the output side rather than in the clock itself, as you said. But we shouldn't lose sight of the findings of Swaab et al (1985) of reduced cell counts at post mortem in the SCN of patients with Alzheimer's disease. Is there a problem with that report?

Moore: We have replicated that study, but found no change whatsoever in cell numbers in the SCN of 10 well-characterized Alzheimer's disease patients. If anything, there are slightly more neurons in these SCN. However, we are continuing to work on this problem.

Mrosovsky: Perhaps some of your cases were healthier than those studied by Swaab et al (1985).

Lewy: It is more complicated than that. The polygraphic sleep recordings of elderly people are different from those of younger people, in general, but elderly people do not complain to the same extent. It is possible that, just as patients with winter depression grow accustomed to each winter, ageing is a 'disorder' that one learns to live with. Just because an elderly person's sleep record is different from a younger person's doesn't mean that it is pathological.

Daan: We are dealing here with amplitude on the output side. It has long been known that in the course of life the activity of small rodents gradually declines. If there is a decline in total activity there is bound to be a decline in amplitude, which thus doesn't in itself need to have anything to do with a change in the underlying circadian system. We should be careful before assuming that output amplitude tells us something about what the pacemaker is doing.

Czeisler: You are thinking about activity in terms of wheel running. Dr Lewy was talking about a decrease in the amplitude of the melatonin rhythm. Our studies of body temperature rhythms are with young and older humans whose activity we restrict to the same, sedentary level throughout the constant routine, because we recognize that activity has to be controlled if we are to evaluate whether there's a change in the amplitude of the hormonal or thermoregulatory signal.

Lewy: I'm not convinced that the decrease in melatonin production reflects a decrease in the amplitude of its underlying pacemaker.

Czeisler: I'm just saying that there are multiple, output rhythms that appear to be changing in their amplitude with advancing age. It is true that they are all output rhythms, but they can't all be dismissed as being related to the motor activity of the individual.

Turek: For years we ignored amplitude, because we all knew that the amplitude could be affected by something downstream from the clock. However, we should at least consider the possibility that amplitude changes seen in output rhythms could be a function of the circadian clock system. We have focused too much on phase and period.

Daan: We should not ignore amplitude, but should study it, as you have done, by looking at responses to an external signal which tell us something about the amplitude of the pacemaker.

Kronauer: If some mean level, such as activity, is decreasing, you could perhaps compensate for that by dealing with the modulation index, a normalization in which the modulation is divided by the mean. That may provide a better way of handling activity as an output variable, but doesn't do away with the problem of it not being a fundamental variable.

Daan: There are other aspects to the regulation of sleep and wakefulness. For example, reduced activity might lead to reduced sleepiness in humans; if general activity decreased, the whole rhythm of sleep and wakefulness would be flattened out and the amplitude thereby reduced.

Kronauer: I was talking about animal activity records.

Daan: Furthermore, in activity records the daily minimum is zero, for both young and old animals. Dividing the modulation by the mean thus gives you about the same relative amplitude for young and old.

Armstrong: Another important aspect of ageing is that as people get older the probability of disease and illness increases and they take medication, often chronically, which could affect the circadian clock. For example, β-blockers are administered for high blood pressure and some of these, such as propranolol, would affect the pineal and melatonin release. Many elderly people take benzodiazepines, such as triazolam, for sleeping problems, and if humans respond to triazolam like Fred Turek's hamsters they will be phase shifting all over the place. A variety of medications given for a variety of age-associated disorders could be affecting the clock. People don't know that and when they feel lousy they don't realize why. They could be artificially inducing jet lag in themselves all the time. Medications for a traditional disorder such as hypertension could be causing a circadian rhythm disorder of which both the patient and the administering physician are totally unaware.

Moore: Ageing is a complex problem, obviously. The one sure way of preventing the outcome of ageing in rodents is to reduce body weight to about 75% of normal. One mouse maintained like this has just reached four years of age, which is an extraordinary age for a mouse, and is still going strong. It would be interesting to see if long-term reduction of body weight changed circadian functioning in hamsters.

Turek: We plan to do such experiments.

References

Czeisler CA, Kronauer RE, Johnson MP, Allan JS, Dumont M 1989 Action of light on the human circadian pacemaker: treatment of patients with circadian rhythm sleep disorders. In: Horne J (ed) Sleep '88. Proceedings of the 8th European Congress on Sleep Research. Gustav Fischer Verlag, Stuttgart, p 42–47

Czeisler CA, Dumont M, Duffy JF et al 1992 Association of sleep–wake habits in older people with changes in output of circadian pacemaker. Lancet 340:933–936

Dumont M, Richardson GS, Czeisler CA 1990 Endogenous circadian phase and amplitude in elderly patients with a complaint of insomnia. Sleep Res 19:217

Edmunds LN Jr 1994 Clocks, cell cycles, cancer, and aging: role of the adenylate cyclase–cyclic AMP–phosphodiesterase axis in signal transduction between circadian oscillator and cell division cycle. Ann NY Acad Sci 719:77–96

Foster RG 1993 Photoreceptors and circadian systems. Curr Dir Psychiatr Sci 2:34–39

Highkin HR 1960 The effect of constant temperature environments and of continuous light on the growth and development of pea plants. Cold Spring Harbor Symp Quant Biol 25:231–238

Monk TH, Reynolds CF, Buysse DJ et al 1991 Circadian characteristics of healthy 80 year olds and their relationship to objectively recorded sleep. J Gerontol 46:M171–175

Natelson BH, Tapp WN, Drastel S, Gross J, Ottenweller JE 1992 Constant light extends life in hamsters with heart disease. Proc Soc Exp Biol Med 202:69–74

Okawa M, Mishima K, Hishikawa Y, Hozumi S, Hori H, Takahashi K 1991 Circadian rhythm disorders in sleep, waking and body temperature in elderly patients with dementia and their treatment. Sleep 14:478–483

Provencio I, Wong SY, Lederman AB, Argamaso SM, Foster RG 1994 Visual and circadian responses to light in aged retinally degenerate (*rd*) mice. Vision Res 34:1799–1806

Sack RL, Lewy AJ, Erb DL, Vollmer WM, Singer CM 1986 Human melatonin production decreases with age. J Pineal Res 3:379–388

Satinoff E, Hua L, Tcheng TK et al 1993 Do the suprachiasmatic nuclei oscillate in old rats as they do in young ones? Am J Physiol 265:R1216–R1222

Swaab DF, Fliers E, Partiman TS 1985 The suprachiasmatic nucleus of the human brain in relation to sex, age and senile dementia. Brain Res 342:37–44

Witting W, Mirmiran M, Bos NPA, Swaab DF 1993 Effect of light intensity on diurnal sleep–wake distribution in young and old rats. Brain Res Bull 30:157–162

Clinical chronopharmacology: the importance of time in drug treatment

Björn Lemmer

Zentrum der Pharmakologie, J. W. Goethe-Universität, Theodor-Stern-Kai 7, D-60590 Frankfurt/M, Germany

Abstract. Nearly all functions of the body, including those influencing pharmacokinetic parameters such as drug absorption and distribution, drug metabolism and renal elimination, show significant daily variations: these include liver metabolism, hepatic blood flow and the first-pass effect; glomerular filtration, renal plasma flow and urine volume and pH; blood pressure, heart rate and organ perfusion rates; acid secretion in the gastro-intestinal tract and gastric emptying time. The onset and symptoms of diseases such as asthma attacks, coronary infarction, angina pectoris, stroke and ventricular tachycardia are circadian phase dependent. In humans, variations during the 24 h day in pharmacokinetics (chronopharmacokinetics) have been shown for cardiovascularly active drugs (propranolol, nifedipine, verapamil, enalapril, isosorbide 5-mononitrate and digoxin), antiasthmatics (theophylline and terbutaline), anticancer drugs, psychotropics, analgesics, local anaesthetics and antibiotics, to mention but a few. Even more drugs have been shown to display significant variations in their effects throughout the day (chronopharmacodynamics and chronotoxicology) even after chronic application or constant infusion. Moreover, there is clear evidence that even dose/concentration–response relationships can be significantly modified by the time of day. Thus, circadian time has to be taken into account as an important variable influencing a drug's pharmacokinetics and its effects or side-effects.

1995 Circadian clocks and their adjustment. Wiley, Chichester (Ciba Foundation Symposium 183) p 235–253

There is sound evidence demonstrating that living systems, humans included, are organized not only in space but also in time. Endogenous clocks keep time for them and it has been assumed that the rhythmicity inherent to living systems allows them to adapt and survive more easily under changing environmental conditions during the 24 h of a day as well as during the varying conditions of the changing seasons. In humans, however, most rhythmic fluctuations could not, and still cannot, be studied under free-running conditions, leaving open the question of whether or not they are really 'circadian'. Nevertheless, under 'real life' conditions, nearly all functions of the human body as well as the onset and symptoms of various diseases have been shown to be circadian phase dependent (see Reinberg & Halberg 1971, Minors & Waterhouse 1981, Reinberg & Smolensky 1983, Lemmer 1989, Arendt et al 1989, Touitou & Haus 1992).

Rhythms in health and disease: implications for clinical pharmacology

In clinical pharmacology pharmacokinetic parameters often are considered not to be influenced by time of day, and, with regard to drug concentration-versus-time profiles, 'the flatter the better' is also a common aim in drug targeting. However, this paradigm can be held no longer (Lemmer 1991a), because it is now well established that nearly all functions of the body, including those influencing pharmacokinetic parameters, show significant variations throughout the day. Circadian, 24 h or daily rhythms have been shown in heart rate, body temperature, blood pressure, blood flow, stroke volume, peripheral resistance, parameters of electrocardiographic (ECG) recordings, plasma concentrations of hormones, neurotransmitters and second messengers (e.g., cortisol, melatonin, insulin, prolactin, atrial natriuretic hormone, noradrenaline, cyclic AMP), in the renin–angiotensin–aldosterone system, in blood viscosity, aggregability and fibrinolytic activity, in the plasma concentrations of glucose, electrolytes, plasma proteins and enzymes, and in the number of circulating red and white blood cells and and blood platelets. Moreover, various functions of the lung (such as minute volume, peak flow, forced expiratory volume and dynamic compliance), of the liver (metabolism, estimated hepatic blood flow and the first-pass effect) and of the kidneys (glomerular filtration, renal plasma flow, urine pH, urine volume and electrolyte excretion) vary with the time of day. Also, gastric acid secretion has a pronounced circadian variation, with peak values in the late afternoon in normal subjects and in patients suffering from peptic ulcers.

The importance of time in humans can also be seen in certain diseases or states of disease in which the onset and symptoms do not occur randomly within the 24 h of a day. Asthma attacks happen more often during night hours than at other times of day, as observed nearly 300 years ago by John Floyer (1698). Similarly, the occurrence of coronary infarction, angina pectoris attacks and pathological ECG recordings are unevenly distributed over the 24 h span of the day, with a predominant peak between eight and 12 o'clock in the morning. Moreover, subtypes of a particular disease, such as forms of vasospastic and stable angina pectoris or of primary and secondary hypertension, may have markedly different 24 h symptom patterns (see Lemmer 1989, 1993a).

Chronopharmacology

Bearing in mind the organization in time of living systems, one can see easily that the right amount of the right substance must be at the right place, but at the right *time* (Halberg 1960). This is more important when a living subject has to act or react under environmental conditions which are themselves highly periodic. Thus, it is easy to understand that exogenous compounds, including drugs, may challenge the individual differently depending on the time of exposure. Chronopharmacology, a relatively new field in biological sciences, combines chronobiology and pharmacology (for reviews see Lemmer 1989, Bélanger 1993).

Over the last decade numerous clinical studies have provided convincing evidence that a drug's pharmacokinetics or its effects/side effects can be modified by the circadian time or the timing of its application within the 24 h of a day. Tables 1 and 2 list drugs which have been found to show significant variation around the clock in pharmacokinetics or effects in humans. I shall demonstrate the importance of the rapidly expanding field of clinical chronopharmacology with reference to three areas—peptic ulcer, asthma bronchiale and hypertension—in which the 'time of day' of drug application to patients is now taken into account by the Commission on Drugs of the German Medical Association (Arzneimittelkommission der deutschen Ärzteschaft 1992).

Chronopharmacology of peptic ulcer

The H_2 histamine receptor blockers are now the drugs of first choice in the treatment of peptic ulcers. Moore & Englert (1970) first described the circadian rhythm in gastric acid secretion in humans, finding peak values around 20.00–23.00 h. This finding led to the unanimous recommendation that H_2 blockers (ranitidine, cimetidine, famotidine, roxatidine and nizatidine) should be taken once a day after the last meal in the evening when acid secretion is increasing, regardless of whether the compound has a short or long half-life.

It has been generally assumed that constant drug infusion results in constant drug concentrations, which in turn should lead to constant drug effects. However, constant infusion of ranitidine over 24 h does *not* lead to a constant effect: the increase in gastric pH caused by ranitidine was less during the night than the daytime hours of infusion (Fig. 1), indicating that there might be a partial nocturnal resistance to H_2-receptor blockade (Sanders et al 1988). This interesting finding not only calls for further investigation of the regulatory mechanisms of gastric acid secretion, but could also mean that drugs with a different mechanism of action could be added to the treatment with H_2 blockers during night hours. In this context, however, it is interesting to note that a breakthrough drop in intragastric pH occurred in response to a dose of the proton pump inhibitor omeprazole alone and in combination with ranitidine (Sanders et al 1992). In addition, variations over 24 h in pharmacokinetics were reported for both the H_2 blocker cimetidine and omeprazole, with a higher peak drug concentration (C_{max}) and a shorter time to peak concentration (t_{max}) after morning than evening dosing (Maleev et al 1990, Prichard et al 1985).

TABLE 1 Chronopharmacology: drugs whose effects in humans vary according to the time of day of treatment

Type of drug	Examples
Cardiovascularly active drugs	
β-Blockers	Acebutolol, atenolol, bevantolol, bopindolol, labetolol, mepindolol, bisoprolol, nebivolol, metoprolol, nadolol, oxprenolol, propranolol, sotalol, carvedilol, timolol (IOP)
β-Agonists	Xamoterol, terbutaline (IOP), adrenaline (IOP)
Calcium channel blockers	Amlodipine, nifedipine, nisoldipine, diltiazem, nitrendipine, verapamil, lacidipine, isradipine
Angiotensin-converting enzyme inhibitors	Captopril, quinapril, perindopril, enalapril, lisinopril, spirapril, benazepril
Diuretics	Hydrochlorothiazide, indapamide, xipamide, piretanide, torasemide
Organic nitrates	Glyceryl trinitrate, isosorbide dinitrate, isosorbide 5-mononitrate
Others	Clonidine, indoramine, potassium chloride, sodium nitroprusside, prazosin, phentolamine
Anticancer drugs	Cisplatin, tetrahydropyranyl adriamycin, folinic acid, doxorubicin, floxuridine, oxaliplatin, methotrexate, combinations
Anti-asthmatic drugs	Theophylline, aminophylline, orciprenaline, terbutaline, metacholine, methylprednisolone, dexamethasone, terbutaline + budesonide, adrenaline, isoprenaline
Psychotropic drugs	Diazepam, haloperidol, phenylpropanolamine
H_1 antihistamines	Clemastine, cyproheptadine, mequitazine, terfenadine
Ophthalmological drugs (IOP)	Adrenaline, isoprenaline, terbutaline, timolol
Non-steroidal anti-inflammatory drugs, general and local anaesthetics, and opioids	Acetylsalicylic acid, flurbiprofen, ketoprofen, indomethacin, pranoprofene, paracetamol, tenoxicam, piroxicam, metamizol, carticaine, lidocaine, mepivacaine, morphine, halothane
Endocrinological/ gastrointestinal agents	Prednisone, adrenocorticotropic hormone, methylprednisolone, insulin, tolbutamide, glucose, bezafibrate, clofibrate, omeprazole, lansoprazole, H_2 blockers (cimetidine, nizatidine, roxatidine, famotidine, ranitidine)
Miscellaneous	Tuberculine, ethanol, bright light, placebo

IOP, drugs used to reduce intraocular pressure.

TABLE 2 Chronopharmacokinetics: drugs whose kinetics in humans vary according to the time of day of treatment

Type of drug	Examples
Cardiovascularly active drugs	
β-Blockers	Propranolol
Calcium channel blockers	Diltiazem, nifedipine, verapamil
Angiotensin-converting enzyme inhibitors	Enalapril
Organic nitrates	Isosorbide dinitrate, isosorbide 5-mononitrate
Others	Digoxin, methyldigoxin, potassium chloride, dipyridamol
Anticancer drugs	Cisplatin, doxorubicin, 5-fluorouracil, cyclosporine, methotrexate, busulphan
Anti-asthmatic drugs	Aminophylline, theophylline, terbutaline, prednisolone
Psychotropic drugs	
Benzodiazepines	Diazepam, midazolam, lorazepam, temazepam
Others	Melatonin, hexobarbitone, amitriptyline, nortriptyline, lithium, haloperidol, carbamazepine, diphenylhydantoin, valproic acid
Non-steroidal anti-inflammatory drugs, local anaesthetics	Acetylsalicyclic acid, indomethacin, ketoprofen, phenacetin, paracetamol, lidocaine, bubivacaine
Gastrointestinal agents	Cimetidine, omeprazole
Antibiotics	Ampicillin, gentamicin, griseofulvin, sulphasymazine, sulphisomodine, amikacin
Miscellaneous	Ethanol, mequitazine, 5-methoxypsoralene

Chronopharmacology of nocturnal asthma

Because asthma attacks occur predominantly during the night in asthmatic patients, it is not surprising that anti-asthmatic drugs have also been studied in relation to time of day. Theophylline was one of the first drugs for which variations in pharmacokinetics according to the time of day were reported and more than 49 studies have now been published (Smolensky 1989) covering different theophylline preparations in different galenic formulations. These data in general demonstrated that C_{max} was lower and t_{max} was longer after evening than after morning application of theophylline, as demonstrated in one example in Fig. 2. More recent results have suggested that higher doses of theophylline

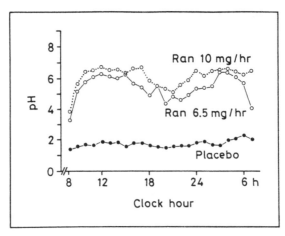

FIG. 1. Effect of constant infusions of the H_2 histamine receptor blocker ranitidine (Ran) (6.25 [——○——] or 10 [- - -○- - -] mg/h) or placebo (——●——) on gastric pH in 15 fasted patients with duodenal ulcers. (Redrawn from Sanders et al 1988.)

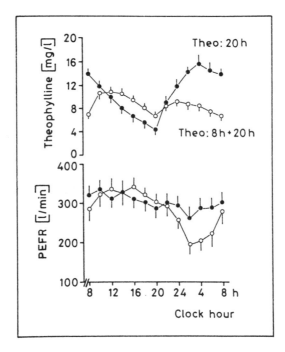

FIG. 2. Serum theophylline (Theo) concentrations and effects on peak expiratory flow rate (PEFR) in two groups of nine asthmatic patients after two doses of theophylline, one at 08.00 h and the other at 20.00 h (○), or after a single dose given at 20.00 h (●). Error bars show SEM. (Redrawn from Neuenkirchen et al 1985.)

should be given during the night than during the day, or even that a single evening dose might be used, in order to overcome the nocturnal decrease in pulmonary function adequately (Neuenkirchen et al 1985). In contrast to the general belief about drug concentration-versus-time profiles—the flatter the better—it seems, therefore, that allowing greater fluctuations in drug concentrations over the 24 h of a day will lead to better treatment of asthmatic patients.

β_2-sympathomimetic drugs are the remedy of first choice in asthma. The pharmacokinetics of the β_2-sympathomimetic terbutaline and its effects on peak expiratory flow are circadian phase dependent (Postma et al 1985, Jonkman et al 1988). After subchronic oral treatment with terbutaline twice daily (7.5 mg at 07.30 and 19.30 h) C_{max} was higher after morning than evening administration, the two with t_{max} values being 3.5 and 6.2 h, respectively (Jonkman et al 1988); this picture resembles the variations observed with theophylline. Moreover, unequal dosing, giving two-thirds of the daily oral dose of terbutaline in the evening, improves control of the nocturnal fall in peak flow in asthmatics (Postma et al 1985). These studies are not only evidence for circadian phase-dependent variations in the pharmacokinetics and effects of oral terbutaline, but also provide further evidence for the circadian phase dependency of the dose–response relationship. Such clinical data demonstrate that higher doses of antiasthmatic drugs such as theophylline and β-sympathomimetics should be given in the evening than during the daytime when the asthma attacks are predominantly nocturnal, a recommendation recently put forward by the Commission on Drugs of the German Medical Association (Arzneimittelkommission der deutschen Ärzteschaft 1992).

Chronopharmacology of hypertension

The development of ambulatory blood pressure monitoring devices to continuously monitor human blood pressure and heart rate helped chronobiological observations make their mark in cardiovascular medicine. Such measurements over 24 h demonstrated that blood pressure in normotensive and in hypertensive patients is clearly dependent on the time of day (see Fig. 3). Moreover, different circadian patterns may be found with different forms of hypertension: in normal conditions and in primary hypertension there is, in general, a nightly drop in blood pressure (Lemmer 1993b, Fig. 3), whereas in secondary hypertension due to, for example, renal disease, hyperthyroidism, pregnancy or Cushing's disease, the rhythm in blood pressure is lost in about 70% of cases, or even reversed such that the highest values are at night.

In cardiovascular research animal experiments are still needed to provide a detailed picture of regulatory mechanisms from overt rhythms, such as those in blood pressure and heart rate, down to the cellular and subcellular level of signal transduction processes. Telemetric studies in various normotensive and hypertensive strains of rat revealed strain-dependent 24 h patterns in blood

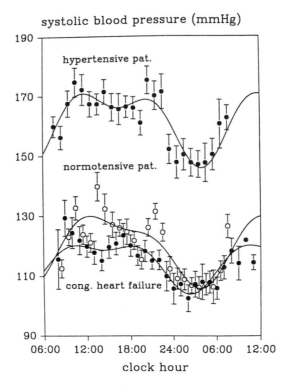

FIG. 3. Circadian rhythm in systolic blood pressure in normotensive (○) and primary hypertensive (●, *top*) patients and in patients suffering from congestive heart failure (●, *bottom*), as determined by ambulatory blood pressure recording. Mean hourly data ± SEM from 12–17 subjects are shown. (Reproduced from Lemmer 1993a, with permission.)

pressure and heart rate (Lemmer et al 1993a). Spontaneously hypertensive rats (SHR) show circadian blood pressure and heart rate rhythms with peak values in the rats' activity span (Fig. 4) similar to those found in humans with essential/primary hypertension (see Fig. 3). Thus, we were able to confirm the suitability of the SHR strain as a model for human primary hypertension. We were the first to describe the transgenic TGR(mRen-2)27 hypertensive rat, in which blood pressure peaks during the resting phase while the heart rate rhythm remains undisturbed (Fig. 4); we proposed this rat strain as a model with which to study regulatory mechanisms in human secondary hypertension (Lemmer et al 1993a). These animal models for primary (SHR) and secondary (TGR) hypertension may be helpful in the further evaluation of modes of action, efficacy and circadian phase dependency of antihypertensive drugs (Lemmer et al 1993b, 1994) for different subtypes of the disease in humans. These chronobiological findings in hypertension led to the recommendation (Arzneimittelkommission der deutschen Ärzteschaft 1992) that in primary

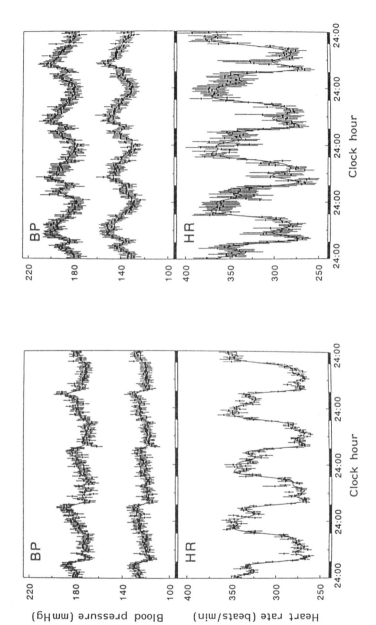

FIG. 4. 24 hour rhythms in systolic and diastolic blood pressure (BP) and in heart rate (HR) in (*left*) spontaneously hypertensive (SHR) (*right*) TGR(mRen-2)27 transgenic hypertensive (TGR) rats as determined by telemetry. Mean values ± SEM of measurements made at 30-minute intervals in 11 spontaneously hypertensive rats and nine transgenic hypertensive rats are shown. (Reproduced from Lemmer et al 1993a with permission. Copyright 1993 American Heart Association.)

hypertension antihypertensive drugs should be taken at early morning hours, whereas in secondary hypertension an additional evening dose may be necessary. Thus, the time of day has become an important factor in the treatment of hypertension.

The chronopharmacokinetics of cardiovascularly active drugs have been reviewed recently (Lemmer 1989, 1991b, Lemmer & Bruguerolle 1994). Our studies in normotensive subjects and hypertensive patients have shown that various compounds which affect the cardiovascular system, including propranolol, oral nitrates, the calcium channel blocker nifedipine and the angiotensin-converting enzyme inhibitor enalapril, showed higher C_{max} values and/or shorter t_{max} values after morning (oral) administration than after evening administration, at least when immediate-release formulations were used (Table 3). Sustained-release formulations of IS-5-MN and nifedipine, however, showed no pharmacokinetic circadian phase dependency (Table 3) (for details see Langner & Lemmer 1988, Lemmer et al 1991a,b,c, Lemmer 1989, 1991b, Witte et al 1993). A faster gastric emptying time in the morning (Goo et al 1987) and, more importantly, a higher gastro-intestinal perfusion in the morning than in the evening, are assumed to be involved in the underlying mechanism responsible for the chronokinetic behaviour of these lipophilic compounds (Lemmer & Nold 1991, Fig. 5).

From our clinical studies, it appears that the circadian phase dependency of components of the cardiovascular system—due to variations in regulatory mechanisms (see Witte & Lemmer 1991)—seems to be of greater importance for drug efficacy than daily variations in pharmacokinetics.

TABLE 3 Chronopharmacokinetic parameters of drugs affecting the human cardiovascular system

Drug	C_{max}^a (ng/ml) Time of treatment		t_{max}^b (h) Time of treatment		Reference
	Morning	Evening	Morning	Evening	
Propranolol	38.6	26.2*	2.5	3.0	Langner & Lemmer 1988
IS-5-MN i.r.	1605.0	1588.0	0.9	2.1*	Lemmer et al 1991b
IS-5-MN s.r.	509.0	530.0	5.2	4.9	Lemmer et al 1991a
Nifedipine i.r.	82.0	45.7*	0.4	0.6*	Lemmer et al 1991c
Nifedipine s.r.	48.5	50.1	2.3	2.8	Lemmer et al 1991c
Enalaprilat (Enalapril 21 days)	46.7	53.5	3.5	5.6*	Witte et al 1993

$^a C_{max}$, peak drug serum concentration.
$^b t_{max}$, time to peak concentration.
*Evening treatment significantly different from morning treatment, $P<0.05$.
i.r., immediate release; s.r., sustained release.

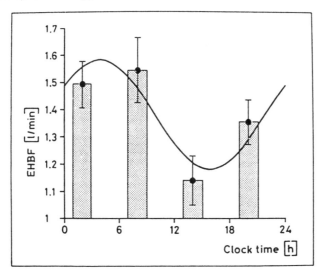

FIG. 5. Circadian rhythm in estimated hepatic blood flow (EHBF) in 10 healthy supine subjects ($\pm$ SEM) as determined by the indocyanine green method at different times of the day. (Redrawn from Lemmer & Nold 1991.)

Conclusions

Various parameters controlling functions of the gastro-intestinal tract, the lungs and the cardiovascular system, as well as the onset and the symptoms of diseases of these systems, are clearly circadian phase dependent in humans. Experiments in rats in which blood pressure, heart rate and motility were monitored by telemetry gave the first evidence that these rhythms persist under free-running conditions (Lemmer & Witte 1993). To what a degree these functions are regulated by internal clocks has yet to be fully established. However, there is sound evidence that the rhythmic patterns within these systems can be disturbed by particular diseases and are modified by drugs. Thus the time of day has become an important factor in the diagnosis and drug treatment of various diseases. Both the pharmacokinetics and the effects of drugs can be dependent on the circadian phase, but the relationship might not be causal. Moreover, clear evidence that drug dose–response relationships can vary markedly with circadian time has led to circadian time-specified recommendations for drug dosing.

References

Arendt J, Minors DS, Waterhouse JM (eds) 1989 Biological rhythms in clinical practice. Wright, London

Arzneimittelkommission der deutschen Ärzteschaft 1992 Arzneiverordnungen, 17th edn. Deutscher Ärzte-Verlag, Köln

Bélanger P 1993 Chronopharmacology in drug research and therapy. Adv Drug Res 24:1-80

Floyer J 1698 A treatise of asthma. R. Wilkins & W. Innis, London

Goo RH, Moore JG, Greenberg E, Alazraki NP 1987 Circadian variation in gastric emptying of meals in humans. Gastroenterology 93:515-518

Halberg F 1960 The 24-hour scale: a time dimension of adaptive functional organization. Perspect Biol Med 3:491-527

Jonkman JHG, Borgström L, van der Boon WJV, de Noord OE 1988 Theophylline–terbutaline, a steady state study on possible pharmacokinetic interactions with special reference to chronopharmacokinetic aspects. Br J Clin Pharmacol 26:285-293

Langner B, Lemmer B 1988 Circadian changes in the pharmacokinetics and cardiovascular effects of oral propranolol in healthy subjects. Eur J Clin Pharmacol 33:619-624

Lemmer B (ed) 1989 Chronopharmacology—cellular and biochemical interactions. Marcel Dekker, New York

Lemmer B 1991a Implications of chronopharmacokinetics for drug delivery: antiasthmatics—H_2-blockers—cardiovascular active drugs. Adv Drug Delivery Rev 6:83-100

Lemmer B 1991b The cardiovascular system and daily variation in response to antihypertensive and antianginal drugs: recent advances. Pharmacol & Ther 51:269-274

Lemmer B 1993a Why are so many biological systems periodic? In: Gurny R, Junginger HE, Peppas NA (eds) Pulsatile drug delivery—current applications and future trends. Wissenschäftliche Verlagsgesellschaft, Stuttgart, p 11-24

Lemmer B 1993b Chronopharmacology: time—a key in drug treatment. Ann Biol Clin 51:179-181

Lemmer B, Nold G 1991 Circadian changes in estimated hepatic blood in healthy subjects. Br J Clin Pharmacol 32:627-629

Lemmer B, Bruguerolle B 1994 Chronopharmacokinetics—are they clinically relevant? Clin Pharmacokinet 26:419-427

Lemmer B, Nold G, Behne S, Kaiser R 1991a Chronopharmacokinetics and cardiovascular effects of nifedipine. Chronobiol Int 8:485-494

Lemmer B, Scheidel B, Behne S 1991b Chronopharmacokinetics and chronopharmacodynamics of cardiovascular active drugs: propranolol, organic nitrates, nifedipine. Ann NY Acad Sci 618:166-181

Lemmer B, Scheidel B, Blume H, Becker H-J 1991c Clinical chronopharmacology of oral sustained-release isosorbide-5-mononitrate in healthy subjects. Eur J Clin Pharmacol 40:71-75

Lemmer B, Mattes A, Böhm M, Ganten D 1993a Circadian blood pressure variation in transgenic hypertensive rats. Hypertension 22:97-101

Lemmer B, Boese S, Mattes A 1993b Wirkung von Doxazosin auf circadiane Rhythmen im Blutdruck normotensiver und hypertensiver Ratten. Z Kardiol 82(suppl 1):139 (abstr)

Lemmer B, Witte K, Makabe T, Ganten D, Mattes A 1994 Effects of enalaprilat on circadian profiles in blood pressure and heart rate of spontaneously and transgenic hypertensive rats. J Cardiovasc Pharmacol 23:311-314

Maleev A, Terziivanov D, Kostova N, Skrinska E, Belovezhdov N, Budevsky O 1990 Chronopharmacokinetics of the preparation Biomet[400] (cimetidine). In: Lemmer B, Hüller H (eds) Clinical pharmacology, vol 6: Clinical chronopharmacology. Zuckschwerdt Verlag, Munich, p 155-159

Minors DS, Waterhouse JM 1981 Circadian rhythms and the human. Wright, London

Moore JG, Englert E 1970 Circadian rhythm of gastric acid secretion in man. Nature 226:1261–1262

Neuenkirchen H, Wilkens JH, Oellerich M, Sybrecht GW 1985 Nocturnal asthma: effect of a once per evening dose of sustained release theophylline. Eur J Respir Dis 66:196–204

Postma DS, Koëter GH, van de Mark TW, Reig RP, Sluiter HJ 1985 The effects of oral slow-release terbutaline on the circadian variation in spirometry and arterial blood gas levels in patients with chronic airflow obstruction. Chest 87:653–657

Prichard PJ, Yeomans ND, Mihaly GW et al 1985 Omeprazole: a study of its inhibition of gastric pH and oral pharmacokinetics after morning or evening dosage. Gastroenterology 88:64–69

Reinberg A, Halberg F 1971 Circadian chronopharmacology. Annu Rev Pharmacol 11:455–492

Reinberg A, Smolensky MH (eds) 1983 Biological rhythms and medicine. Springer-Verlag, New York

Sanders SW, Moore JG, Büchi KN, Bishop AL 1988 Circadian variation in the pharmacodynamic effect of intravenous ranitidine. Annu Rev Chronopharmacol 5:335–338

Sanders SW, Moore JG, Day GM, Tolman KG 1992 Circadian differences in pharmacological blockade of meal-stimulated gastric acid secretion. Aliment Pharmacol & Ther 6:187–193

Smolensky MH 1989 Chronopharmacology of theophylline and beta-sympathomimetics. In: Lemmer B (ed) Chronopharmacology—cellular and biochemical interactions. Marcel Dekker, New York, p 65–113

Touitou Y, Haus E (eds) 1992 Biological rhythms in clinical and laboratory medicine. Springer-Verlag, Berlin

Witte K, Lemmer B 1991 Rhythms in second messenger mechanisms. Pharmacol & Ther 51:231–237

Witte K, Lemmer B 1993 Free-running rhythms in blood pressure and heart rate in normotensive Sprague–Dawley rats. J Interdiscipl Cycle Res 24:328–334

Witte K, Weisser K, Neubeck M et al 1993 Cardiovascular effects, pharmacokinetics, and converting enzyme inhibition of enalapril after morning versus evening administration. Clin Pharmacol & Ther 54:177–186

DISCUSSION

Turek: Many of the effects of drug treatment at different times of day may be related to gastric emptying. How did you control for the meals that people had? Did people have small meals over a 24 h period to make sure that the effect is not a result of them having an empty digestive system after having slept all night?

Lemmer: There are two possible approaches with such studies in humans. One is to apply drugs in the normal setting of meal times, in real life. The other, which we chose, is to use an unnatural situation. The subjects were not allowed to take any meals 8–10 h before drug dosing, whether the drug was taken in the morning or in the evening. The timing of the meal can influence gastric emptying, drug absorption and so on. Thus, we have to take this into account.

Armstrong: You have gone to a lot of trouble to ensure you have good control over the situations in which you administer the drugs you are testing, controlling for meals, gut emptying, time of day etc. Do you then check back to see whether you still see the differences in the natural situation, where all these things are confounded? The differences that you find under clean experimental conditions may be totally swamped if someone has a big breakfast, a small lunch, a few snacks and naps and alcoholic drinks.

Lemmer: This depends on the type of compound. Some compounds are not influenced by meals whereas others are influenced greatly. When setting up a drug study design we know in advance whether we should have a rigid experimental design or not. Nakano et al (1984) have shown, by comparing people who have had a heavy meal with those who have not, that taking diazepam together with a meal can reduce the day–night differences in peak plasma concentrations but does not abolish the rhythm in the pharmacokinetics of diazepam.

Armstrong: The attitude of many people in the pharmaceutical industry is that any rhythms will be completely swamped under normal conditions so that pills might as well be taken three times a day.

Lemmer: That attitude is clearly wrong. For several compounds we know that the dose–response relationship after morning administration is completely different from that after evening application.

Turek: Why has the clinical community paid so little attention over the last 20–30 years to this kind of work? There are clearly timing variations in the effects of cancer drugs, really important drugs given in life-threatening situations, yet it is not normal practice in US hospitals to be concerned about this.

Lemmer: I always have to argue with clinicians who tell me they have been following the same drug treatment regimes for the last 20 years without problems and see no reason to change. We have to provide the clinicians with good data from cross-over studies that demonstrate clearly the benefits of giving a drug at a particular time. Things are improving, though. As I said, The Commission on Drugs of the German Medical Association has recently recommended that drugs should be given in the morning for primary hypertension with a second dose in the evening for secondary hypertension such as in renal disease, endocrine disorders or gestational hypertension. Time of the day is gradually becoming a consideration for clinicians.

We are now making progress in introducing 'time of day' into clinical medicine because of the availability and use of new devices allowing constant monitoring, e.g., of blood pressure and electrical activity of the heart. Now that we can monitor these things constantly over 24 h no one can deny that there are physiological rhythms in humans which can be disturbed in disease.

Redfern: The pharmaceutical industry is also increasingly using this kind of apparatus to monitor blood pressure in experimental animals. However, routine recording of blood pressure over 24 h does present a second problem, how to handle the mass of data that is produced.

Waterhouse: Is there epidemiological evidence that it's advantageous to prescribe drugs in the way you recommend?

Lemmer: No, not to my knowledge. We always have to rule out either time of day or real life in such clinical studies. I prefer to work in an artificial system in order to generate more data purely on the 'time-of-day effect'; such data are really lacking.

Armstrong: I agree with you, but the other approach will be needed to convince sceptical clinicians.

Mrosovsky: Are activity patterns altered in the transgenic rats with extra renin? I wonder whether there are any analogies to the Zugunruhe in birds.

Lemmer: We were asked this of course by the referees of *Hypertension*, in which our paper was published (Lemmer et al 1993). With the telemetric system we additionally monitored motor activity. The transgenic TGR(mRen-2)27 rats are still nocturnal, which confirms an important point, that the level of activity does *not* determine the blood pressure. Activity is of course linked to heart rate, but not to blood pressure.

Redfern: I understand that you have found that after chronic treatment with enalapril there is effectively a shift in the endogenous 24 h blood pressure pattern. We have shown a similar effect of antidepressants on 5-HT turnover. Such effects obviously make it much more difficult to interpret the experimental results. At the same time this is in my view a more realistic model of what to expect clinically, where of course chronic drug administration is the rule rather than the exception. It is easy to design neat experiments involving administration of a single dose of drug at different time points. The results obtained don't necessarily give any indication of what to expect clinically.

Lemmer: This is one of the problems we have to face. It's very difficult to correlate acute as well as chronic effects with kinetics. In our studies with various cardiovascularly active drugs there was evidence that mainly during night there was a bad or no correlation between effects and the drugs' pharmacokinetics.

Waterhouse: In a meeting such as this there is always the possibility of a chiasm between possible benefits to humankind on the one hand and the mechanisms that act at the cellular and molecular levels on the other. You drew attention to the clear dissociation between the kinetics and the response which was observed. What's happening at the cellular level to cause that?

Lemmer: We are now using different types of drugs—α-blockers, β-blockers, Ca^{2+} channel blockers—in animal experiments at different times of day in order to find out the relative importance of these receptor systems for the regulation of the blood pressure rhythm. Also, we have begun to investigate the signal transduction processes in cardiac tissue and in the vascular bed.

This cannot be done in humans other than in a restricted way, but we can do these studies in rats.

The regulatory mechanism controlling blood pressure in humans and in rats is much more complex than had been assumed. β-Receptor-mediated processes may dominate during the daytime and α-receptor-mediated processes at night or during the early morning hours. We can study signal transduction processes much better in animals using ligand binding techniques to determine receptor numbers, studying the basal and drug-induced (*in vitro*) activities of, for example, adenylate cyclase and guanylate cyclase in different tissues and at different circadian times. We have already observed changes in the concentration of angiotensin-converting enzyme in target tissue such as the heart, and in the activity of the cyclases mentioned above. Telemetry combined with biochemical studies should provide more insights into the regulatory mechanisms, which are not yet understood either in the rat or in humans.

Dunlap: Are you saying that the pharmacological treatment hasn't changed the oscillator, but has somehow changed the set points for the oscillation, so that the maximum and minimum will be different?

Lemmer: To my knowledge, no one has demonstrated the degree to which the blood pressure rhythm is endogenous. There have been no studies in humans under free-running conditions. We have addressed this point by telemetry in rats kept completely undisturbed for 64 days in constant darkness, and we could observe a free-running rhythm of systolic/diastolic blood pressure and heart rate, with a period of slightly longer than 24 h (Witte & Lemmer 1993). When we put the rats in constant light, rhythmicity broke down. When they were returned to a 12:12 h light–dark cycle the rhythms reappeared. When the light–dark cycle was shifted by six hours (advancing the dark span), it took 8–11 days for the animals' rhythms in systolic and diastolic blood pressure, heart rate and locomotor activity to become resynchronized to the shifted cycle. I do not know what the masking effect of motor activity on blood pressure is, but from our work with the transgenic rats we can assume that motor activity is at least not mainly responsible for the blood pressure rhythmic pattern, though it may be of importance for heart rate rhythm. This is the first study I know of demonstrating a prominent endogenous component to the rhythm of blood pressure.

Cassone: We have measured free-running rhythms in heart rate in rats, which are abolished by lesions of the SCN (Warren et al 1994). We can block the rhythm of heart rate with sympathetic blockade without affecting the rhythms of body temperature or activity. You can dissociate the two.

Lemmer: There is no reliable experimental data telling us to what degree activity contributes to the heart rate rhythm. Even in our transgenic animals heart rate coincides with locomotor activity, but not with blood pressure. Thus, motor activity could contribute to the heart rate rhythm.

Takahashi: One of the parameters that we gloss over and ignore in animal experiments, in which we might inject a drug such as triazolam and plot a phase response curve, is the pharmacokinetics of that agent. Have the pharmacokinetic characteristics of the agents that you listed in Table 2 been tested in animals? If there have been good studies, we could ask what the real phase dependence of agents on the clock system is.

Lemmer: There are quite a lot of data on anticancer drugs, and we've studied β-blockers and antidepressants in rats (Lemmer et al 1985, Lemmer & Holle 1991). It is difficult to predict whether a drug will show pharmacokinetic variations. The advantage of animal experiments is that we can determine the concentration of a drug in target tissues. In humans you can determine only plasma concentration. When phase response curves for different drugs have been set up the differences in pharmacokinetics between different drugs have not adequately been taken into account. For example, time to peak drug concentration (t_{max}), which can vary enormously between compounds, must have an impact on the phase response curve. Lipophilic compounds normally have a shorter t_{max} than others; this could confound the data. In addition, a drug's half-life, the duration of its effect, can vary enormously between different compounds, and may also confound the data in phase response curves. Triazolam, for example, is a short-acting benzodiazepine, with a short t_{max} and a short half-life, whereas other benzodiazepines such as diazepam have a very long half-life, and can have significant after effects. Thus, if the interval between applying a drug and measuring its effects is the same for drugs with different pharmacokinetics, the resulting phase response curves may be misleading.

Takahashi: You mentioned the importance of formulation. Are there formulations you might recommend us to use in our animal experiments so as to minimize the pharmacokinetic effects?

Lemmer: If you want to minimize differences you should use intravenous injection, which produces peak concentrations instantaneously. The oral route is not used often in animal experiments and the variability in the extent and time of absorption may be great. After intraperitoneal injections there is only a short delay, and a slightly greater one after subcutaneous injection, before peak concentration is reached. In rats, at least, intravenous or intraperitoneal injections would be the first choice.

Mrosovsky: Do you think the pharmacokinetics of triazolam are really a problem when we are using doses at the high end of the dose response curve? The drug should become effective rapidly.

Takahashi: The pharmacokinetics would certainly affect the drug's duration, its clearance rate.

Mrosovsky: Would an hour or so make a huge difference with such a short-acting agent?

Lemmer: It could make a huge difference. Triazolam affects the benzodiazepine receptors immediately in humans and there are pronounced after effects; it binds to and dissociates from the receptor rapidly and leaves the body. This pattern may be one reason for triazolam's prominent after effects. The original assumption that a compound with a short half-life would be an ideal sleeping pill has not held.

Roenneberg: You have discussed the chronopharmacology of drugs. Isn't there another side to chronopharmacology, that of the receptor? If one doesn't see a rhythm it might be because there are two rhythms—drug uptake and receptor abundance or affinity—out of phase.

Lemmer: That's true. The differing dose–response relationships at different times of day are partly due to the drug's kinetics and partly to the changing sensitivity of the target organ. We have demonstrated this in cardiac tissue and brain tissue, starting with the signal transduction processes via β-adrenergic receptors in which a pronounced rhythm in cylic AMP, adenylate cyclase and phosphodiesterase could be demonstrated (Lemmer et al 1987a,b), and we are now looking at other systems such as nitric oxide-mediated processes. The dissociation between kinetics and effects can be explained mainly in terms of daily variations at the level of the receptor or the enzymes involved in the signal transduction process, at least within the cardiovascular system.

Miller: Earlier on we were discussing differences between drugs' effects *in vivo* and *in vitro* in the suprachiasmatic nucleus slice. To what extent might those differences be due to the pharmacokinetic and pharmacodynamic variables that you have talked about, particularly with respect to indoleamines such as melatonin? Is anything known about the chronopharmacology of indoleamines?

Arendt: I don't know of any reports showing changes in the pharmacokinetics of orally administered melatonin with time of day in sheep or human plasma. In sheep there were no differences between morning (09.30) and afternoon (15.30) administration. Some elegant work by B. Claustrat (presented at the Philippe Laudat Conference on the Neurobiology of Circadian and Seasonal Rhythms: Animal and Clinical Studies, 1991) giving infusions to humans doesn't appear to show this either.

Lemmer: One additional problem in pharmacokinetic/pharmacodynamic studies is linked to the lipophilicity of drugs. What you measure in plasma is generally only a small amount of the total amount of the drug applied. Lipophilic drugs accumulate in brain, muscle, lung, and heart tissue. Thus, drug concentration at the target tissue can be much higher than would be assumed from plasma values. The advantage of animal experiments clearly is that you can measure drug concentrations in target tissues.

References

Lemmer B, Winkler H, Ohm T, Fink M 1985 Chronopharmacokinetics of β-receptor blocking drugs of different lipophilicity (propranolol, metoprolol, sotalol, atenolol) in plasma and tissues after single and multiple dosing in the rat. Naunyn Schmiedebergs Arch Pharmacol 330:42–49

Lemmer B, Holle L 1991 Chronopharmacokinetics of imipramine and desipramine in rat forebrain and plasma after single and chronic treatment with imipramine. Chronobiol Int 8:176–185

Lemmer B, Bärmeier H, Schmidt S, Lang P-H 1987a On the daily variation in the beta-receptor–adenylate cyclase–cAMP–phosphodiesterase system in rat forebrain. Chronobiol Int 4:469–475

Lemmer B, Lang P-H, Schmidt S, Bärmeier H 1987b Evidence for circadian rhythmicity of the β-adrenoceptor–adenylate cyclase–cAMP–phosphodiesterase system in the rat. J Cardiovasc Pharmacol 10(suppl 4):S138–S140

Lemmer B, Mattes A, Böhm M, Ganten D 1993 Circadian blood pressure variation in transgenic hypertensive rats. Hypertension 22:97–101

Nakano S, Watanabe H, Nagai K, Ogawa N 1984 Circadian stage-dependent changes in diazepam kinetics. Clin Pharmacol & Ther 36:271–277

Warren WS, Champney TH, Cassone VM 1994 The suprachiasmatic nucleus controls the circadian rhythm of heart rate via the sympathetic nervous system. Physiol Behav 55:1091–1099

Witte K, Weisser K, Neubeck M et al 1993 Cardiovascular effects, pharmacokinetics, and converting enzyme inhibition of enalapril after morning versus evening administration. Clin Pharmacol & Ther 54:177–186

The effect of light on the human circadian pacemaker

Charles A. Czeisler

Section on Sleep Disorders/Circadian Medicine, Division of Endocrinology, Department of Medicine, Harvard Medical School, Brigham and Women's Hospital, 221 Longwood Avenue, Boston, Massachusetts 02115, USA

Abstract. The periodic light–dark cycle provides the primary signal by which the human circadian pacemaker is synchronized to the 24 h day. Earlier reports that social contacts were more effective than light in the entrainment of human circadian rhythms have not been supported by more recent studies. In fact, we have found that exposure to a cyclic light stimulus can induce strong (type 0) resetting of the human circadian pacemaker, indicating that exposure to light affects the pacemaker's amplitude of oscillation as well as its phase. These findings support Winfree's long-standing prediction, based on his pioneering recognition of the importance of amplitude in the analysis of circadian clocks, that strong (type 0) resetting would prove to be a common property of circadian resetting responses to light across a wide array of species, from algae to humans. Research on humans has shown, for the first time, that the response of the circadian pacemaker to light depends not only on the timing, intensity and duration of light exposure, but also on the number of consecutive daily light exposures. Exposure to light of a critical strength at a critical phase can even drive the human circadian pacemaker to its region of singularity, akin to temporarily 'stopping' the human circadian clock. These findings have important implications for the treatment of circadian rhythm sleep disorders, because properly timed exposure to light can reset the human clock to any desired hour within one to three days.

1995 Circadian clocks and their adjustment. Wiley, Chichester (Ciba Foundation Symposium 183) p 254–302

In nearly all eukaryotic organisms studied, periodic environmental stimuli normally synchronize (or entrain) the non-24 h intrinsic period of endogenous circadian rhythms to a period of exactly 24 h. In mammals, several environmental and behavioural stimuli have been shown to act as circadian synchronizers. These include the timing of food availability and activity itself (Moore-Ede et al 1982, Mrosovsky & Salmon 1987, Van Reeth et al 1988). However, it is generally accepted that the daily alternation of light and darkness is the most important periodic environmental stimulus for entraining circadian rhythms (Rusak 1979). In the late 1950s, it was first demonstrated that single light exposures could reset circadian phase (Pittendrigh & Bruce 1957) and that the magnitude and direction of these phase shifts were dependent on the intensity and the duration of the light exposure (described by a dose response curve [DRC]) as well as on the circadian phase of the light exposure (described by a phase response curve [PRC]) (Hastings & Sweeney 1958). Circadian DRCs, which have also been reported in nocturnal mammals, indicate that the resetting response to light is enhanced by increasing either the intensity or the duration of the light stimulus (Hastings & Sweeney 1958, Takahashi et al 1984). Studies in nocturnal mammals indicate that the resetting response also varies with the wavelength of the light, with the peak response at about 500 nm (Takahashi et al 1984).

Circadian PRCs produced in response to light have been described in nearly every eukaryotic organism studied (Johnson 1990). The resetting response to light described by the PRC has been found to share the following properties across a wide array of species, from unicellular algae to primates (Johnson 1990): (1) light stimuli early in the subjective night* induce phase delay shifts; (2) light stimuli late in the subjective night induce phase advance shifts; (3) light stimuli during the subjective day induce minimal phase shifts.

In fact, the principal difference between light-induced PRCs in different species seems to be the magnitude or strength of the resetting response to a given stimulus. Phase response curves of two general types have been reported: a low amplitude PRC (with maximal phase shifts of only a few hours) and a high amplitude PRC (with phase shifts as great as ± 12 h). Winfree has designated these two classes of PRC as weak type 1 and strong type 0 phase resetting, respectively (Winfree 1980), on the basis of the topology of their resetting contours. It was initially thought that lower organisms, such as unicellular algae, plants and insects, were uniquely capable of strong (Winfree's type 0) phase resetting, whereas higher organisms, such as mammals, were reportedly limited to weak (Winfree's type 1) phase resetting (Moore-Ede et al 1982).

Consistent with the notion of such decreasing circadian light sensitivity across species were the reports in 1970 and 1971 that in humans—by self-definition the highest organism of all—social interaction, rather than environmental light, was the principal synchronizer of the circadian pacemaker to the 24 h day (Wever 1970, Aschoff et al 1971, 1975, Aschoff & Wever 1976, 1981). Wever and Aschoff, in a reversal of their earlier position that exposure to the light–dark cycle was sufficient to entrain human circadian rhythms (Aschoff et al 1969), reported that the gong signal system they had used to schedule the collection of subjects' urine samples had been viewed by the experimental subjects as a call or a command from the experimenters, and

---

*In the absence of environmental time cues, the subjective night is defined as the circadian phase at which an organism behaves as it would during the sidereal night. Noctural animals are most active during their subjective night and sleep during their subjective day, whereas the reverse is true for diurnal animals.

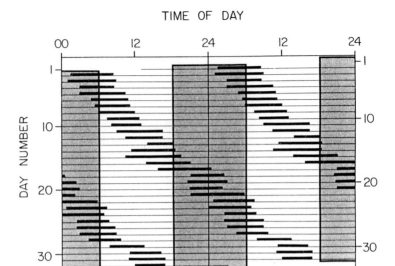

FIG. 1. Rest-activity cycle of a human subject living in an underground bunker in Erling-Andechs, Germany, shielded from periodic influences from the outside world. The data are double-plotted in a raster format, with successive days shown both next to and beneath each other. Solid bars indicate reported sleep episodes and thin horizontal lines reported waking hours. Overhead ceiling lights were switched off each day at the times indicated by stippling, but the subject had free access to kitchen lights, bedside and desk lamps and bathroom lights at self-selected times throughout the study. Reproduced from Czeisler et al 1981, as redrawn from Wever 1970.

that such 'social contact' was primarily responsible for the entrainment by light–dark cycles they had reported earlier (Wever 1970, Aschoff et al 1975). When their gong signal system accidentally failed, they reported that exposure to the light–dark cycle alone—in the absence of the scheduled gong signals—was insufficient to entrain human circadian rhythms (Fig. 1) (Wever 1970, Aschoff et al 1975). In fact, Wever (1970) reported that weak electrical fields had more of an impact on human circadian rhythms than light; he concluded that the light–dark cycle was almost without effect on humans under laboratory conditions. He inferred that the gong signals represented a form of social contact between the experimenters (who lived and worked on a 24 h schedule) and the subjects living in temporal isolation, and that exposure to these social cues was sufficient to entrain human circadian rhythms. In the following year, Aschoff et al (1971) reported that in constant darkness, human circadian rhythms could indeed be entrained by social cues alone, in the absence of the light–dark cycle. These two results led Aschoff and Wever (Aschoff et al 1975, Wever 1979) to the general conclusion that the light–dark cycle was unimportant in the entrainment of human circadian rhythms and that periodic social cues were necessary for our synchronization to the 24 h period of the earth's rotation (Aschoff & Wever 1976, 1981).

It thus appeared that in humans, in contrast to lower animals, circadian rhythm synchronization was primarily dependent on higher cortical functioning. Because of the pioneering role that Aschoff and Wever played in the establishment of the modern field of human circadian rhythm research, their concept that social interaction was the principal synchronizer of human circadian rhythms became the framework on which other results were interpreted. For example, on the basis of that concept, it was presumed that maladaptation to night shift work was due to continued social interaction between night-active shift workers and the day-oriented society (Aschoff et al 1975). Similarly, it was presumed that jet lag could be more rapidly overcome by intensifying social interaction with the indigenous population on arrival in a new time zone (Klein & Wegmann 1974, Aschoff et al 1975).

Unfortunately, in both of the key experiments that provided the basis for the conclusion that social cues were the principal synchronizer of human circadian rhythms, there were unrecognized flaws. Wever's protocol was confounded because experimental subjects, studied in the temporal isolation facility at the Max-Planck-Institut für Verhaltensphysiologie in Erling-Andechs, Germany, had access to and controlled different sources of auxiliary lighting during scheduled episodes of 'darkness'. These included bedside lamps, desk lamps and kitchen and bathroom lights (Aschoff et al 1969). Only the overhead lights were controlled by the experimenters, through an elaborate electrical panel outside each subject's suite. The subjects were free to use the auxiliary lighting at any time, which was more than sufficient to allow them to carry on ordinary daily activities in the absence of the overhead lights. These conditions were not comparable to the conditions of entrainment experiments in plants and lower animals, in which there was no access to auxiliary lighting during scheduled dark episodes. In fact, the conditions were more similar to those in experiments on animals living with a self-selected light–dark cycle. Under such conditions, both birds and mammals will free run (Wahlström 1965, Warden 1978). Thus, no conclusion could be drawn from Wever's 1970 experiment as to whether or not humans were different from other species in their response to an imposed light–dark cycle. The second experiment by Aschoff et al (1971), intended to be a direct test of the synchronizing effect of social cues in humans, was confounded by the effects of scheduled activity, which exerts powerful masking effects on all of the observed rhythms (Czeisler 1978, Czeisler et al 1986, Minors & Waterhouse 1989, Lee et al 1992, Klein et al 1993) and which is now known to exert a synchronizing effect on the circadian pacemaker in mammals (Mrosovsky & Salmon 1987, Van Reeth & Turek 1989, Turek 1989, Edgar & Dement 1991). Furthermore, the four-day experiment was far too short to evaluate circadian entrainment in humans and therefore did not justify the

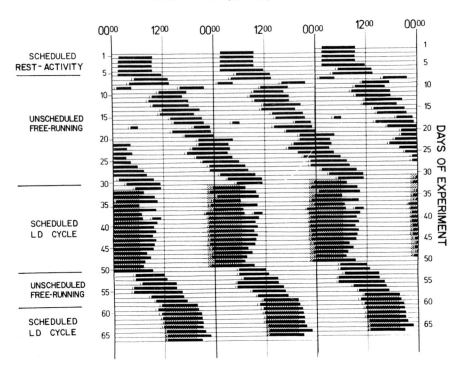

FIG. 2. Rest–activity cycle of a 24-year-old man living in an environment free of time cues atop the Van Cortlandt Pavilion of the Montefiore Hospital and Medical Center in New York. Data are triple-plotted in a raster format; the symbols are as in Fig. 1, except that stippling indicates times when the subject was exposed to complete darkness. Thin vertical hatch marks indicate the beginning of bedtime preparation before unscheduled bedrest episodes. As shown on the left, a strict light–dark (LD) cycle was imposed on two separate occasions. Light and dark episodes were separated by a one-hour dawn–dusk (twilight) transition during the scheduled light–dark cycle. Subjects were not disturbed or visited by staff during chosen or scheduled bedrest episodes unless the subject called for aid (e.g., for nocturnal micturition). Reproduced from Czeisler et al 1981.

authors' conclusion that human circadian rhythms could be entrained by social cues in constant darkness. Thus, the conclusion that humans were different from nearly all other species studied in the mechanism by which their circadian rhythms were synchronized to the 24 h day was not justified on the basis of these two experiments.

We therefore repeated the first of those two experiments to evaluate whether or not human circadian rhythms could be entrained by a light–dark cycle, using a strict light–dark cycle that was comparable to that used in entrainment studies of plants and lower animals. We found that 16 h of exposure to ordinary room light alternating with eight hours' exposure to total darkness could entrain both the rest–activity and body temperature cycles of human subjects to a 24 h period (Fig. 2), as reported in animal studies (Czeisler 1978, Czeisler et al 1981). However, as we indicated at that time, our initial studies of entrainment in human subjects could not distinguish whether synchronization occurred: (a) directly, through an action of light on the endogenous circadian pacemaker, or (b) indirectly, by an influence of lighting conditions on the subject's self-selected scheduling of their behavioural rest–activity/sleep–wake cycle which might, in turn, have affected the endogenous circadian pacemaker (Czeisler et al 1981). That same uncertainty about mechanism applies, of course, to studies of entrainment by light–dark cycles in all animals, especially photophobic nocturnal rodents, whose activity levels are acutely affected by light. We therefore concluded that these results did not support the claim that humans are different from nearly all other species; in fact, when conditions were comparable, we found that a cycle of ordinary indoor room light alternating with darkness was capable of entraining the human circadian system.

Although Aschoff and Wever still argued that any such apparent entrainment effects of an absolute light–dark cycle in humans were behavioural rather than physiological (Aschoff & Wever 1981, Wever et al 1983, Wever 1989), other lines of evidence supported the notion that light is a synchronizer of the human circadian pacemaker. First, studies of the neuroanatomy of the circadian pacemaker in mammals indicated that the 'wiring' needed to mediate circadian phase resetting by light was in place in mammals (reviewed in Moore 1995, this volume). A monosynaptic retinohypothalamic tract (RHT), discovered by Moore in 1972, was found to convey photic information from specialized retinal ganglion cells to the central circadian pacemaker in the suprachiasmatic nucleus (SCN) of the hypothalamus. Furthermore, it was shown that the RHT is able to mediate circadian phase resetting by light even after the induction of behavioural blindness by bilateral transection of the primary optic tracts (Rusak 1979). Finally, neuroanatomical structures analogous to those subserving circadian rhythmicity (the SCN) and photic entrainment (the RHT) in other mammals were documented in humans (Lydic et al 1980, Sadun et al 1984, Stopa et al 1984).

Second, Orth and his colleagues had shown that the circadian rhythm of plasma cortisol concentration—one of the most robust endogenous hormonal oscillations then known—could be phase delayed in healthy subjects by entraining them for two weeks to a schedule in which their nocturnal exposure to darkness was extended from 08.00 until noon (Orth et al 1967, Orth & Island 1969, Osterman 1974). The three- to four-hour phase delay in their cortisol rhythm occurred despite the fact that the subjects continued to awaken at 08.00 throughout their two weeks of delayed 'dawn' (Fig. 3). Although

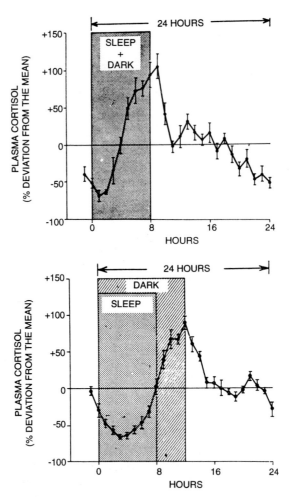

FIG. 3. *Top:* Average (±SEM) plasma cortisol levels from eight studies in six subjects who slept for eight hours in the dark for several days before and during the period of sampling. *Bottom:* Average (±SEM) plasma cortisol levels from nine days of study in three subjects after 14 days of adaptation to a schedule in which sleep was restricted to eight hours, but darkness was imposed for 12 h, i.e., extending for four hours after awakening. In both panels hour 0 represents the onset of the habitual sleep episode. Reproduced from Orth & Island 1969.

Aschoff et al (1975) suggested that this phase delay was due to a change in habits rather than to a direct effect of light, Orth's provocative endocrinological result could not be so readily dismissed as a behavioural effect.

Third, Miles et al (1977) and Orth et al (1979) showed that the sleep tendency and cortisol rhythms, respectively, were not entrained to the 24 h day in two different blind humans living on a 24 h schedule, despite the imposition of a strict daily schedule of activity and full social interaction with individuals synchronized to the 24 h day. The results of these two extensively documented case reports, which have since been confirmed and extended (Sack et al 1992, Klein et al 1993), suggested that entrainment was impaired in the absence of light input to the human circadian pacemaker. Although it could still be argued that relative social isolation of the blind, rather than a loss of light input to the pacemaker, was responsible for their loss of synchronization to the 24 h day, the strength of that argument was weakening.

Fourth, Lewy et al (1980) reported that in humans a different and direct biological action of light—the inhibition of pineal melatonin secretion—occurred only if light intensity was much brighter than ordinary indoor room light (Lewy et al 1980). The subsequent success that they and others had using phototherapy for the treatment of depression (Lewy et al 1982, Kripke et al 1983, Ralph & Menaker 1986, Wirz-Justice et al 1986, Terman 1988) led Lewy and colleagues to hypothesize that light intensity had to be 'sufficiently bright' to exceed a melatonin suppression threshold (Lewy 1983, Lewy & Sack 1986) to affect human circadian rhythms (Lewy 1983, Lewy et al 1984, Lewy et al 1987, Lewy et al 1991), and ultimately led them to suggest that human circadian rhythms are 'unperturbed by the use of ordinary indoor light' (Lewy et al 1984, Lewy et al 1987). Wever et al (1983) then reported that exposure to bright light (~ 4000 lux), presumably together with the gong signals, greatly expanded the range of entrainment of human circadian rhythms, leading Wever (1989) to concur with Lewy's suggestion that only light exceeding the melatonin suppression threshold could entrain human circadian rhythms. However, the study of Wever et al (1983) was confounded by the masking effects of activity on the observed rhythms and by the use of a protocol involving a continuously changing schedule in which the imposed day length increased each day, introducing a systematic bias that could lead to a misestimation of the range of entrainment by more than 50% (Kronauer et al 1983). Therefore, although these studies were supportive of a potential role of light in the synchronization of human circadian rhythms, they defined neither the magnitude nor the phase dependency of the phase-resetting effect of light on the human circadian pacemaker (Lewy et al 1984). Wever and colleagues now claimed that only light above an intensity of 1500 lux—greater than had initially been used at their laboratory in Erling-Andechs—affected human circadian rhythms, and continued to maintain that light below an intensity of 1500 lux did not have any direct physiological effect on human circadian rhythms (Wever et al 1983, Wever 1989).

Strategy for evaluating the light sensitivity of the human circadian pacemaker

To reconcile these differences, we set out to develop a strategy for assessing quantitatively the resetting response of the human circadian pacemaker to light. Evaluation of the resetting capacity of the human circadian pacemaker required a reliable assessment of circadian phase before and after the administration of a putative stimulus. Experimental variables used to mark the output of the circadian pacemaker differ widely across species. In the fruit fly *Drosophila*, eclosion from the pupal casing is used as a reliable marker of circadian phase and phase resetting (Pittendrigh & Bruce 1957), just as leaf movements are used in the *Phaseolus* plant (Moore-Ede et al 1982), the luminescence rhythm in the marine dinoflagellate *Gonyaulax polyedra* (Hastings & Sweeney 1958) and activity onset in small mammals such as the golden hamster *Mesocricetus auratus* (Mrosovsky & Salmon 1987, Takahashi et al 1984, Van Reeth & Turek 1989).

Choice of an appropriate and reliable variable to mark the output of the circadian pacemaker is thus key to the evaluation of the pacemaker's resetting responses. Humans, like nearly all other species studied, continue to behave in a periodic manner when studied in conditions free of environmental time cues (Aschoff & Wever 1976, Czeisler 1978, Wever 1979, Aschoff & Wever 1981). However, in a self-selected light–dark cycle in the absence of external time cues, human subjects exhibit a free-running period of the rest–activity cycle that averages about 25 h (Wever 1979). It is important to realize that this approximately 25 h period actually represents a compromise between two oscillatory processes: the intrinsic period of the endogenous circadian pacemaker* and the behavioural rest–activity (and hence self-selected light–dark) cycles (Czeisler 1978, Wever 1979, Kronauer et al 1982). Most commonly, these two processes remain synchronized with each other, a condition known as *internal* synchronization, even though the subjects are *externally* desynchronized from the 24 h day. During such internally synchronized free runs, there is a four- to six-hour change in the phase relationship between the human activity rhythm and physiological markers of the output of the human circadian pacemaker such as the body temperature, cortisol and presumably the melatonin rhythms (Wever 1973, Czeisler 1978, Czeisler & Jewett 1990). Thus, in humans, the activity rhythm, widely used as a marker of circadian phase in small mammals, does not maintain a fixed phase relationship to the output of the circadian pacemaker under entrained versus free-running conditions, and is therefore not a suitable, reliable marker of circadian phase.

Furthermore, in nearly all long-term (more than two months) free-running studies in humans (and in about a quarter of the short-term studies) (Czeisler 1978), the rest–activity cycle adopts a period very different from the near-24 h

---

*Various authors modelling its interaction with the periodic, behavioural rest–activity cycle in humans have referred to this oscillator as the deep (X) oscillator (Kronauer et al 1982, Kronauer 1987, Kronauer 1990), the strong (Group I) oscillator (Wever 1973, 1979), the C (circadian) oscillator (Daan et al 1984) and, more generally, as the endogenous circadian pacemaker (Czeisler & Jewett 1990), the name which will be used hereafter in this text.

period of the endogenous circadian pacemaker. Humans do not always choose to sleep and wake in synchrony with their endogenous circadian rhythms. Thus, the human activity rhythm is often desynchronized from the output of the human circadian pacemaker during temporal isolation studies, just as it is when shift workers decide to work at night and attempt to sleep by day or when transmeridian travellers propel themselves rapidly across time zones and immediately attempt to schedule their activity rhythm to local time.

This dissociation between the sleep–wake cycle and body temperature cycle was recognized to occur in night-shift workers by Benedict at the turn of the last century (Czeisler et al 1990). Some 50 years later, Kleitman was the first to monitor the body temperature cycle throughout a complete cycle of desynchrony from the sleep–wake cycle in studies of human subjects scheduled to non-24 h routines of living (Kleitman & Kleitman 1953, Kleitman 1963). Aschoff and colleagues (Wever 1973, Aschoff & Wever 1976, Wever 1979, Aschoff & Wever 1981) later termed this phenomenon 'spontaneous *internal* desynchronization' when it occurred under free-running conditions, or 'forced desynchrony' when it occurred as a result of scheduling subjects to a non-24 h day length, as Kleitman had. Most physiological rhythms, including those of body temperature, cortisol and melatonin secretion, sleep duration, sleep and wake propensity, rapid eye movement sleep propensity, urinary potassium excretion, alertness and cognitive performance continue to oscillate with a near-24 h period during such internal desynchronization (Aschoff & Wever 1976, Czeisler 1978, Wever 1979, Czeisler et al 1980, Czeisler & Guilleminault 1980, Zulley et al 1981, Aschoff & Wever 1981), consistent with the hypothesis that such physiological rhythms reflect the output of a hypothalamic pacemaker located in the SCN (Klein et al 1991) and are not driven passively by the rest–activity cycle. In contrast, there is a group of other rhythms, such as blood pressure, heart rate, growth hormone secretion and skin temperature, which appear to respond predominantly to the changes in posture, activity level and sleep–wake state inherent in the behavioural rest–activity cycle, but which fail to persist if activity and posture are held constant. During this state of internal desynchrony, these physiological markers of the circadian pacemaker oscillate with a remarkably stable period, whereas the period of the behavioural rest–activity rhythm is labile and shows marked day-to-day variations. Thus, the activity rhythm, which is an inconsistent marker of endogenous circadian phase during internally synchronized free runs, is altogether unreliable during internal desynchrony. This confirms that the free-running activity rhythm in humans is unfortunately not a sufficiently reliable marker of endogenous circadian phase for assessment of the resetting response of the pacemaker to a light stimulus.

Our group therefore refined and validated an alternative method by which to assess the phase and amplitude of the output of the endogenous circadian pacemaker. As described above, during internal desynchrony, there are several physiological rhythms which maintain a persistent, near-24 h oscillation even in the absence of synchrony with the behavioural rest–activity cycle. Unfortunately, the output of the circadian pacemaker, as characterized by the phase and amplitude of those rhythms, is often obscured by the effects of activity, sleep, meals or changes in posture (masking effects) superimposed on the observed rhythms. During long-term studies in which behavioural activity is desynchronized from the output of the endogenous circadian pacemaker, the masking effects of activity are distributed across a variety of circadian phases, allowing the average phase of the circadian pacemaker to be extracted from a three- to four-week segment of data. However, even during desynchrony, masking effects obscure the contribution of the pacemaker to any *single* day of recording, such that lengthy segments of data must be collected during desynchrony to allow extraction of the endogenous circadian component from the observed data. However, we have found that it is possible to assess both endogenous circadian phase and endogenous circadian amplitude much more rapidly (within one or two days) by monitoring circadian rhythms (core body temperature and hormonal secretory patterns) under constant environmental and behavioural conditions, using a modification (Czeisler et al 1986) of the constant routine technique first proposed by Mills et al (1978). This constant routine procedure consists of 30–40 h of enforced semi-recumbent wakefulness (with the head of the bed raised to a 45° angle) in constant indoor room light (~150 lux) with hourly snacks distributed throughout day and night (Czeisler et al 1990). Using this methodology, we have made endogenous circadian phase and amplitude (ECPA) assessments in a control group of 29 young male subjects (age $\pm$ SD = 20.9 $\pm$ 3.1 years) whose sleep–wake cycles were stably entrained to the 24 h day, using a set of statistical methods based on a harmonic-regression-plus-correlated-noise model for estimating the phases and the amplitude of the endogenous circadian pacemaker from core temperature data collected under constant routine (Brown & Czeisler 1992). We have found that the fitted minimum of the body temperature cycle (the most widely accepted and best characterized circadian marker) during the constant routine in these normal young subjects occurred at 06.42 $\pm$ 15 min, an average of 1.68 $\pm$ 1.01 h before each subject's habitual waking time; the average amplitude of the fitted waveform was 0.38 $\pm$ 0.02 °C. Repeated ECPA estimates in eight subjects demonstrated that the technique yields reproducible results (Pearson's product-moment correlation coefficient of repeated estimates = 0.998, $P<0.001$). We have recently compared the entrained phase of the endogenous circadian temperature cycle in young and older subjects using this technique (Czeisler et al 1992).

We have validated this method of assessing the endogenous component of circadian rhythms in both normal and extreme cases by comparing the results it gives with those obtained with standard techniques during internal desynchronization. We studied five young male subjects (mean age 20.2 years) and two elderly subjects (a 66-year-old female and an 85-year-old male) during

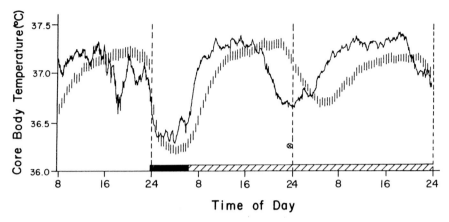

FIG. 4. Core body temperature cycle (solid line) of a 66-year-old woman under baseline and constant routine (hatched horizontal bar) conditions; her sleep episode is indicated by the solid bar. Data from this subject are superimposed on average ($\pm$ SEM) temperature data (vertical lines) from 29 young, healthy men living on the same protocol. Hour 24 represents the onset of the habitual sleep episode. The encircled cross indicates the minimum of a harmonic regression model fitted to her temperature data (Brown & Czeisler 1992). Note that in contrast to the healthy young men, the subject's temperature minimum occurs nearly seven hours before her habitual waketime—a difference which is not apparent on the baseline day when the data are masked by the timing of the rest–activity cycle. The phase of her cortisol rhythm was similarly advanced. Reproduced from Czeisler et al 1986.

197 days of temporal isolation. Assessments of ECPA were made before, during and after study segments in which each of the primary conditions observed in studies of temporal isolation—synchrony, spontaneous desynchrony and forced desynchrony between the rest–activity cycle and the endogenous rhythms driven by the circadian pacemaker—were evident. Independent assessments of the period and phase of the temperature cycle made using the ECPA technique agreed with those estimated by the standard techniques of non-parametric spectral analysis and waveform eduction using continuous temperature recordings during desynchrony (Pearson's product-moment correlation coefficient $= 0.995$, $P < 0.001$). These studies demonstrated that the ECPA technique could be used to assess the phase of the endogenous circadian temperature cycle accurately within one to two days, instead of the one to two months required to assess endogenous phase during desynchrony. The availability of this rapid phase assessment technique made it feasible to determine the acute phase-shifting effect of a discrete stimulus, such as exposure to environmental light, on the endogenous circadian pacemaker in humans.

In order to evaluate the effect of light on the endogenous circadian pacemaker, we monitored subjects in constant routine procedures before and after exposure to a bright light stimulus. We reasoned that it would be inherently difficult to

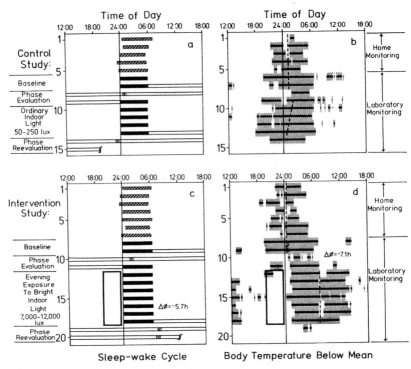

FIG. 5. Evening exposure to bright indoor light reset the circadian pacemaker by six or more hours in this phase-advanced, 66-year-old woman (see Fig. 4), even when the timing of her rest–activity cycle was held constant. (a) The entrainment control study (24 h) involved exposure to ordinary room light (~150 lux). The encircled cross represents the fitted temperature minimum, as in Fig. 4. Hatched bars represent bedrest episodes reported from home sleep–wake diaries during ambulatory monitoring; black bars represent scheduled bedrest episodes in the laboratory. The endogenous circadian phase assessments obtained from the initial and final constant routines suggest a small cumulative advance of the circadian pacemaker in this woman, whose intrinsic circadian period was shorter than 24 h (Czeisler et al 1986). This finding is consistent with the observation that subjects in the laboratory tend to drift in the direction of their intrinsic period, presumably because the laboratory provides weaker synchronizing cues than the external environment. (b) Raster plot of body temperature troughs (horizontal black bars indicate the specific times and days when values were below the baseline, entrained body temperature mean of 36.83 °C). An apparent shortening of the average temperature-cycle period in the laboratory is indicated by the dashed diagonal line. (c) Ten months later, the same subject returned for the entrainment study with the addition of bright indoor light (7000–12 000 lux, comparable in intensity to natural outdoor sunlight around twilight). The initial and final constant routine phase assessments showed that evening exposure to bright light caused a 5.7 h phase delay ($-\Delta\phi$) of the endogenous circadian temperature rhythm. The subject was exposed to a bank of wide-spectrum fluorescent lights between 19.40 and 23.40 h each day for seven days (open vertical box). Light intensity was measured in the direction of gaze with a photometer placed on her forehead. Symbols as in a. (d) Raster plot of body temperature troughs (temperature below entrained

demonstrate a direct physiological action of light as a circadian synchronizer, beyond its potentially indirect influence via behaviour, because the circadian phase of maximal light sensitivity normally occurs during the night in day-active species. Therefore, we sought individuals whose endogenous circadian phase, as revealed by the timing of the endogenous circadian temperature minimum measured during a constant routine, was markedly displaced with respect to their habitual sleep–wake and light–dark schedules. Given our finding of a shorter period of the free-running temperature rhythm during synchrony among older people (Weitzman et al 1982), we reasoned that the circadian system in some older people might show an internal phase advance with respect to the habitual sleep–wake schedule. In 1986, after screening a group of healthy elderly subjects with the constant routine, we identified a woman with a marked internal phase advance of her endogenous circadian temperature cycle with respect to her habitual sleep–wake schedule (Fig. 4) (Czeisler et al 1986). Using pre- and post-stimulus constant routines, we demonstrated that seven consecutive evening exposures to bright light (7000–12 000 lux, comparable to ambient outdoor light intensity at dawn) could in this phase-advanced subject rapidly reset the advanced phase of the circadian timing system (as marked by both the endogenous components of the body temperature rhythm and the cortisol secretory pattern) by six hours, even when the timing of the sleep–wake schedule and social contacts remained unchanged (Fig. 5) (Czeisler et al 1986). A raster plot of the temperature troughs before and during the intervention study suggested that the six-hour phase delay was evident one to two days after the start of the intervention (Fig. 5d). These results indicated that light could exert a direct biological effect on the human circadian pacemaker driving the endogenous component of the body temperature rhythm, above and beyond any effects mediated by the timing of the behavioural rest–activity cycle (Czeisler et al 1986). The unexpected magnitude, rapidity and stability of the shift challenged then-existing concepts regarding the relative insensitivity of the human circadian pacemaker to light, and suggested that exposure to environmental light could indeed reset the human circadian pacemaker, independently of changes in the sleep–wake schedule or in the timing of social interaction. The results of this case study have been greatly strengthened by a series of related papers: Dijk et al (1987, 1989) reported that three consecutive mornings of exposure to bright light advanced the body temperature,

---

mean of 36.69 °C) before and during the intervention study shown in c, confirming the magnitude (about six hours) and unexpected rapidity (one to two days) of the phase delay. Although a masking effect of light could explain the disappearance of the trough between 19.00 and 24.00 h, the extension of the trough from 08.00 to 14.00 h cannot be similarly explained. Symbols as in (b). Estimates of the phase shift were based on spectral analysis of data sets disjoint from those used to derive the estimates of the endogenous circadian temperature minima shown in (c). Reproduced from Czeisler et al 1986.

melatonin and sleep propensity rhythms by an hour, again while the timing of the sleep–wake cycle was held constant; Broadway et al (1987) reported a similar approximately two-hour phase advance of the melatonin rhythm following six weeks of exposure to bright light in the morning and evening during the Antarctic winter. Clodoré et al (1990) and Foret et al (1993) have reported that in addition, the rhythms of cortisol, alertness and performance can be phase advanced by repeated morning exposure to bright light. Lewy et al (1987) and Drennan et al (1989) found that the onset of nocturnal melatonin secretion, and the body temperature cycle, respectively, could be delayed by two to three hours after three to seven consecutive evenings of bright light exposure, even when the timing of sleep was held constant. Honma et al (1987a,b) also reported entrainment of human circadian rhythms by the light–dark cycle and phase-dependent shifts of human circadian rhythms in response to single bright light pulses (Honma & Honma 1988), although the responses they observed were limited largely to phase advances, leading them to question whether or not the human circadian pacemaker could achieve phase delays in normal subjects (Honma et al 1987b, Honma & Honma 1988).

On the basis of the results of our initial case study, we recognized that light acted directly on the endogenous circadian pacemaker in humans (Czeisler et al 1986). We also realized that the apparent role of the rest–activity cycle as an intermediary in conveying a synchronizing input from the external environment to the endogenous circadian pacemaker, that we had previously concluded from model simulations based on temporal isolation studies conducted in ordinary indoor light (Kronauer et al 1982), was evident only because the subjects in those studies were exposed to light only when they were awake and were exposed to darkness whenever they were asleep (Czeisler et al 1986). We thus proposed a new hierarchical model in which the external light–dark cycle synchronizes the endogenous circadian pacemaker *directly* (Czeisler et al 1986, Kronauer 1987, Kronauer 1990), in turn affecting the timing of the sleep–wake cycle (Czeisler et al 1980). We concluded that our results were incompatible with alternative models in which the behavioural sleep–wake cycle played a primary or essential intermediary role in the entrainment of physiological rhythms under ordinary conditions (Czeisler et al 1986). Our hierarchical model, in turn, explained the observed coupling influence (Wever 1973, Wever 1979, Kronauer et al 1982) of the free-running behavioural rest–activity cycle (and hence light–dark cycle) on the period of the free-running circadian temperature cycle during internal synchrony versus internal desynchrony, as was suggested by Daan et al (1984), simulated by Beersma et al (1987) and substantiated by Klerman et al (1992). This hierarchical model postulating a direct action of light—even ordinary indoor light—on the pacemaker also explained a previously paradoxical observation. Imposition of a relative light–dark cycle during internal desynchrony synchronizes the body temperature cycle to a 24 h day, even when the behavioural rest–activity cycle has a period longer than

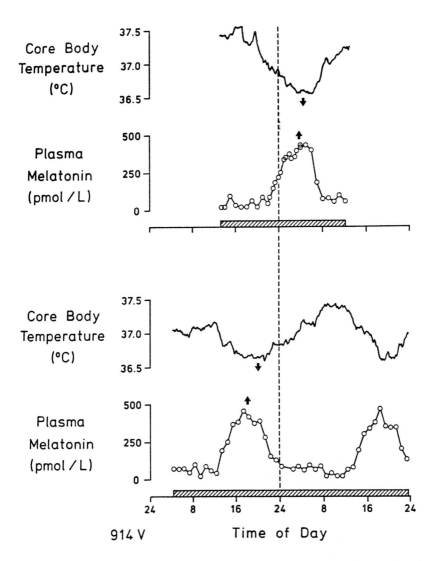

FIG. 6. The melatonin and temperature rhythms from a 22-year-old man observed under constant routine conditions (~150 lux; hatched bars) before (*upper panel*) and after (*lower panel*) a three-cycle bright light stimulus (five hours, ~9500 lux) designed to reset the circadian pacemaker. The timing of the fitted maximum of the plasma melatonin rhythm (up arrow) and that of the fitted minimum of the core body temperature rhythm (down arrow) were shifted by +9.8 and +8.4 h, respectively. Reproduced from Shanahan & Czeisler 1991.

30 h (Wever 1979), suggesting the possibility of a direct action of light on the pacemaker even at that time (Aschoff & Wever 1981, Daan et al 1984).

In many ways, recognition of this hierarchy reconciles the initial conclusions of Aschoff and Wever with our own: we can now agree that in our society, behaviour plays a far greater role in the synchronization of circadian rhythms than it does in most other animals. We disagree only about the mechanism by which human behaviour exerts such a strong synchronizing influence on the circadian pacemaker. In our view, the primary and most important pathway by which the behaviour of modern humans exerts this powerful synchronizing effect is through our voluntary control—unique within the animal kingdom— of the ordinary electric light switch and all other behaviours that determine the timing and duration of our retinal exposure to light of *all* intensities each day. As noted above, studies of blind people whose circadian rhythms are not entrained to the 24 h day despite their living, working and eating on a 24 h schedule suggest that non-photic synchronizers have a much less powerful influence on the human circadian pacemaker (Orth et al 1979, Sack et al 1992, Klein et al 1993), although entrainment by weaker non-photic synchronizers (Mrosovsky & Salmon 1987, Turek 1989, Van Cauter et al 1993) may occur in some cases, even when there is a complete absence of light input to the SCN (Czeisler et al 1994).

The results of our initial case study (Czeisler et al 1986) formed the basis of Kronauer's mathematical model of the effect of light on the human circadian pacemaker (Kronauer 1987, 1990, Kronauer & Frangioni 1987, Kronauer & Czeisler 1993, Kronauer et al 1993). We have subsequently found that similarly large phase shifts can be achieved in normal subjects after only three exposures to bright light (Czeisler & Allan 1987, Czeisler et al 1987, 1989a). As shown in Fig. 6, bright light can rapidly shift the phase of the human circadian pacemaker, as marked by both the endogenous circadian temperature cycle and the melatonin cycle (Shanahan & Czeisler 1991). Furthermore, our data indicate that this rapid phase shifting is not observed if sleep is similarly displaced but subjects are exposed to darkness instead of bright light (Fig. 7). Thus, we found that a three-cycle stimulus produces phase shifts of similar magnitude to the seven-cycle stimulus used in our initial case study.

Kronauer's model has successfully reproduced and predicted the results of most of our subsequent studies with light, including the suppression of amplitude by a two-cycle stimulus (see below). Use of the model has allowed us to extrapolate beyond our experimental data in order to construct protocols designed to achieve specific scientific goals. Furthermore, although several authors have suggested (on the basis, for example, of melatonin suppression data) that ordinary room light does not have a direct effect on human circadian rhythms (Lewy et al 1984, 1987, Wever 1989), both the mathematical model and our experimental results suggest that the timing of exposure to room light can modify both the magnitude and the direction of phase shifts induced by

Light and the human circadian pacemaker 271

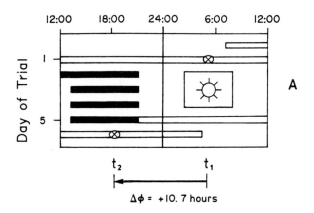

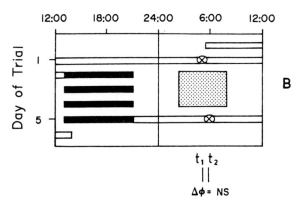

FIG. 7. Response of the circadian pacemaker to inversion of the sleep–wake schedule in an 18-year-old man (subject 720) exposed to either bright light (A) or darkness (B) centred around the initial endogenous circadian temperature minimum (encircled cross). The constant routine procedure (open bars) was imposed at the beginning and end of each study. The time scale is defined relative to the initial fitted minimum of the body temperature cycle that is assigned a reference value of 05.00 h (the approximate clock hour at which the fitted temperature minimum might occur in a healthy young man living at home who sleeps from about 23.00 to 07.00 h). In A, the intervention (open box with sun) consisted of three cycles of daily exposure to five hours of bright light (averaging 9843 lux) centred around the initial fitted temperature minimum ($t_1 = 05.00$ h). The bright light stimulus substantially phase advanced ($\Delta\phi = +10.7$ h) the fitted temperature minimum ($t_2 = 18.18$ h). B shows the control study for the same subject, consisting of an intervention of three cycles of daily exposure to five hours of darkness (<0.02 lux) centred around the initial fitted temperature minimum ($t_1 = 05.00$ h). Despite inversion of his sleep–wake schedule, there was no significant difference (NS) between the subject's final fitted temperature minimum ($t_2 = 05.45$ h) and the initial fitted temperature minimum ($t_1 = 05.00$ h). Reproduced from Czeisler et al 1989a.

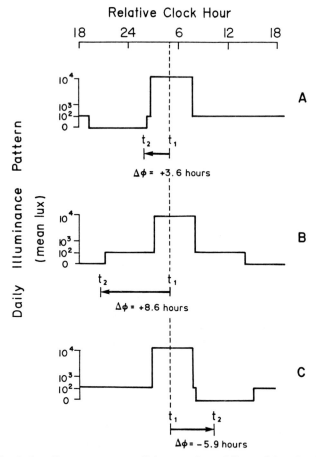

FIG. 8. The timing of exposure to room light can substantially modulate the phase-shifting effect of bright light when the midpoint of the bright light exposure, t_{BL}, occurs at the most light-sensitive phase. Daily illuminance patterns and resulting phase shifts are shown in three different trials of light exposure in a 22-year-old man (subject 713). In each of these trials, t_{BL} occurred at approximately the same initial temperature minimum (t_1 indicated by a vertical dashed line), whereas the timing of exposure to room light (and therefore darkness or sleep) was varied. In (A), exposure to room light occurred predominantly after the exposure to bright light, whereas in (C), most of the exposure to room light was before that to bright light. In (B), the midpoint of the exposure to room light (t_{RL}) was the same as that of the bright light exposure (t_{BL}). Whereas t_{BL} occurred at a relative clock hour of 05.20 h ($\pm$ 15 min) in all three cases, the relative clock hours at which the midpoints of the overall light exposures occurred (T_L, which is a brightness-weighted average of t_{BL} and t_{RL} [Czeisler et al 1989a]) were 06.36, 05.34 and 03.43 h for (A), (B) and (C), respectively. These T_L values correspond to initial circadian phases at which the stimuli occurred (ϕ_i) of 1.6, 0.6 and 22.7, respectively. These differences in ϕ_i were associated with marked differences in the magnitude and direction of the resetting response to the light stimulus ($\Delta\phi$ for A = +3.6 h; for B = +8.6 h and for C = −5.9 h), consistent with the results shown in Fig. 12. Reproduced from Czeisler et al 1989a.

Light and the human circadian pacemaker 273

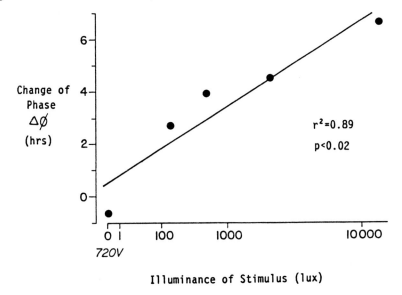

FIG. 9. The magnitude of the change in circadian phase following a three-cycle intervention stimulus is plotted against the illuminance level of the intervention stimulus on a cube-root scale. As can be seen, the phase-shifting ability of bright light increases with increasing dose (measured as illuminance) in a non-linear manner. We have found that a cube-root model approximates ($r^2 = 0.89$, $P < 0.02$) the non-linearity observed in this type of phase-shifting protocol (Allan et al 1988), as predicted by Kronauer's mathematical model (see text).

exposure to bright light (Czeisler et al 1989a). When the phase of bright light administration is centred at the endogenous circadian temperature minimum and held constant, the timing of exposure to ordinary room light (~ 150 lux) in relation to darkness/sleep can affect both the magnitude and direction of phase shifts induced by the bright light stimulus in this critical region, as shown in Fig. 8.

These results are consistent with: (1) our initial report of entrainment of human circadian rhythms by an imposed cycle of ordinary room light (~ 150 lux) versus darkness (Czeisler 1978, Czeisler et al 1981); (2) the prediction by Kronauer's mathematical model of a non-linear relationship between the illuminence level of the light exposure and its strength as a phase-shifting stimulus (Kronauer 1990, Kronauer & Czeisler 1993, Kronauer et al 1993); (3) our preliminary results of a dose-dependent resetting response to increasing light intensity (Fig. 9) (Allan et al 1988, Boivin et al 1994); and (4) Laakso's recent finding that one hour's exposure to 500 lux before bedtime induces a 30 min delay in the human melatonin rhythm (Laakso et al 1993). Taken together, these findings do not support the currently prevailing view that bright light ($\geqslant 500$ lux) is

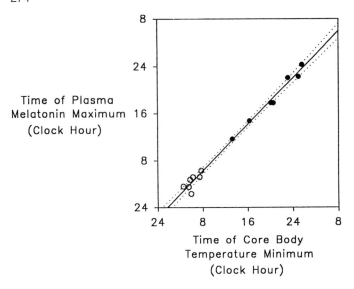

FIG. 10. Correlation ($n = 14$, $r = 0.995$, $P < 0.0001$) between the timing of the fitted maximum of the melatonin rhythm and the fitted minimum of the temperature rhythm in seven young males before (○, initial constant routine) and after (●, final constant routine) a three-cycle bright light stimulus (five hours, ~9500 lux) designed to reset the circadian pacemaker. The fitted maximum of the melatonin rhythm occurs 1.5–2 h before the fitted minimum of the temperature rhythm. The 95% confidence interval (dotted lines) is plotted for the regression line ($y = 0.99x - 1.64$). Reproduced from Shanahan & Czeisler 1991.

required to phase shift or entrain the circadian pacemaker in humans (Lewy et al 1984, 1985a,b, 1987, Wever 1989, Lewy et al 1991, Laakso et al 1993). These findings instead support the conclusion noted above that exposure to ordinary indoor room light exerts a direct physiological effect on the human circadian pacemaker which is sufficient to achieve the small daily phase shifts necessary to entrain the human circadian pacemaker to the 24 h day.

Having demonstrated that light exposure could reliably shift endogenous circadian temperature phase, we next sought to determine whether or not this shift in endogenous circadian temperature phase was reflected in the output of other endogenous circadian markers. By monitoring the melatonin and temperature rhythms before and after light-induced phase shifts, we found that exposure to bright light induced substantial and equivalent phase shifts in the melatonin and temperature rhythms (mean ± SEM difference in the phase-shifting response $= 0.03 \pm 0.32$ h) (Fig. 10), which maintained their usual phase relationship even after light-induced circadian phase inversion (Shanahan & Czeisler 1991) (Fig. 6). We therefore evaluated the timing of the release of the hormones thyroid-stimulating hormone and cortisol, as well as the timing of daily variations in alertness, cognitive performance and urine production in constant routines before and after large phase shifts induced by bright light. For all variables, we found that the average phase relationship between the endogenous components of each of these rhythms and the body temperature cycle remained fixed after 4–12 h phase shifts, indicating that the phase of each of these rhythms shifted by an equivalent amount in response to the bright light stimulus (Czeisler et al 1989a, Allan & Czeisler 1994, Czeisler et al 1990, Shanahan & Czeisler 1991). On the basis of these results, we have concluded that the bright light stimulus shifts the phase of a master circadian pacemaker, probably located in the hypothalamic SCN, which drives the endogenous component of each of these rhythms. The results of our studies, in which the endogenous components of each of these rhythms were assessed during constant routines before and after exposure to a three-cycle bright light stimulus, thus do not support the hypothesis that each of these physiological and behavioural rhythms is driven by a separate oscillator, with the outputs of these oscillators being transiently dissociated from each other after a phase shift (Hauty & Adams 1965, Aschoff 1969, 1973, Elliott et al 1972, Aschoff et al 1975, Rutenfranz et al 1975, DeRoshia et al 1976, Moore-Ede et al 1977, Knauth et al 1978, Winget et al 1978), and with different rhythms resynchronizing at different rates. Although it remains possible that some dissociation may have occurred during the three days between the initial and final constant routines (Jewett et al 1994), our results suggest that earlier findings of such dissociation between rhythm markers persisting for a week or longer after a phase shift were probably due largely to the confounding effect of masking on the observed rhythms.

The light PRC in humans: type 0 resetting to a three-cycle stimulus

As discussed above, the circadian response to a given light stimulus is usually described in terms of a PRC which depicts graphically the relationship between the phase of administration of the light stimulus and the resultant magnitude and direction of the induced phase shift (Hastings & Sweeney 1958). We conducted 45 resetting trials in 14 subjects in which the bright light stimulus was administered across different circadian phases, thereby constructing the first light PRC in humans (Czeisler et al 1989, Johnson 1990). We found that both the magnitude and the direction of the resetting response to three cycles of exposure to five-hour episodes of bright light were dependent on the timing of the bright light in relation to the initial endogenous circadian temperature minimum (i.e., the initial circadian phase). The bright light stimulus, which consisted of three cycles of exposure to a repeated daily illuminance pattern, included, per cycle, five waking hours' direct exposure to bright light (~ 7000–$12\,000$ lux) and another 11 waking hours' exposure to room light (~ 150 lux), together with eight hours' exposure to darkness (< 0.02 lux) during scheduled sleep episodes. By repeating this protocol,

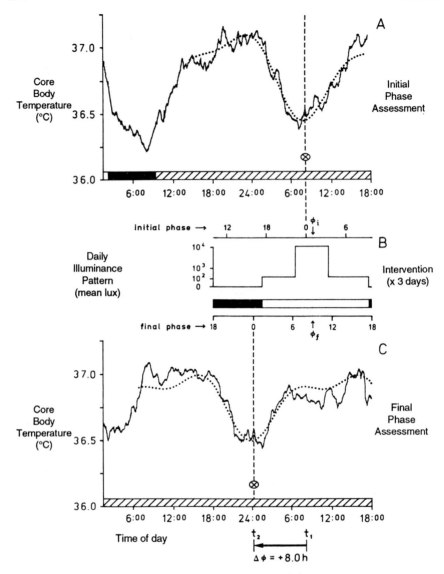

FIG. 11. Circadian phase resetting by bright light in a 20-year-old male (subject 706). The initial sleep episode (black bar in A) was at his habitual sleeping time. Phase assessments were made using the constant routine procedure (open hatched bar, ~150 lux) before (A) and after (C) the light stimulus (B). Endogenous circadian phase (ECP) was assessed by fitting a two-harmonic regression model (dotted line) to the constant routine core body temperature data (solid line); the clock hour of the initial (08.10 h) fitted temperature minimum (encircled cross and a vertical dashed line) is shown in (A). The light stimulus (B) consisted of three consecutive days of exposure to the following daily pattern of illuminance: ordinary indoor room light (100–200 lux) during scheduled

illustrated in Figs 7A and 11, exposing subjects to light at different phases, we have constructed a PRC to this cyclic stimulus of bright light and darkness. As can be seen in Fig. 12(A), the largest phase shifts were observed when the bright light episodes were centred near the initial temperature minimum, whereas smaller shifts were observed when the bright light episodes were centred 180° away from the initial temperature minimum (Czeisler et al 1989). We found that advances to an earlier phase occurred when the bright light episodes were centred after the minimum of the body temperature cycle, whereas there were delays to a later phase when light pulses were given before the temperature minimum. Statistical analysis by E. N. Brown (Czeisler et al 1989a) indicated that these data fitted well with Kronauer's model of the effect of light on the human circadian pacemaker (Kronauer 1990, Kronauer & Czeisler 1993), and suggested that an important factor contributing to the observed PRC is a circadian variation in the pacemaker's sensitivity to light (Czeisler et al 1989a). The latter finding is consistent with the reported circadian variation in retinal input to the SCN in mammals (Tierstein et al 1980, Terman & Terman 1985, Terman et al 1991).

Although the dependency of the resetting response to light on the initial circadian phase of stimulus application that we observed shares several similarities with PRC data reported in other species, our findings differed in two critical respects from predictions based on similar studies in other mammals. First, we found that the so-called break-point—when resetting responses to a light stimulus switch from delays to advances—occurs closer to dawn than to midnight in humans, four to six hours later than Daan & Lewy (1984) had surmised from PRC data derived from nocturnal rodents. This difference has major implications for the practical application of bright light phototherapy intended to shift circadian phase (Kronauer et al 1991). For example, our results revealed that for the average person, exposure to bright light from midnight to 04.00 h will actually induce a substantial phase delay (required for westward travel) rather than the phase advance (required for eastward travel) which Daan & Lewy (1984) had expected. Second, the large magnitude of the phase shifts

---

times awake, five-hour bright light episodes (7000–12 000 lux), and darkness (<0.02 lux) throughout scheduled sleep episodes. The average daily illuminance pattern (solid line) is plotted on a cube-root scale and the initial temperature minimum (t_1) has been assigned a reference value of 0 with ϕ_i indicating the initial circadian phase at which the midpoint of the overall light exposure occurred (one hour after the initial temperature minimum). On the final phase scale, the final temperature minimum (t_2), observed after the intervention, has been assigned a reference value of 0; ϕ_f indicates the circadian phase of the light stimulus with respect to the final phase scale, where $\Delta\phi$ is the phase shift and $\phi_f = \phi_i + \Delta\phi$. For this subject, ϕ_f was 9.0 h after the final temperature minimum. After the intervention, the final temperature minimum (t_2), was at 00.11 h, 8.0 h earlier than the initial temperature minimum (t_1), indicating that an eight-hour phase-advance shift ($\Delta\phi = +8.0$ h) was induced by the three-cycle light stimulus protocol. Reproduced from Czeisler et al 1989a.

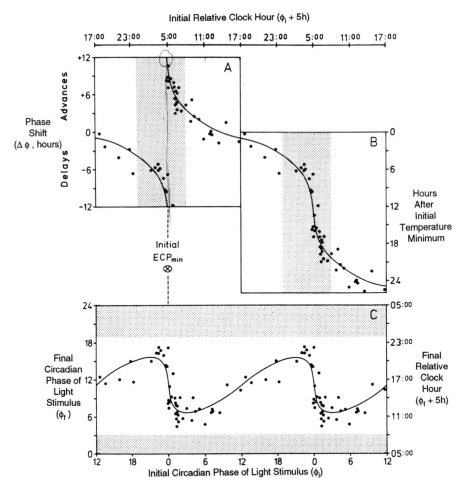

FIG. 12. The human circadian pacemaker exhibits strong type 0 resetting by light in a phase-dependent manner after three cycles of exposure to bright light. Phase shifts are plotted with respect to the initial circadian phase at which the light stimulus occurred (ϕ_i). The initial fitted temperature minimum (ECP_{min}) is defined as 05.00 h on the upper abscissa (relative clock hour scale) and as 0 on the lower abscissa (initial circadian phase scale). (A) The human phase response curve induced by exposure to the three-cycle light stimulus with the initial subjective night indicated by the stippled area. An artifactual discontinuity, or 'break-point', in the curve during the subjective night, when small differences in the initial circadian phase of the light stimulus appear to reverse the direction of the resulting phase shift, is produced when these data are plotted in the standard format. (B) The data from A are replotted with an ordinate which represents the number of hours between the temperature minimum after and before intervention, $t_2 - t_1$ ($= -\Delta\phi$), showing there is not a true discontinuity in the phase response curve data from A. (C) The human phase transition curve to the three-cycle light stimulus, with data plotted in the format introduced by Winfree (1980). The subjective night with respect to the final

observed, the shape of the PRC shown in Fig. 12(A) and the average slope of zero in the phase transition curve shown in Fig. 12(C) indicate that our three-cycle stimulus had induced a type 0 (strong) resetting response in the human (Czeisler et al 1989a, Kronauer 1990, Strogatz 1990, Kronauer & Czeisler 1993). These results—which remain controversial to this day (see Beersma & Daan 1993, Kronauer et al 1993, Lakin-Thomas 1993)—challenged the long-held belief that only lower organisms, such as unicellular algae, plants and insects, are sufficiently sensitive to light to be capable of strong (Winfree's type 0) phase resetting. Higher organisms, such as mammals, were believed to be less sensitive to the resetting effects of light and therefore restricted to weak (Winfree's type 1) resetting (Moore-Ede et al 1982). In fact, the belief that lower organisms are much more sensitive to the phase-resetting effects of light than vertebrates may stem from Pittendrigh's classical studies of the circadian rhythm of pupal emergence in *Drosophila pseudoobscura* (Pittendrigh 1960), as described previously (Czeisler et al 1989a). The circadian rhythm of eclosion in those pupae is so exquisitely sensitive to the phase-resetting effects of light that less than a minute's exposure to dim blue light (1 W/m^2) can induce a 12 h phase shift of that critically timed event (Winfree 1974), resulting from the induction of type 0 resetting of the *Drosophila* circadian pacemaker. In contrast, adult flies of the same species require 12 h exposure to bright light (~ 7000 lux) to achieve comparable type 0 resetting of the circadian pacemaker driving their activity rhythm (Engelmann & Mack 1978). In fact, the cockroach *Leucophaea maderae* requires even more light (80 000 lux for 12 h) to achieve strong type 0 resetting of its locomotor activity rhythm (Wiedenmann 1977). It thus appears from our studies that the sensitivity of the human circadian pacemaker to the phase-shifting effects of light is well within the range of sensitivity observed among insects.

Only Winfree had predicted that humans might be capable of type 0 resetting (Winfree 1987), in which large phase shifts are achieved through an antecedent

---

phase (stippled area) is included as part of a 10 h excluded zone during which the final circadian phase (ϕ_f) never occurs, indicating that the light stimulus was strong enough to transform the initial phase of the bright light stimulus (ϕ_i) into the subjective day, regardless of whether the stimulus was applied during the subjective day or subjective night. A mathematical representation based on Kronauer's model (1987, 1990) of the effect of light on the human circadian pacemaker was fitted to the data in C by non-linear least squares analysis and then projected onto A and B (solid lines). Estimated parameters and standard errors are: $\alpha = 11.06 \pm 0.3$; $\phi_c = 0.05 \pm 0.01$; $a = 0.70 \pm 0.1$; $b = 0.45 \pm 0.13$; covariance $(a,b) = -0.01$. Linear transformations of these variables yield the following estimates of the light sensitivity parameters: $(a+b) = 1.15 \pm 0.06$ (maximal effectiveness); $(a-b) = 0.25 \pm 0.23$ (minimal effectiveness). The standard error of the model (the root-mean-square deviations of the points [ϕ_f] from the model curve) is 1.99 h. On the basis of the curve in C, the maximal width of the 95% confidence interval for the mean of ϕ_f is 1.5 h. Reproduced from Czeisler et al 1989a.

suppression of the amplitude of oscillation of the circadian pacemaker. Oscillator theory predicts that if a strong stimulus induces type 0 resetting and a weak stimulus induces type 1 resetting, there should exist a stimulus intermediate in strength which, if given at the exact circadian phase between that which induces phase advances and that which induces phase delays, will bring the oscillating system to a theoretical point of singularity where the amplitude is reduced to a level indistinguishable from zero and phase is undefined. Experiments in single-cell organisms (Walz & Sweeny 1979, Malinowski et al 1985), plants (Engelmann & Johnsson 1978) and insects (Winfree 1973, Chandrashekaran & Engelmann 1976, Saunders 1978, Peterson 1980) have verified that in type 0 resetting, with a light stimulus of specific timing, duration and intensity, it is possible to dampen circadian amplitude such that the organism appears to be near the theoretical singularity point.

To verify whether such amplitude suppression was occurring in humans when the largest phase shifts were induced by our three-cycle stimulus, we did a series of studies involving interruption of the three-cycle stimulus after two of the cycles for an intermediate assessment of circadian phase and amplitude (Czeisler et al 1988, Jewett et al 1991). The two-cycle stimulus did indeed suppress endogenous circadian temperature amplitude to an average of 42% of the initial amplitude. In four of the trials, the amplitude was so markedly reduced (i.e., by up to 84%) that a meaningful estimate of circadian phase could not be obtained from the core body temperature data. Likewise, there was a nearly complete loss of the circadian rhythm of cortisol secretion in some of the subjects (Jewett et al 1991). These results were predicted by and are consistent with our mathematical model of the effects of light on the circadian pacemaker (Kronauer 1990, Kronauer & Czeisler 1993, Kronauer et al 1993) and support our hypothesis that the three-cycle stimulus induces a type 0 resetting response in humans.

Our data on the effects of one and two cycles of light exposure on endogenous circadian temperature phase do not support the hypothesis, put forth by Beersma & Daan (1993), that the large phase shifts (e.g., 8–12 h) which we have observed after three exposures to bright light represent a linear accumulation of smaller responses (e.g., 3–4 hours to each of the three cycles of exposure). We have found that a one-cycle stimulus does not have as marked an effect on phase as would be the case if these large phase shifts represented a linear accumulation of smaller responses. This is supported by the findings of Kennaway et al (1987), Burešová et al (1991), Minors et al (1991), Dawson et al (1993) and Van Cauter et al (1993), who report more modest phase-shifting responses to single episodes of exposure to bright light. In fact, as we have already noted (Kronauer et al 1993), Beersma & Daan's attempt to model our data using a simple oscillator limited to weak (type 1) resetting requires that they postulate that a single exposure to bright light induces all but 20–25% of the very largest (10–12 h) phase shifts achieved after three cycles of exposure to bright light—a postulate that does not match with existing data (Minors et al 1991, Van Cauter et al 1993).

We have found that a two-cycle bright light stimulus centred at the temperature minimum has variable effects on the circadian timing system, resulting in either marked reductions in the amplitude of the circadian temperature cycle or a phase shift similar to that seen following three exposures (Jewett et al 1991). However, we have consistently observed the largest (8–12 h) phase shifts only after three cycles of exposure to five-hour episodes of bright light. Whether or not the pacemaker itself is completely shifted following one or two exposures, with another one or two cycles of transients occurring before a steady-state phase shift is achieved, cannot be determined with certainty from such studies. However, our preliminary data suggest that this is the case, because subjects with reduced circadian amplitude (after two light exposures) appear to be more sensitive to the phase-shifting effect of a single subsequent light exposure. These studies emphasize the importance of an often neglected variable in circadian rhythm research—endogenous circadian amplitude (Winfree 1973, Kronauer & Czeisler 1988, Pittendrigh et al 1991)—in evaluating resetting responses of the human circadian pacemaker to light.

The fact that humans are much more sensitive to the phase-resetting effects of light than previously recognized has several practical implications for the treatment of circadian rhythm sleep disorders. For example, Eastman & Miescke (1990) initially reported that exposure to bright light could facilitate adaptation among shift workers to a 26 h sleep–wake schedule, although Gallo & Eastman (1993) have subsequently cast doubt on the generality of that finding. We have, however, demonstrated that maladaptation to night-shift work can be treated successfully with properly timed exposure to bright light during night-shift work and darkness during daytime sleep (Czeisler et al 1990), despite the continued exposure of the shift workers to the conflicting synchronizers present in the environment. This finding was later confirmed by Eastman (1992). Similar approaches can be used to facilitate adaptation to transmeridian travel (Czeisler & Allan 1987). In addition, there is mounting evidence that persistent circadian rhythm sleep disorders can be treated effectively with bright light, including the midwinter insomnia that occurs north of the Arctic Circle during the 'dark period' (Hansen et al 1987), the sleep maintenance insomnia of advancing age (Czeisler et al 1989b, Campbell et al 1993), the non-24 h sleep–wake schedule disorder (Hoban et al 1989, Guilleminault et al 1993) and delayed sleep phase syndrome (Czeisler et al 1989b, Rosenthal et al 1990). Continuous ambulatory monitoring of ambient light indicates that average urban dwellers are ordinarily minimally exposed to bright light, even in cities such as San Diego, largely because of self-induced shielding from natural light in modern living and working quarters (Okudaira et al 1983, Savides et al 1986). Conversely, the near-universal availability of artificial electric light enables us to self-select light exposure during the night-time hours. It thus appears that an entire class of circadian rhythm sleep disorders may be largely the result of these fundamental behaviourally induced changes in the periodic environment to which we, as a species, are ordinarily exposed.

Understanding the mechanism by which light synchronizes the mammalian circadian pacemaker may help to refine treatment approaches for these self-inflicted maladies. Several areas are ripe for further investigation. First, additional experiments are needed to explore the importance of circadian amplitude. Second, the role of melatonin in the regulation and entrainment of mammalian circadian rhythms must be evaluated, particularly in light of the claim that oral administration of melatonin every day for a week can shift circadian phase by one to two hours (Lewy et al 1992). These findings must be reconciled with reports that daily melatonin administration cannot synchronize the body temperature, cortisol or melatonin rhythms in most blind patients (Folkard et al 1990, Sack et al 1991) and does not appear to expand the range of entrainment of human circadian rhythms in temporal isolation (Wever 1989). Acute suppression of the melatonin rhythm by light (Lewy et al 1980) does not appear to be necessary for the induction of circadian phase shifts, as has been suggested (Laakso et al 1993), because bright light can induce substantial phase shifts in individuals with no detectable melatonin rhythm (Shanahan & Czeisler 1991), just as melatonin suppression does not appear to be causally involved in the antidepressant effects of bright light phototherapy for affective illness (Dietzel et al 1985, Wehr et al 1986). However, given recent findings that melatonin has hypnotic properties, even in the physiological range (Dollins et al 1993), it may be that melatonin suppression is involved in some of the immediate effects of bright light on sleep tendency and alertness which have been reported (Campbell & Dawson 1990, Dawson & Campbell 1991). This could also be the case for the reportedly causal role of melatonin suppression in bright light-induced elevation of body temperature (Strassman et al 1991, Myers & Badia 1993). Third, the photoreceptors in the mammalian retina responsible for mediating photic entrainment of the circadian system must be identified and characterized. Evidence in animals (Foster et al 1991, Cooper et al 1993) and humans (Martens et al 1992, Czeisler et al 1994) indicates that under certain conditions photic input to the circadian system may be preserved even in the absence of light perception. Furthermore, it is believed that a circadian variation in retinal sensitivity to light (Terman et al 1991) may be an important factor influencing responses to light (and hence the PRC) in mammals. Finally, further progress is needed in the understanding of the molecular events by which light exposure induces circadian phase shifts. Glutamate appears to be the neurotransmitter by which neurons in the RHT transmit photic information from the retina to the hypothalamus (Moore 1995, this volume), leading to the induction of the immediate-early gene c-*fos* (Rea et al 1993) and the phosphorylation of the transcription factor cyclic adenosine monophosphate response element-binding protein (CREB) (Ginty et al 1993) in the ventrolateral neurons of the SCN. Unravelling the molecular events by which light shifts the circadian clock could provide important clues as to the fundamental mechanism by which the circadian pacemaker keeps time.

Acknowledgements

The work described in this paper was supported in part by grants NIMH-1-R01-MH45130 from the National Institute of Mental Health; NAG9-524 from the National Aeronautics and Space Administration; NIA-1-R01-AG06072 and NIA-P01-AG09975 from the National Institute on Aging; by General Clinical Research Center grant NCRR-GCRC-M01-RR02635 from the National Center for Research Resources; and by the National Geographic Society and the Brigham and Women's Hospital.

References

Allan JS, Czeisler CA 1994 Persistence of the circadian thyrotropin rhythm under constant conditions and after light-induced shifts of circadian phase. J Clin Endocrinol & Metab 79:508–512

Allan JS, Czeisler CA, Duffy JF, Kronauer RE 1988 Non-linear dose response of the human circadian pacemaker to light. In: Abstracts of the 154th Annual AAAS Meeting (AAAS Publ No 897-30 101)

Aschoff J 1969 Desynchronization and resynchronization of human circadian rhythms. Aerospace Med 40:844–849

Aschoff J 1973 Internal dissociation and desynchronization of circadian systems. In: Proceedings of the 21st International Congress on Aviation and Space Medicine, p 255 (abstr)

Aschoff J, Wever R 1976 Human circadian rhythms: a multioscillatory system. Fed Proc 35:2326–2332

Aschoff J, Wever R 1981 The circadian system of man. In: Aschoff J (ed) Handbook of behavioral neurobiology, vol 4: Biological rhythms. Plenum, New York, p 311–331

Aschoff J, Fatranská M, Giedke H, Doerr P, Stamm D, Wisser H 1971 Human circadian rhythms in continuous darkness: entrainment by social cues. Science 171:213–215

Aschoff J, Hoffmann K, Pohl H, Wever R 1975 Re-entrainment of circadian rhythms after phase-shifts of the zeitgeber. Chronobiologia 2:23–78

Aschoff J, Pöppel E, Wever R 1969 Circadiane Periodik des Menschen unter dem Einfluss von Licht-Dunkel-Wechseln unterschiedlicher Periode. Eur J Physiol 306:58–70

Beersma DM, Daan S 1993 Strong or weak phase resetting by light pulses in humans? J Biol Rhythms 8:340–347

Beersma DGM, Daan S, Dijk DJ 1987 Sleep intensity and timing: a model for their circadian control. Lect Math Life Sci 19:39–62

Boivin DB, Duffy JF, Kronauer RE, Czeisler CA 1994 Sensitivity of the human circadian pacemaker to moderately bright light. J Biol Rhythms, submitted

Broadway J, Arendt J, Folkard S 1987 Bright light phase shifts the human melatonin rhythm during the Antarctic winter. Neurosci Lett 79:185–189

Brown EN, Czeisler CA 1992 The statistical analysis of circadian phase and amplitude in constant routine core temperature data. J Biol Rhythms 7:177–202

Burešová M, Dvořáková M, Zvolský P, Illnerová H 1991 Early morning bright light phase advances the human circadian pacemaker within one day. Neurosci Lett 121:47–50

Campbell SS, Dawson D 1990 Enhancement of nighttime alertness and performance with bright ambient light. Physiol Behav 48:317–320

Campbell SS, Dawson D, Anderson MW 1993 Alleviation of sleep maintenance insomnia with timed exposure to bright light. J Am Geriatr Soc 41:829–836

Chandrashekaran MK, Engelmann W 1976 Amplitude attenuation of the circadian rhythm in *Drosophila* with light pulses of varying irradiance and duration. Int J Chronobiol 3:231–240

Clodoré M, Foret J, Benoit O et al 1990 Psychophysiological effects of early morning bright light exposure in young adults. Psychoneuroendocrinology 15:193–205

Cooper HM, Herbin M, Nevo E 1993 Ocular regression conceals adaptive progression of the visual system in a blind subterranean mammal. Nature 361:156–159

Czeisler CA 1978 Internal organization of temperature, sleep–wake, and neuroendocrine rhythms monitored in an environment free of time cues. PhD thesis, Stanford University, Stanford, CA

Czeisler CA, Allan JS 1987 Acute circadian phase reversal in man via bright light exposure: application to jet-lag. Sleep Res 16:605

Czeisler CA, Guilleminault C 1980 REM sleep: its temporal distribution. Raven Press, New York

Czeisler CA, Jewett ME 1990 Human circadian physiology: interaction of the behavioral rest–activity cycle with the output of the endogenous circadian pacemaker. In: Thorpy MJ (ed) Handbook of sleep disorders. Marcel Dekker, New York, p 117–137

Czeisler CA, Weitzman ED, Moore-Ede MC, Zimmerman JC, Knauer RS 1980 Human sleep: its duration and organization depend on its circadian phase. Science 210:1264–1267

Czeisler CA, Richardson GS, Zimmerman JC, Moore-Ede MC, Weitzman ED 1981 Entrainment of human circadian rhythms by light–dark cycles: a reassessment. Photochem Photobiol 34:239–247

Czeisler CA, Allan JS, Strogatz SH et al 1986 Bright light resets the human circadian pacemaker independent of the timing of the sleep–wake cycle. Science 233:667–671

Czeisler CA, Allan JS, Kronauer RE 1987 Rapid manipulation of phase and amplitude of the human circadian pacemaker with light–dark cycles. In: Proceedings of the 5th International Congress of Sleep Research, Copenhagen, S11(abstr)

Czeisler CA, Allan JS, Kronauer RE, Duffy JF 1988 Strong circadian phase resetting in man is effected by bright light suppression of circadian amplitude. Sleep Res 17:367

Czeisler CA, Kronauer RE, Allan JS et al 1989a Bright light induction of strong (type 0) resetting of the human circadian pacemaker. Science 244:1328–1333

Czeisler CA, Kronauer RE, Johnson MP, Allan JS, Johnson TS, Dumont M 1989b Action of light on the human circadian pacemaker: treatment of patients with circadian rhythm sleep disorders. In: Horne J (ed) Sleep '88. Proceedings of the 8th European Congress on Sleep Research. Gustav Fischer Verlag, Stuttgart, p 42–47

Czeisler CA, Johnson MP, Duffy JF, Brown EN, Ronda JM, Kronauer RE 1990 Exposure to bright light and darkness to treat physiologic maladaptation to night work. N Engl J Med 322:1253–1259

Czeisler CA, Dumont M, Duffy JF et al 1992 Association of sleep–wake habits in older people with changes in output of circadian pacemaker. Lancet 340:933–936

Czeisler CA, Shanahan TL, Klerman EB et al 1994 Bright light suppression of melatonin concentrations in some blind patients. N Engl J Med, in press

Daan S, Lewy AJ 1984 Scheduled exposure to daylight: a potential strategy to reduce 'jet lag' following transmeridian flight. Psychopharmacol Bull 20:566–568

Daan S, Beersma DGM, Borbély AA 1984 Timing of human sleep: recovery process gated by a circadian pacemaker. Am J Physiol 246:R161–R178

Dawson D, Campbell SS 1991 Timed exposure to bright light improves sleep and alertness during simulated night shifts. Sleep 14:511–516

Dawson D, Lack L, Morris M 1993 Phase resetting of the human circadian pacemaker with use of a single pulse of bright light. Chronobiol Int 10:94–102

DeRoshia CW, Winget CM, Bond GH 1976 Two mechanisms of rephasal of circadian rhythms in response to a 180 degree phase shift (simulated 12-hr time zone change). J Interdiscip Cycle Res 7:279–286

Dietzel M, Waldhauser F, Lesch OM, Masalek M, Walter H 1985 Bright light treatment success not explained by melatonin. J Interdiscip Cycle Res 16:165

Dijk DJ, Visscher CA, Bloem GM, Beersma DGM, Daan S 1987 Reduction of human sleep duration after bright light exposure in the morning. Neurosci Lett 73:181–186

Dijk DJ, Beersma DGM, Daan S, Lewy AJ 1989 Bright morning light advances the human circadian system without affecting NREM sleep homeostasis. Am J Physiol 256:R106–R111

Dollins AB, Zhdanova IV, Wurtman RJ, Lynch HJ, Deng MH 1994 Effect of inducing nocturnal serum melatonin concentrations in daytime on sleep, mood, body temperature, and performance. Proc Natl Acad Sci USA 91:1824–1828

Drennan M, Kripke DF, Gillin JC 1989 Bright light can delay human temperature rhythm independent of sleep. Am J Physiol 257:R136–R141

Eastman CI 1992 High-intensity light for circadian adaptation to a 12-h shift of the sleep schedule. Am J Physiol 263:R428–R436

Eastman CI, Miescke KJ 1990 Entrainment of circadian rhythms with 26-h bright light and sleep-wake schedules. Am J Physiol 259:R1189–R1197

Edgar DM, Dement WC 1991 Regularly scheduled voluntary exercise synchronizes the mouse circadian clock. Am J Physiol 261:R928–R933

Elliott AL, Mills JN, Minors DS, Waterhouse JM 1972 The effect of real and simulated time-zone shifts upon the circadian rhythms of body temperature, plasma 11-hydroxycorticosteroids, and renal excretion in human subjects. J Physiol 221:227–257

Engelmann W, Johnsson A 1978 Attenuation of the petal movement rhythm in *Kalanchoë* with light pulses. Physiol Plant 43:68–76

Engelmann W, Mack J 1978 Different oscillators control the circadian rhythm of eclosion and activity in *Drosophila*. J Comp Physiol 127:229–237

Folkard S, Arendt J, Aldhous M, Kennett H 1990 Melatonin stabilises sleep onset time in a blind man without entrainment of cortisol or temperature rhythms. Neurosci Lett 113:193–198

Foret J, Aguirre A, Touitou Y, Clodoré M, Benoit O 1993 Effect of morning bright light on body temperature, plasma cortisol and wrist motility measured during 24 hours of constant conditions. Neurosci Lett 155:155–158

Foster RG, Provencio I, Hudson D, Fiske S, De Grip W, Menaker M 1991 Circadian photoreception in the retinally degenerate mouse (*rd/rd*). J Comp Physiol A Sens Neural Behav Physiol 169:39–50

Gallo LC, Eastman CI 1993 Circadian rhythms during gradually delaying and advancing sleep and light schedules. Physiol Behav 53:119–126

Ginty DD, Kornhauser JM, Thompson MA et al 1993 Regulation of CREB phosphorylation in the suprachiasmatic nucleus by light and a circadian clock. Science 260:238–241

Guilleminault C, McCann CC, Quera-Salva MA, Cetel M 1993 Light therapy as treatment of dyschronosis in brain impaired children. Eur J Pediatr 152:754–759

Hansen T, Bratlid T, Lingjärde O, Brenn T 1987 Midwinter insomnia in the subarctic region: evening levels of serum melatonin and cortisol before and after treatment with bright artificial light. Acta Psychiatr Scand 75:428–434

Hastings JW, Sweeney BM 1958 A persistent diurnal rhythm of luminescence in *Gonyaulax polyedra*. Biol Bull (Woods Hole) 115:440–458

Hauty GT, Adams T 1965 Phase shifting of the human circadian system. In: Aschoff J (ed) Circadian clocks. North-Holland, Amsterdam, p 413–425

Hoban TM, Sack RL, Lewy AJ, Miller LS, Singer CM 1989 Entrainment of a free-running human with bright light. Chronobiol Int 6:347–353

Honma K, Honma S 1988 A human phase response curve for bright light pulses. Jpn J Psychiatry Neurol 42:167–168

Honma K, Honma S, Wada T 1987a Entrainment of human circadian rhythms by artificial bright light cycles. Experientia 43:572–574

Honma K, Honma S, Wada T 1987b Phase-dependent shift of free-running human circadian rhythms in response to a single bright light pulse. Experientia 43:1205–1207

Jewett ME, Kronauer RE, Czeisler CA 1991 Light-induced suppression of endogenous circadian amplitude in humans. Nature 350:59–62

Johnson CH 1990 An atlas of phase response curves for circadian and circatidal rhythms. Department of Biology, Vanderbilt University, Nashville, TN

Kennaway DJ, Earl CR, Shaw PF, Royles P, Carbone F, Webb H 1987 Phase delay of the rhythm of 6-sulphatoxy melatonin excretion by artificial light. J Pineal Res 4:315–320

Klein DC, Moore RY, Reppert SM (eds) 1991 The suprachiasmatic nucleus: the mind's clock. Oxford University Press, New York

Klein KE, Wegmann H-M 1974 The resynchronization of human circadian rhythms after transmeridian flights as a result of flight direction and mode of activity. In: Scheving LE, Halberg F, Pauly JE (eds) Chronobiology. Igaku Shoin, Tokyo, p 564–570

Klein T, Martens H, Dijk DJ, Kronauer RE, Seely EW, Czeisler CA 1993 Circadian sleep regulation in the absence of light perception: chronic non-24-hour circadian rhythm sleep disorder in a blind man with a regular 24-h sleep–wake schedule. Sleep 16:333–343

Kleitman N 1963 Sleep and wakefulness. University of Chicago Press, Chicago, IL

Kleitman N, Kleitman E 1953 Effect of non-twenty-four-hour routines of living in oral temperature and heart rate. J Appl Physiol 6:283–291

Klerman EB, Dijk DJ, Czeisler CA, Kronauer RE 1992 Simulations using self-selected light–dark cycles from 'free-running' protocols in humans result in an apparent *tau* significantly longer than the intrinsic *tau*. In: Proceedings of the 3rd meeting of the Society for Research on Biological Rhythms, p 110(abstr)

Knauth P, Rutenfranz J, Herrmann G, Poeppl SJ 1978 Re-entrainment of body temperature in experimental shift-work studies. Ergonomics 21:775–783

Kripke DF, Risch SC, Janowsky DS 1983 Bright white light alleviates depression. Psychiatry Res 10:105–112

Kronauer RE 1987 A model for the effect of light on the human 'deep' circadian pacemaker. Sleep Res 16:621

Kronauer RE 1990 A quantitative model for the effects of light on the amplitude and phase of the deep circadian pacemaker, based on human data. In: Horne J (ed) Sleep '90. Proceedings of the 10th European Congress on Sleep Research. Pontenagel Press, Dusseldorf, p 306–309

Kronauer RE, Czeisler CA 1988 The significance of endogenous amplitude in circadian rhythms. In: Proceedings of the 1st meeting of the Society for Research on Biological Rhythms. Wild Dunes, SC, p 28(abstr)

Kronauer RE, Czeisler CA 1993 Understanding the use of light to control the circadian pacemaker in humans. In: Wetterberg L (ed) Light and biological rhythms in man. Pergamon Press, Oxford, p 217–236

Kronauer RE, Frangioni JV 1987 Modeling laboratory bright-light protocols. Sleep Res 16:622

Kronauer RE, Czeisler CA, Pilato SF, Moore-Ede MC, Weitzman ED 1982 Mathematical model of the human circadian system with two interacting oscillators. Am J Physiol 242:R3–R17

Kronauer RE, Gander P, Moore-Ede M, Czeisler CA 1983 A two-oscillator model derived from free-running human circadian rhythms accurately predicts range of zeitgeber entrainment. Sleep Res 12:368

Kronauer RE, Jewett ME, Czeisler CA 1991 Human circadian rhythms. Nature 351:193
Kronauer RE, Jewett ME, Czeisler CA 1993 The human circadian response to light: strong *and* weak resetting. J Biol Rhythms 8:351-360
Laakso ML, Hätönen T, Stenberg D, Alila A, Smith S 1993 One-hour exposure to moderate illuminance (500 lux) shifts the human melatonin rhythm. J Pineal Res 15:21-26
Lakin-Thomas PL 1993 Commentary: strong or weak phase resetting by light pulses in humans? J Biol Rhythms 8:348-350
Lee YS, Klerman EB, Czeisler CA, Kronauer RE 1992 Can ambulatory temperature data be used to assess endogenous circadian phase? Sleep Res 21:380(abstr)
Lewy AJ 1983 Effects of light on human melatonin production and the human circadian system. Prog Neuropsychopharmacol Biol Psychiatry 7:551-556
Lewy AJ, Sack RL 1986 Minireview: light therapy and psychiatry. Proc Soc Exp Biol Med 183:11-18
Lewy AJ, Wehr TA, Goodwin FK, Newsome DA, Markey SP 1980 Light suppresses melatonin secretion in humans. Science 210:1267-1269
Lewy AJ, Kern HA, Rosenthal NE, Wehr TA 1982 Bright artificial light treatment of a manic-depressive patient with a seasonal mood cycle. Am J Psychiatry 139:1496-1498
Lewy AJ, Sack RA, Singer CL 1984 Assessment and treatment of chronobiologic disorders using plasma melatonin levels and bright light exposure: the clock-gate model and the phase response curve. Psychopharmacol Bull 20:561-565
Lewy AJ, Sack RL, Singer CM 1985a Immediate and delayed effects of bright light on human melatonin production: shifting 'dawn' and 'dusk' shifts the dim light melatonin onset (DLMO). In: Wurtman RJ, Baum MJ, Potts JT (eds) The medical and biological effects of light. New York Academy of Science, New York, p 253-259
Lewy AJ, Sack RL, Singer CM 1985b Treating phase typed chronobiologic sleep and mood disorders using appropriately timed bright artificial light. Psychopharmacol Bull 21:368-372
Lewy AJ, Sack RL, Miller LS, Hoban TM 1987 Antidepressant and circadian phase-shifting effects of light. Science 235:352-354
Lewy AJ, Sack RL, Latham JM 1991 Melatonin and the acute suppressant effect of light may help regulate circadian rhythms in humans. In: Arendt J, Pévet P (eds) Advances in pineal research. John Libbey, London, p 285-293
Lewy AJ, Saeeduddin A, Jackson JML, Sack RL 1992 Melatonin shifts human circadian rhythms according to a phase-response curve. Chronobiol Int 9:380-392
Lydic R, Schoene WC, Czeisler CA, Moore-Ede MC 1980 Suprachiasmatic region of the human hypothalamus: homolog to the primate circadian pacemaker? Sleep 2:355-361
Malinowski JR, Laval-Martin DL, Edmunds LN Jr 1985 Circadian oscillators, cell cycles, and singularities: light perturbations of the free-running rhythm of cell division in *Euglena*. J Comp Physiol B Biochem Syst Environ Physiol 155:257-267
Martens H, Klein T, Rizzo JF III, Shanahan TL, Czeisler CA 1992 Light-induced melatonin suppression in a blind man. In: Proceedings of the 3rd meeting of the Society for Research on Biological Rhythms, p 58(abstr)
Miles LEM, Raynal DM, Wilson MA 1977 Blind man living in normal society has circadian rhythms of 24.9 hours. Science 198:421-423
Mills JN, Minors DS, Waterhouse JM 1978 Adaptation to abrupt time shifts of the oscillator(s) controlling human circadian rhythms. J Physiol 285:455-470
Minors DS, Waterhouse JM 1989 Masking in humans: the problem and some attempts to solve it. Chronobiol Int 6:29-53
Minors DS, Waterhouse JM, Wirz-Justice A 1991 A human phase-response curve to light. Neurosci Lett 133:36-40

Moore RY 1995 Organization of the mammalian circadian system. In: Circadian clocks and their adjustment. Wiley, Chichester (Ciba Found Symp 183) p 88–106
Moore-Ede MC, Kass DA, Herd JA 1977 Transient circadian internal desynchronization after light–dark phase shift in monkeys. Am J Physiol 232:R31–R37
Moore-Ede MC, Sulzman FM, Fuller CA 1982 The clocks that time us: physiology of the circadian timing system. Harvard University Press, Cambridge, MA
Mrosovsky N, Salmon PA 1987 A behavioural method for accelerating re-entrainment of rhythms to new light–dark cycles. Nature 330:372–373
Myers BL, Badia P 1993 Immediate effects of different light intensities on body temperature and alertness. Physiol Behav 54:199–202
Okudaira N, Kripke DF, Webster JB 1983 Naturalistic studies of human light exposure. Am J Physiol 245:R613–R615
Orth DN, Island DP 1969 Light synchronization of the circadian rhythm in plasma cortisol (17-OHCS) concentration in man. J Clin Endocrinol 29:479–486
Orth DN, Island DP, Liddle GW 1967 Experimental alteration of the circadian rhythm in plasma cortisol (17-OHCS) concentration in man. J Clin Endocrinol 27:549–555
Orth DN, Besser GM, King PH, Nicholson WE 1979 Free-running circadian plasma cortisol rhythm in a blind human subject. Clin Endocrinol 10:603–617
Osterman PO 1974 Light synchronization of the circadian rhythm of plasma 11-hydroxycorticosteroids in man. Acta Endocrinol 77:128–134
Peterson EL 1980 Phase-resetting a mosquito circadian oscillator. I. Phase-resetting surface. J Comp Physiol A Sens Neural Behav Physiol 138:201–211
Pittendrigh CS 1960 Circadian rhythms and the circadian organization of living systems. Cold Spring Harbor Symp Quant Biol 25:159–184
Pittendrigh CS, Bruce VG 1957 An oscillator model for biological clocks. In: Rudnick D (ed) Rhythmic and synthetic processes in growth. Princeton University Press, Princeton, NJ, p 75–109
Pittendrigh CS, Kyner WT, Takamura T 1991 The amplitude of circadian oscillations: temperature dependence, latitudinal clines, and the photoperiodic time measurement. J Biol Rhythms 6:299–313
Ralph MR, Menaker M 1986 Effects of diazepam on circadian phase advances and delays. Brain Res 372:405–408
Rea MA, Michel AM, Lutton LM 1993 Is *fos* expression necessary and sufficient to mediate light-induced phase advances of the suprachiasmatic circadian oscillator? J Biol Rhythms 8:S59–S64
Rosenthal NE, Joseph-Vanderpool JR, Levendosky AA et al 1990 Phase-shifting effects of bright morning light as treatment for delayed sleep phase syndrome. Sleep 13:354–361
Rusak B 1979 Neural mechanisms for entrainment and generation of mammalian circadian rhythms. Fed Proc 38:2589–2595
Rutenfranz J, Klimmer F, Knauth P 1975 Desynchronization of different physiological functions during three weeks of experimental nightshift with limited and unlimited sleep. In: Colquhoun P, Folkard S, Knauth P, Rutenfranz J (eds) Experimental studies of shiftwork. Westdeutscher Verlag, Leverkusen, p 74–77
Sack RL, Lewy AJ, Blood ML, Stevenson J, Keith LD 1991 Melatonin administration to blind people: phase advances and entrainment. J Biol Rhythms 6:249–261
Sack RL, Lewy AJ, Blood ML, Keith LD, Nakagawa H 1992 Circadian rhythm abnormalities in totally blind people: incidence and clinical significance. J Clin Endocrinol & Metab 75:127–134
Sadun AA, Schaechter JD, Smith LEH 1984 A retinohypothalamic pathway in man: light mediation of circadian rhythms. Brain Res 302:371–377

Saunders DS 1978 An experimental and theoretical analysis of photoperiodic induction in the flesh-fly, *Sarcophaga argyrostoma*. J Comp Physiol A Sens Neural Behav Physiol 124:75–95

Savides TJ, Messin S, Senger C, Kripke DF 1986 Natural light exposure of young adults. Physiol Behav 38:571–574

Shanahan TL, Czeisler CA 1991 Light exposure induces equivalent phase shifts of the endogenous circadian rhythms of circulating plasma melatonin and core body temperature in men. J Clin Endocrinol & Metab 73:227–235

Stopa EG, King JC, Lydic R, Schoene WC 1984 Human brain contains vasopressin and vasoactive intestinal polypeptide neuronal subpopulations in the suprachiasmatic region. Brain Res 297:159–163

Strassman RJ, Qualls CR, Lisansky EJ, Peake GT 1991 Elevated rectal temperature produced by all-night bright light is reversed by melatonin infusion in men. J Appl Physiol 71:2178–2182

Strogatz SH 1990 Interpreting the human phase response curve to multiple bright-light exposures. J Biol Rhythms 5:169–174

Takahashi JS, DeCoursey PJ, Bauman L, Menaker M 1984 Spectral sensitivity of a novel photoreceptive system mediating entrainment of mammalian circadian rhythms. Nature 308:186–188

Terman M 1988 On the question of mechanism in phototherapy for seasonal affective disorder: considerations of clinical efficacy and epidemiology. J Biol Rhythms 3:155–172

Terman M, Terman J 1985 A circadian pacemaker for visual sensitivity? Ann NY Acad Sci 453:147–161

Terman M, Remé CE, Wirz-Justice A 1991 The visual input stage of the mammalian circadian pacemaking system. II. The effect of light and drugs on retinal function. J Biol Rhythms 6:31–48

Tierstein PS, Goldman AI, O'Brien PJ 1980 Evidence for both local and central regulation of rat rod outer segment disc shedding. Invest Ophthalmol & Visual Sci 19:1268–1273

Turek FW 1989 Effects of stimulated physical activity on the circadian pacemaker of vertebrates. J Biol Rhythms 4:135–147

Van Cauter E, Sturis J, Byrne MM et al 1993 Preliminary studies on the immediate phase-shifting effects of light and exercise on the human circadian clock. J Biol Rhythms 8:S99–S108

Van Reeth O, Turek FW 1989 Stimulated activity mediates phase shifts in the hamster circadian clock induced by dark pulses or benzodiazepines. Nature 339:49–51

Van Reeth O, Vanderhaeghen JJ, Turek FW 1988 Phase advancing effects of benzodiazepine on the hamster circadian clock can be blocked by preventing the associated hyperactivity. In: Proceedings of the 1st Meeting of the Society for Research on Biological Rhythms, Charleston, NC, p 29(abstr)

Wahlström G 1965 Experimental modifications of the internal clock in the canary, studied by self-selection of light and darkness. In: Aschoff J (ed) Circadian Clocks. North-Holland, Amsterdam, p 324–328

Walz B, Sweeney BM 1979 Kinetics of the cycloheximide-induced phase changes in the biological clock in *Gonyaulax*. Proc Natl Acad Sci USA 76:6443–6447

Warden AW 1978 Circadian rhythms of self-selected lighting in golden hamsters: relation to gonadal condition. Chronobiologia 5:28–38

Wehr TA, Jacobsen FM, Sack DA, Arendt J, Tamarkin L, Rosenthal NE 1986 Phototherapy of seasonal affective disorder: time of day and suppression of melatonin are not critical for antidepressant effects. Arch Gen Psychiatry 43:870–875

Weitzman ED, Moline ML, Czeisler CA, Zimmerman JC 1982 Chronobiology of aging: temperature, sleep–wake rhythms and entrainment. Neurobiol Aging 3:299–309

Wever R 1970 Zur Zeitgeber-Stärke eines Licht-Dunkel-Wechsels für die circadiane Periodik des Menschen. Eur J Physiol 321:133–142

Wever R 1973 Internal phase-angle differences in human circadian rhythms: causes for changes and problems of determinations. Int J Chronobiol 1:371–390

Wever RA 1979 The circadian system of man: results of experiments under temporal isolation. Springer-Verlag, New York

Wever RA 1989 Light effects on human circadian rhythms: a review of recent Andechs experiments. J Biol Rhythms 4:161–185

Wever RA, Polášek J, Wildgruber CM 1983 Bright light affects human circadian rhythms. Pflügers Arch Eur J Physiol 396:85–87

Wiedenmann G 1977 Weak and strong phase shifting in the activity rhythm of *Leucophaea maderae* (*blaberidae*) after light pulses of high intensity. Z Naturforsch Sect C Biosci 32:464–465

Winfree AT 1973 Resetting the amplitude of *Drosophila*'s circadian chronometer. J Comp Physiol 85:105–140

Winfree AT 1974 Suppressing *Drosophila* circadian rhythm with dim light. Science 183:970–972

Winfree AT 1980 The geometry of biological time. Springer-Verlag, New York

Winfree AT 1987 The timing of biological clocks. Scientific American, New York

Winget CM, Hughes L, LaDou J 1978 Physiological effects of rotational work shifting: a review. J Occup Med 20:204–210

Wirz-Justice A, Bucheli C, Graw P, Kielholz P, Fisch HU, Woggon B 1986 Light treatment of seasonal affective disorder in Switzerland. Acta Psychiatr Scand 74:193–204

Zulley J, Wever R, Aschoff J 1981 The dependence of onset and duration of sleep on the circadian rhythm of rectal temperature. Pflügers Arch Eur J Physiol 391:314–318

DISCUSSION

Lewy: I was pleased to see in Shanahan & Czeisler (1991a) that there is an excellent correlation between the melatonin rhythm and the fitted temperature minimum. However, Fig. 2 of that paper, which is Fig. 6 in the paper you have just presented, appears to be in error. The original data appeared in Theresa Shanahan's thesis (Shanahan 1990), where the mark for zero melatonin is at the level of the abscissa, indicating higher melatonin concentrations than you report in Shanahan & Czeisler (1991a), where the zero mark is somewhat above the abscissa. That is, in Shanahan & Czeisler (1991a) the lowest levels appear to be close to zero, when they were in fact about 30 pmol/l; similarly, the average baseline levels in Shanahan & Czeisler (1991a) are higher than apparent in Shanahan (1990).

Czeisler: Both the temperature and the melatonin axis are correctly labelled. Daytime levels of melatonin were quite low, which is normal. Note that the first tick on the y-axis is at 250 pmol/l.

Lewy: The point I'm trying to make is that you have stated that the dim light melatonin onset (DLMO) is not as reliable a phase marker as the fitted melatonin maximum, when in fact there is a considerable amount of noise in your melatonin assay that confounds daytime levels (needed to determine the DLMO) more than night-time levels (needed for the maximum). In light of this, do you still think that the DLMO is not as reliable as the fitted melatonin maximum?

Czeisler: Yes. Wehr's data (Wehr 1991, Wehr et al 1993) make me even more concerned about using the DLMO as a marker. Although melatonin onset is an excellent marker of phase and is highly correlated with the fitted maximum, exclusive reliance on melatonin onset as a phase reference point assumes that the waveform of the melatonin rhythm remains fixed across conditions. However, when we have monitored melatonin profiles around the clock in our research studies, we have found that some subjects have broader or narrower melatonin patterns than others. Wehr found that the duration of melatonin secretion varied within the same subject under different photoperiodic conditions. However, I would agree with you that measurement of the onset of melatonin secretion is more convenient than assessment of the entire waveform; melatonin onset can be measured when people are awake, without the need to collect data throughout the night. One question that Theresa Shanahan addressed in her undergraduate thesis (1990) was whether the onset of the nocturnal rise in melatonin levels or the fitted peak of the rhythm was better correlated with the endogenous phase of the temperature cycle. She found that the fitted peak was somewhat better correlated with temperature than was the onset of melatonin secretion. As Professor Richard Kronauer once put it, you can't judge from the foot of a mountain exactly where the peak will be. We continue to look not only at melatonin onsets but also at the whole curve whenever possible, because if the shape of the curve and the duration of melatonin secretion vary, we feel that the entire rhythm must be evaluated in order to characterize fully its response to a stimulus.

Arendt: The onset and the decline can be affected differentially depending on the circumstances of the experiment, but, in gross terms, the calculated peak time and the onset usually correlate closely.

Czeisler: I agree; we are quibbling over details. Theresa Shanahan showed that there was a very high correlation, 0.965, between the melatonin onset time and peak time under normal conditions.

Lewy: We measured some samples for Theresa Shanahan so that she could see how the assay she had used compared with our gas chromatography–mass spectrometry assay (Lewy & Markey 1978). As indicated in her thesis (Shanahan 1990, Fig. 2 *top*, p 62), the y intercept was not at the origin but at 80 pmol/l, indicating substantial interference (most probably due to cross-reactivity) in her radioimmunoassay. The same assay run in Josephine Arendt's laboratory (Shanahan 1990, Fig. 2 *bottom*, p 62) had 40 pmol/l less interference. Melatonin assays must have less noise when one is using the DLMO as a phase marker rather than using the entire night-time curve.

Czeisler: Theresa Shanahan (1990) used a variety of levels to assess melatonin onset. Even at 150 pmol/l, well above the noise of the radioimmunoassay, her results were consistent, indicating that the peak was better correlated with the temperature minimum than the onset of secretion, however defined.

Lewy: It's true that one way to get around the problem of noise in assays is to set the threshold for the DLMO higher; however, the higher it is, the further away it is from the SCN's melatonin onset signal.

Waterhouse: There are two issues here. The first is one of analysis, which though it's important for the research worker might not be of great clinical significance. The second issue is that you are measuring different aspects of the output from the oscillator—the onset of secretion and the overall output. Illnerová's work shows that there is apparently quite a sophisticated model that can be constructed by considering the onset and termination of secretion separately.

Czeisler: I see no reason to throw away the rest of the curve if the data are available—that's not to say that the DLMO is not a pragmatic way of assessing phase in the clinical situation.

Takahashi: As an 'impartial' observer, it seems to me that the DLMO is more sensitive to assay variation, and therefore is not as good a phase marker, because it won't work in as many conditions.

Lewy: But other melatonin assays (including radioimmunoassays), and even the assay in question as done in other laboratories, are less noisy and therefore more useful for determining the DLMO.

Takahashi: I don't want to get into the details. It seems to me that if it is necessary to get into arguments about baseline values and zero points for your particular analytical method, that actually compromises that phase marker, and makes it a less useful one.

Waterhouse: Dr Czeisler, what is the correlation between the different markers you use in your constant routine?

Czeisler: The fitted body temperature minimum correlated more strongly with the fitted maximum of the melatonin rhythm ($r^2 = 0.99$) than with melatonin onset ($r^2 = 0.96$), although both were highly correlated. In addition to the Arendt assay we are now using an assay distributed by Elias USA, Inc. We shall be able to validate these two different assays and see whether Theresa Shanahan's finding is consistent using the new assay.

Lewy: Even with the noise in Dr Czeisler's melatonin assay, the DLMO correlates well ($r^2 = 0.963$, $P < 0.0001$) with the fitted temperature minimum; this correlation is no different from that between the latter and the fitted melatonin maximum ($r^2 = 0.990$, $P < 0.0001$) (Shanahan 1990). You stated in Shanahan & Czeisler (1991a) that there is no significant difference in the average duration of the melatonin peak before (7.5 ± 0.6 h) and after (7.8 ± 0.3 h) light-induced phase shifts. Therefore, the DLMO appears to be an accurate marker for the circadian phase of the entire melatonin curve as it responds to the phase-shifting effects of light.

Czeisler: Shanahan's (1990) data were collected before and after exposure to light. Her more recent data collected throughout the light-resetting protocol suggest that the fitted melatonin maximum is the more robust marker of phase when melatonin amplitude declines. Also, you need to consider data collected under a variety of conditions. In Wehr's experiments, for example, I would be interested to know which component of the melatonin rhythm proves to be the best marker of phase. When the duration of melatonin release greatly expands in response to a shortened photoperiod, I would be interested to know where the true phase of the circadian system is. Is it still linked to the onset of melatonin release, or, as Illnerová et al (1989) suggest, does this increased duration reflect interaction of a morning and evening oscillator, or is the phase of the melatonin rhythm best marked by the peak of the rhythm?

Waterhouse: You could take advantage of the differences to see whether there are some circumstances in which one method has an advantage over another. Many of the data obtained in this field are heavily masked (that is, distorted by other factors such as activity and sleep), so one tries to unmask them. In the field, constant routines and DLMO methods are not applicable. Therefore, less sophisticated but more practicable methods must be used instead (see, for example, Folkard 1989, Minors & Waterhouse 1992). There might have to be a compromise between what's desirable in one sense and what's achievable in another. It's up to the various protagonists to assess the value of the different methods in specific cases.

Armstrong: What is the minimum light level needed to entrain the human circadian system? We did an experiment in the mid 1980s in which even 350 lux of light significantly reduced melatonin levels but 200 lux of light reduced melatonin but not significantly (McIntyre et al 1989). Dr Czeisler thought that Fig. 2 in his paper (p 258, this volume; Czeisler et al 1981) showed entrainment to a dim light–dark cycle, but it looked to me as though the subject was still free running. If the subject were entrained to the scheduled light–dark cycle at low light intensities during Days 30–50, the free run should start from the start of the dark period on the last day of the light–dark cycle, whereas it starts towards the end of the dark period. There are therefore two explanations other than entrainment which have to be considered. The first is that the subject was never entrained but continued to free run; because the imposed light–dark cycle is mandatory, the subject has no option but to rest in the dark period, so the result can be attributed to masking. If the subject is free running, the phase of the free run in the first few days at the termination of the light–dark cycle should be predictable from the phase in the last few days before the commencement of the light–dark cycle. This is a possible interpretation of the figure. The second possibility is that the subject was truly entrained but that there was a phase jump on the first day of release from the light–dark cycle. This possibility cannot easily be excluded but is unlikely. Examples of phase jumps are found in the literature on rodents but not humans. However, if a phase jump did occur, not only must light have fallen on the maximum delay portion of the phase response curve (PRC) but the sensitivity of the PRC to light must also be far greater than has hitherto been shown to be the case in humans.

Czeisler: I would disagree that the subject was still free-running during exposure to the light–dark cycle. If he were, the phase of the activity rhythm on release from the light–dark cycle would be predictable from the subject's free run before entrainment, which is not the case (see *Note* opposite). Also, your view that a typical human subject goes to sleep on the first day of a free run at the same time as the last entrained day is not supported by the data in the literature. In fact, as I have written in my doctoral thesis (Czeisler 1978, p 119), 'Subjects FR01, FR05, FR06 and FR09 [four of the six who showed synchronized free-running patterns] had a very rapid sleep–wake cycle shift immediately after being released from entrainment. In each case, there was a delay of approximately 3–5 hours in the chosen sleep onset time just after the start of the free-running condition. This occurred within one or two sleep periods after the last entrained night. The period length of the [sleep–wake] rhythm then shortened to about twenty-five hours for the rest of the experiment. That pattern was not observed in FR02 or FR07.' This observation is consistent with both Wever's data (1979) and my observation (Czeisler 1978) of an internal phase delay of the sleep–wake cycle with respect to the body temperature and cortisol rhythms under free-running conditions in comparison with entrained conditions.

Finally, you have offered no alternative explanation for the demonstration that the body temperature cycle was also entrained by the imposed light–dark cycle in the experiments which you cited (Czeisler et al 1981). I must therefore respectfully disagree with your re-analysis of our data and reassert my conclusion that daily exposure to low levels of light, around 150 lux, for 16 h/day is sufficient to entrain the circadian pacemaker to the 24 h day in normal young men.

Kronauer: In fact, the minimum light level needed to entrain the human circadian system depends on the intrinsic τ. If the intrinsic τ is 24.2, 150 lux is enough.

Armstrong: Then why didn't Wever get entrainment in his bunker experiments at Andechs (Wever 1979)?

Kronauer: I think you misunderstood Dr Czeisler's figure (Fig. 2, p 258), where there was strict control over darkness. In the bunker experiments the failure to get entrainment was because the subjects were able to use lights in the bedroom, kitchen and bathroom.

Armstrong: Is a bedside lamp, turned on under the subject's own volition, enough to upset entrainment?

Czeisler: The auxiliary lighting available to the subjects in the bunker at the Max Planck Institute was certainly enough to affect entrainment, even when the overhead lights controlled by the experimenters were switched off. Dr Wever

Light and the human circadian pacemaker

Note

After the symposium, the following material was submitted for inclusion in the present volume.

Czeisler: Linear regression through the subject's free-running activity onsets for three weeks *before* exposure to the 24 h cycle of ordinary room light and darkness do not project onto the free-running activity onsets during the eight days *after* release from exposure to the light–dark cycle. In fact, regression analysis indicates that activity onset upon release from the light–dark cycle occurs 17.4 h (not *modulo* 24 h, as Armstrong suggests) before that projected from the prior free run.

Armstrong: Regression lines for free-running rhythms are conventionally plotted using 10–14 days of data.

Waterhouse: The second segment is short and the times of sleep termination are irregular enough for some doubt to remain with regard to the extrapolation of phase.

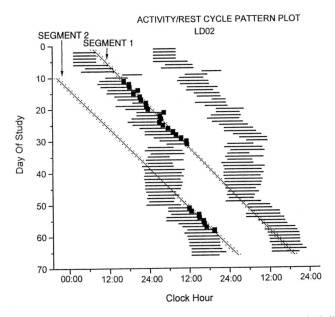

FIG. 1. Free-running activity–rest cycle before and after exposure to a 24 h light–dark cycle consisting of 16 h room light and eight hours' darkness per day (between Days 31 and 50). Horizontal bars represent sleep episodes, as in Fig. 2, p. 258. There is no overlap between the 95% confidence intervals (dashed lines) of regression lines fitted to self-selected activity onset times (■) for the three weeks prior to (Segment 1) and the eight days after (Segment 2) exposure to the light–dark cycle, even when they are repeated at *modulo* 24 h intervals. Projected activity onset upon release from the light–dark cycle to free run (Day 51) is at 05.31 h (95% confidence interval 04.52–06.10 h) using Segment 1 regression data (*before* exposure to the light–dark cycle), whereas activity onset on Day 51 projected using Segment 2 regression data (*after* exposure to the light–dark cycle) is at 12.05 h (95% confidence interval 11.22–12.48 h). In fact, activity onsets from Segment 1 project onto Segment 2 bedtimes. Adapted from Czeisler et al (1981).

took a group of scientists on a tour of the bunker in 1976, which is when I learned about the lighting conditions in those experiments. I asked Dr Wever what it was like when it was dark, and he responded that it was 'dark'. Yet, the light intensity in the bunker at that time, with only the auxiliary lighting controlled by the subjects, was sufficiently bright for Dr Wever to conduct a tour of the facility for an international group of scientists without using the experimenter-controlled overhead lights. To characterize such extensive auxiliary lighting as a mere 'bedside lamp' would be misleading. Certainly, the lighting available to the subjects in the German bunker was sufficient to override entraining effects of an imposed light–dark cycle. Furthermore, we have found that the availability of such relatively low intensity light (around 150 lux) to subjects in the traditional free-running studies feeds back on the circadian pacemaker, resulting in an observed free-running period longer than that of the true intrinsic period of the pacemaker.

We have shown in recent studies in Boston (C. A. Czeisler et al, unpublished work) using a forced desynchrony protocol that the true intrinsic period of the human circadian pacemaker is much closer to 24 h than had previously been estimated, averaging 24.25 h in normal, young, sighted subjects following release from entrainment to the 24 h day (Czeisler et al 1990, Shanahan & Czeisler 1991b). This is roughly a quarter of an hour shorter than the average intrinsic period in totally blind subjects who have not been synchronized to the 24 h day. The shorter intrinsic period in sighted subjects is most probably due to the after-effects of their prior entrainment to the 24 h day.

The next question is, what is the most relevant period? The amount by which we are resetting our circadian clock every day is the difference between the intrinsic period, which on average in young people is 24.25 h, and 24 h. We are not ordinarily exposed to the 100 or 200 days of constant darkness required for the innate period to be revealed. We are always operating under the after-effects of entrainment to a 24 h light–dark cycle. It is certainly consistent with Eskin's data (1969, 1971) that the after-effects of entrainment to the 24 h day would have a long-lasting effect on the free-running period, of a few tenths of an hour. Therefore, we feel that the 24.25 h period we observe in young subjects is the value most relevant to the entrainment of normal young human subjects.

Turek: We have seen after-effects on period of triazolam treatment that last for well over a hundred days (Van Reeth & Turek 1990).

Armstrong: Having taught undergraduates for about 15 years that the τ of the human clock is around 25 h but can be as long as 27 or even 33 h, it's a bit of a shock to find that the τ of your free-running subjects is so short and that the traditionally reported longer τs were due to an after-effect of the previous light cycle. Yet, the τs of blind people in the studies by Sack & Lewy (e.g., Sack et al 1992) are not so short. Are there any other explanations for these discrepant findings? What is the *real* τ of the human clock?

Czeisler: First, let me clarify what I have just said. I said that the after-effects of entrainment to the 24 h day may explain why sighted subjects have an average period about 15 min shorter than the average blind subject. However, both the data from our forced desynchrony studies in sighted subjects (average τ about 24.25 h) and the data from blind subjects (average τ of about 24.5 h) indicate that the intrinsic period of the human circadian pacemaker is considerably shorter than the 25 h figure you have been teaching your undergraduates. We do not think after-effects explain this difference. Rather, we have concluded that the subjects' self-selected exposure to room light in the German bunker led to a misestimation of the τ of the human clock. In fact, Dr E. Klerman has taken Professor Kronauer's model of the circadian pacemaker and run simulations (Klerman et al 1992). She input an intrinsic period of 24.3 h into the Kronauer model equations and then simulated the effect of the timing of the self-selected light–dark cycle from free-running studies on the period of the pacemaker. She found that the period of the oscillator under the influence of feedback from the self-selected light–dark cycle lengthened from 24.3 h to 24.7–24.8 h. If she then added the confounding effects of masking, reversing a procedure similar to that suggested by Minors & Waterhouse (1989, 1992) in which the temperature is increased slightly when the subject is awake and decreased when the subject is asleep, and analysed the simulated, masked 'temperature' data by traditional spectral analysis techniques, she extracted an observed period of 25.1–25.2 h. This was the output of her simulations despite the fact that the intrinsic pacemaker input period was 24.3 h.

Takahashi: You have shown before that the sleep–wake cycle must have a weak coupling effect on temperature oscillators. Is that still true?

Kronauer: Do you mean by means of light?

Takahashi: No, I just mean oscillator interaction.

Kronauer: The data that we reported (Kronauer et al 1993a) are based on a series of experiments in which neither bright light nor normal room light were ever used. The subjects were in either 10–15 lux or darkness. After an initial constant routine, the sleep–wake cycle was acutely inverted and kept inverted for the course of the protocol (about 10 days). An intermediate constant routine was imposed after three cycles of sleep–wake inversion and a third after three more cycles of inversion. Data for 13 subjects showed no change in rhythm amplitude over this time and only a slow phase drift, in the delay sense, of about 0.25 h/day. We ascribe the drift to an endogenous τ of 24.25 h (on average) and consequently find no effect of the inversion of the sleep–wake cycle on either the pacemaker amplitude or phase.

Takahashi: So, is the original model now modified and the oscillator coupling between the sleep–wake cycle and the temperature rhythm weaker?

Kronauer: The drive from the sleep–wake cycle is essentially zero.

Takahashi: If that's true, then experimentally you should be able to drive the sleep–wake cycle to any period and you should always get a body temperature rhythm of 24.2. Have you done control experiments driving with different periods?

Czeisler: Yes, we have done that, using a forced desynchrony protocol on day lengths of 28, 20 and 11 h. We managed to get one subject to come back for three month-long experiments, one at 28, one at 20 and one at 11 h, and his intrinsic period was the same in all three conditions.

We are now doing studies of nine blind subjects. None have any conscious light perception. Five of them are free-running, two still have positive melatonin suppression (Czeisler et al 1994), with light coming into the system in the same manner as it does in retinally degenerate mice (Foster et al 1991, 1993, Foster & Menaker 1993, Provencio et al 1994). We can't say that the other two are free-running; in fact, they appear to be entrained, even though they exhibit negative melatonin suppression and are bilateral enucleated. Either they have a period which is very close to 24 h, or it may be that when the intrinsic period is in the range of 24.0–24.1 h, relatively weaker entraining effects of non-photic synchronizers that can't yet be detected from our paradigms—such as the effects of activity, sleep–wake, exercise, food, etc.—can synchronize the pacemaker to the 24 h day. But the response of the human circadian pacemaker to these non-photic synchronizers is certainly at least an order of magnitude lower than the response to light.

Takahashi: I gather that in the particular study where you give a single bright light pulse you don't get type 0 resetting (Jewett et al 1994). If you were able to do the resetting experiment in which you give a single bright light pulse against a completely dark background you might see type 0 resetting in response to single pulses. This kind of effect is seen in animal experiments. Hermann Pohl varied the background illumination and then measured the PRC to the same light pulse using one-hour light pulses in hamsters. When the illumination background was very dim he got a type 0 curve and when it was raised to about one lux the response became type 1. The background does have effects.

Kronauer: How long was the animal in that restricted background?

Takahashi: A long time, perhaps months, in constant illumination.

Kronauer: One thing we found in our experiments showing a reduced amplitude of output markers (Jewett et al 1991) was that if you expose subjects to between eight and nine consecutive hours of bright light it is possible to get large amplitude reductions in the output markers, comparable to those produced by two five-and-a-half-hour episodes of bright light on successive days.

Daan: I (Beersma & Daan 1993) take a slightly different view from Dr Czeisler and Dr Kronauer (Kronauer et al 1993b) with regard to the interpretation of these data, but I should emphasize that I have great respect for the quality of the data and for the entire work. I do not agree with the conclusion that the resetting data (Czeisler et al 1989) are evidence for strong phase resetting. Certainly these data appear to produce a type 0 PRC, but one has to realize that they have been obtained after three pulses, three cycles of exposure to an entraining agent. We have been sitting in this room for about three cycles of Greenwich time, so, regardless of where we came from, America or Australia,

Light and the human circadian pacemaker

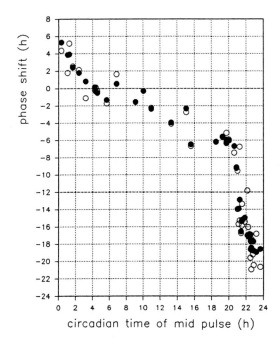

Fig. 2. *(Daan) Upper panel:* Circles show steady-state phase shifts of free-running human circadian rhythms following three-hour, 5000 lux light pulses, assessed by Honma & Honma (1988) and Minors et al (1991). The line connects hourly values (divided by 2.06 to correct for differences in intensity and duration of light in the two studies) of a single-pulse PRC which generates, when applied three times in 24 h intervals, the best fit (smallest sum of squared deviations) to the phase shifts measured by Czeisler et al (1989). *Lower panel:* ○, phase shifts of human circadian rhythms following three-hour, 9500 lux light pulses (from Czeisler et al 1989); ●, phase shifts calculated by applying the single-pulse PRC (line in upper panel) three times. Reproduced from Beersma & Daan 1993, with permission.

we have become nearly entrained to this time zone. New phase becomes restricted to a narrow band after three days. Eventually, any organism, if you expose it long enough to zeitgeber cycles, will show a type 0 PRC. For humans, a single pulse PRC is available from Honma & Honma (1988) and from Dr Waterhouse's group. Single pulses clearly produce a type 1 PRC (Fig. 2, *upper panel*). There are methodological differences, of course, but I don't think they are particularly relevant. Dr D. G. M. Beersma and I have recently calculated the PRC for single pulses (no matter which type, and independent of the data of Honma & Honma [1988] and Minors et al [1991]), which generates the closest possible approximation to Dr Czeisler's phase shift data when the pulses are applied three times in 24 h intervals. The fitted points are shown in the lower panel of the figure. The calculated PRC was reduced by a factor of 2.06 to account for differences in light intensity and duration between the studies (see Beersma & Daan 1993) and is shown in the upper panel. This PRC is clearly type 1. It is, moreover, not distinctly different from the single-pulse PRC from the Waterhouse and Honma groups. This shows that it is possible with type 1 single pulses applied three times to get something which looks like a type 0 PRC.

Kronauer: There are two features of this calculation on which I must comment. The data points in the upper panel of Fig. 2 were for single three-hour stimulus episodes at an inferred light intensity of 5000 lux. In trying to relate these data to our experiments in which intensity was 9500 lux and the duration of the light episodes was five hours, Dr Daan proposes to increase the magnitude of the phase shifts by a factor of 2.06. Subsequently, when the computer was posed the inverse problem, to produce the best fit to our data, Dr Daan chose to reduce the computer solution by the factor of 2.06. That is, the computer result is not as shown by the continuous line in the upper panel of Fig. 2—it is actually 2.06 times larger than shown. Consequently, the maximum and minimum phase shifts attributed by the computer to a *single* five-hour bright light episode are $+8.3$ and -8.0 h, respectively, and are inconsistent with empirical data (Minors et al 1991, Jewett et al 1994, Kronauer et al 1993b). This range of phase shifts encompasses all but seven of our 45 experimental phase shifts, in response to *three* five-hour episodes. Moreover, the computer solution has seven local maxima and seven local minima. This jaggedness arises because the computer was placed under no constraint to produce a smooth, plausible PRC. When this curve is iterated three times the PRC produced is even more jagged, with 21 local maxima and 21 local minima. The selected computer results, the filled circles in the lower panel of Fig. 2, don't begin to show the true excursions and irregularity of the thrice-iterated PRC. For example, in the range of initial stimulus phase between CT 20.1 and CT 20.9 (the critical phase is around CT 20.9) there are three local minima (with phase shifts of -10.2, -11.7 and -9.5 h) and maxima in between (with phase shifts of -6.2 and -4.6 h). Neither the single-episode PRC (*upper panel*) nor the three-episode PRC (*lower panel*) can be thought of as realistic.

Daan: The erratic nature of the calculated curve was as produced by the computer. This may seem to be a problem but, on the other hand, if you do the computation the other way round, and use the smoothed Honma–Minors one-pulse PRC (Honma & Honma 1988, Minors et al 1991) and apply that three times, you end up with a type 0 PRC which is again very close to the Czeisler data. There are no erratic regions in this—it's a smooth curve.

Turek: Let's be clear about this. The published human PRCs do not have much data behind them. Before we get carried away with simulations based on few data, let's get a really good PRC with single pulses of light of different durations and different intensities in humans.

Kronauer: I agree with you, but there is another point central to the exchange Dr Daan and I have been having. The question is whether is it possible to discriminate accurately between type 1 and type 0 resetting in the protocol which we used (Czeisler et al 1989), bearing in mind that phase assessments before and after stimulus intervention are unavoidably noisy. Dr Daan has undertaken to demonstrate that successive application of three single-episode type 1 PRCs can come close to our data, thereby implying that the data do not definitively show type 0 resetting. I have been arguing that the single-episode PRC which he has come up with is unrealistic from two points of view: the maximum phase shifts are very large, and the PRC has a succession of unbelievable peaks and troughs. Furthermore, owing to large negative slope, it cannot represent a simple phase-only oscillator.

Waterhouse: This issue clearly needs further discussion. There appear to be differences in concept as well as methodological approach.

References

Beersma DGM, Daan S 1993 Strong or weak phase resetting by light pulses in humans? J Biol Rhythms 8:340–347
Czeisler CA 1978 Internal organization of temperature, sleep–wake, and neuroendocrine rhythms monitored in an environment free of time cues. PhD thesis, Stanford University, Stanford, CA
Czeisler CA, Richardson GS, Zimmerman JC, Moore-Ede MC, Weitzman ED 1981 Entrainment of human circadian rhythms by light–dark cycles: a reassessment. Photochem Photobiol 34:239–247
Czeisler CA, Kronauer RE, Allan JS et al 1989 Bright light induction of strong (Type 0) resetting of the human circadian pacemaker. Science 244:1328–1333
Czeisler CA, Allan JS, Kronauer RE 1990 A method for assaying the effects of therapeutic agents on the period of the endogenous circadian pacemaker in man. In: Montplaisir J, Godbout R (eds) Sleep and biological rhythms: basic mechanisms and applications to psychiatry. Oxford University Press, New York, p 87–98
Czeisler CA, Shanahan TL, Klerman EB et al 1994 Bright light suppression of melatonin concentrations in some blind patients. N Engl J Med, in press
Eskin A 1969 The sparrow clock: behavior of the free running rhythm and entrainment analysis. PhD thesis, University of Texas, Austin, TX, USA
Eskin A 1971 Some properties of the system controlling the circadian activity rhythm of sparrows. In: Menaker M (ed) Biochronometry. National Academy of Sciences, Washington, DC, p 55–80
Folkard S 1989 The pragmatic approach to masking. Chronobiol Int 6:55–64
Foster RG, Menaker M 1993 Circadian photoreception in mammals and other vertebrates. In: Wetterberg L (ed) Light and biological rhythms in man. Pergamon Press, New York, p 73–91

Foster RG, Provencio I, Hudson D, Fiske S, DeGrip W, Menaker M 1991 Circadian photoreception in the retinally degenerate mouse (*rd/rd*). J Comp Physiol A Sens Neural Behav Physiol 169:39–50

Foster RG, Argamaso S, Coleman S, Colwell CS, Lederman A, Provencio I 1993 Photoreceptors regulating circadian behavior: a mouse model. J Biol Rhythms 8:S17–S23

Honma K, Honma S 1988 A human phase response curve for bright light pulses. Jpn J Psychiatry 42:167–168

Illnerová H, Vaněček J, Hoffmann K 1989 Different mechanisms of phase delays and phase advances of the circadian rhythm in rat pineal *N*-acetyltransferase activity. J Biol Rhythms 4:187–200

Jewett ME, Kronauer RE, Czeisler CA 1991 Light-induced suppression of endogenous circadian amplitude in humans. Nature 350:59–62

Jewett ME, Kronauer RE, Czeisler CA 1994 Phase/amplitude resetting of the human circadian pacemaker. J Biol Rhythyms, submitted

Klerman EB, Dijk DJ, Czeisler CA, Kronauer RE 1992 Simulations using self-selected light–dark cycles from 'free-running' protocols in humans result in an apparent *tau* significantly longer than the intrinsic *tau*. In: Proceedings of the 3rd meeting of the Society for Research on Biological Rhythms, p 110(abstr)

Kronauer RE, Duffy JE, Czeisler CA 1993a Inversion of the sleep–wake cycle in a dim light environment has no significant effect on the circadian pacemaker. Sleep Res 22:626

Kronauer RE, Jewett ME, Czeisler CA 1993b The human circadian response to light: strong *and* weak resetting. J Biol Rhythms 8:351–360

Lewy AJ, Markey SP 1978 Analysis of melatonin in human plasma by gas chromatography negative ionization mass spectrometry. Science 201:741–743

McIntyre IM, Norman TR, Burrows GD, Armstrong SM 1989 Human melatonin suppression by light is intensity dependent. J Pineal Res 6:149–156

Minors DS, Waterhouse JM 1989 Masking in humans: the problem and some attempts to solve it. Chronobiol Int 6:29–53

Minors DS, Waterhouse JM 1992 Investigating the endogenous component of human circadian rhythms: a review of some simple alternatives to constant routines. Chronobiol Int 9:55–78

Minors DS, Waterhouse JM, Wirz-Justice A 1991 A human phase–response curve to light. Neurosci Lett 133:36–40

Provencio I, Wong SY, Lederman AB, Argamaso SM, Foster RG 1994 Visual and circadian responses to light in aged retinally degenerate (*rd*) mice. Vision Res 34:1799–1806

Sack RL, Lewy AJ, Blood ML, Keith LD, Nakagawa H 1992 Circadian rhythm abnormalities in totally blind people: incidence and clinical significance. J Clin Endocrinol & Metab 75:127–134

Shanahan TL 1990 Evaluation of the twenty-four hour plasma melatonin pattern as a marker of the human circadian system. BSc thesis, Boston College, Boston, MA, USA

Shanahan TL, Czeisler CA 1991a Light exposure induces equivalent phase shifts of the endogenous circadian rhythms of circulating plasma melatonin and core body temperature in men. J Clin Endocrinol & Metab 73:227–235

Shanahan TL, Czeisler CA 1991b Intrinsic period of the endogenous circadian rhythm of plasma melatonin is consistent with that of core body temperature during forced desynchrony in a 21-year-old man. Sleep Res 20A:557(abstr)

Van Reeth O, Turek FW 1990 Daily injections of triazolam induce long-term change in hamster circadian period. Am J Physiol 259:514–520

Wehr TA 1991 The durations of human melatonin secretion and sleep respond to changes in daylength (photoperiod). J Clin Endocrinol & Metab 73:1276–1280

Wehr TA, Moul De, Barbato G et al 1993 Conservation of photoperiod-responsive mechanisms in humans. Am J Physiol 265:R846-R857

Wever RA 1979 The circadian system of man: results of experiments under temporal isolation. Springer-Verlag, New York

Melatonin marks circadian phase position and resets the endogenous circadian pacemaker in humans

Alfred J. Lewy, Robert L. Sack, Mary L. Blood, Vance K. Bauer, Neil L. Cutler and Katherine H. Thomas

Sleep and Mood Disorders Laboratory, L-469, Departments of Psychiatry, Ophthalmology and Pharmacology, Oregon Health Sciences University, 3181 SW Sam Jackson Park Road, Portland, OR 97201-3098, USA

>*Abstract.* Measuring the dim light melatonin onset (DLMO) is a useful and practical way to assess circadian phase position in humans. As a marker for the phase and period of the endogenous circadian pacemaker, the DLMO has been shown to advance with exposure to bright light in the morning and to delay with exposure to bright light in the evening. This 'phase response curve' (PRC) to light has been applied in the treatment of winter depression, jet lag and shift work, as well as circadian phase sleep disorders. Exogenous melatonin has phase-shifting effects described by a PRC that is about 12 h out of phase with the PRC to light. That is, melatonin administration in the morning causes phase delays and in the afternoon causes phase advances. All of the circadian phase disorders that have been successfully treated with appropriately timed exposure to bright light can be treated with appropriately scheduled melatonin administration. Melatonin administration is more convenient and therefore may be the preferred treatment.
>
>*1995 Circadian clocks and their adjustment. Wiley, Chichester (Ciba Foundation Symposium 183) p 303–321*

Over the past decade, the use of the melatonin rhythm as a marker for the endogenous circadian pacemaker has gained increasing recognition (Lewy & Sack 1989, Lewy et al 1988). Measuring the onset of melatonin production is a useful and practical way in which to assess the phase position of the endogenous circadian pacemaker. The melatonin rhythm appears to be tightly coupled to the endogenous circadian pacemaker. In totally blind people, there do not appear to be any masking effects on melatonin levels: that is, melatonin is relatively unaffected by influences that could diminish its reliability as a marker for circadian phase. In sighted individuals, melatonin is produced only during nighttime darkness and appears to be significantly affected only by exposure to bright light. Therefore, in sighted people, melatonin is assayed in blood drawn under dim light; we call this the dim light melatonin onset (DLMO).

The DLMO as a marker for circadian phase position in humans

The DLMO is a remarkably good marker for circadian phase position in humans, similar to the onset of locomotor activity in certain rodents. The DLMO offers several advantages over the use of the 24 h melatonin curve or the night-time melatonin profile, particularly when melatonin is measured with an accurate and sensitive technique, such as the gas chromatographic–negative ionization mass spectrometric (GCMS) assay (Lewy & Markey 1978). A smaller volume of blood is required, permitting assessment of the phase of the endogenous circadian pacemaker on frequent occasions. For most situations, subjects need not stay overnight in the laboratory, thus lessening cost and reducing interference with sleep. The DLMO is convenient for subjects in actual field situations, including patients with circadian phase disturbances. Also, for theoretical reasons, the melatonin onset is the most accurate part of the melatonin curve for assessing circadian phase position.

The salivary DLMO is also a good marker for circadian phase position and usually can be obtained before the subject goes to sleep. Because salivary concentrations are about one-third of plasma levels, measurement of the salivary DLMO requires a very sensitive melatonin assay. In some individuals, salivary melatonin can be measured accurately only with the GCMS assay.

There are, however, three potential problems with the DLMO. In most experimental and clinical situations these are not major; when they do occur, they can usually be ameliorated. The first potential problem is that the approximate time of the DLMO must be known for selection of the appropriate window for sample collection; otherwise, samples may have to be collected for up to 24 h. The onset of melatonin production is usually in the evening, at circadian time (CT) 14. (We designate the melatonin onset as CT 14 for a number of reasons, primarily because it usually occurs 14 h after '[bright] lights on' in the morning.) A second potential problem is variability in the shape of the melatonin curve between individuals (within individuals, the melatonin curve appears to vary little). In people who are extremely low melatonin producers the melatonin onset may occur later than CT 14, so for inter-individual comparisons the DLMOs of these people can be adjusted using a mathematical correction according to a formula that correlates DLMO time with melatonin amplitude. A third problem may arise if it is shown that two (loosely coupled) dawn–dusk pacemakers drive the melatonin rhythm (Illnerová & Vanecek 1982, Wehr et al 1993). If this is so, the DLMO will probably most accurately mark the circadian phase position of the evening pacemaker, and time of year may be a factor that influences the DLMO. However, the duration of melatonin production does not appear to change after light-induced phase shifts (Shanahan & Czeisler 1991). Therefore, the DLMO nevertheless accurately reflects phase shifts in the endogenous circadian pacemaker, at least after steady-state entrainment ensues.

Effects of bright light on the human circadian system

The ability to measure melatonin accurately resulted in the discovery that bright light could suppress night-time melatonin production in humans (Fig. 1) (Lewy et al 1980) and that shifting the (bright) light–dark cycle could shift the melatonin rhythm, even when the sleep–wake cycle was held constant (Lewy et al 1984, 1985a). Specifically, exposure to bright light in the morning causes phase advances and in the evening causes phase delays (Lewy et al 1987). This 'phase response curve' (PRC) to light—i.e., the phase shift induced by light at different times—is the basis for treating the two types of circadian phase disorders, the phase-delay type and the phase-advance type. These disorders

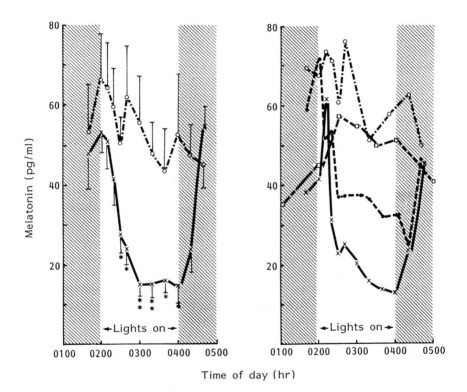

FIG. 1. Effect of light on melatonin secretion. *Left:* Each point represents the mean concentration of melatonin ($\pm$ SE) for six subjects. A paired t-test, comparing exposure to 500 lux (○) with exposure to 2500 lux (×), was performed for each data point. A two-way analysis of variance with repeated measures and the Newman–Keuls statistic for the comparison of means showed significant differences between 02.50 and 04.00 h (*$P<0.05$, **$P<0.01$). *Right:* Effect of different light intensities on melatonin secretion. The averaged values for two subjects are shown. ○, 500 lux; ×, 2500 lux; ●, 1500 lux; and □, asleep in the dark. From Lewy et al 1980, with permission.

include advanced (Lewy et al 1985b) and delayed (Lewy et al 1983) sleep phase syndrome, as well as jet lag (Daan & Lewy 1984) and—more recently—maladaptation to shift work (Czeisler et al 1990, Eastman 1987, 1990).

Bright light was first used in the treatment of affective disorders, such as winter depression (Lewy et al 1982, 1987, Sack et al 1990), where its use remains somewhat controversial. Although the circadian phase-shift hypothesis explaining the antidepressant mechanism of action of bright light in winter depression has gained increasing acceptance, the significant placebo component has to varying extents confounded many studies (Eastman et al 1992). This question should be resolved by the use of a non-photic zeitgeber (melatonin) to induce phase shifts in these patients.

Effects of melatonin administration on the human circadian system

A physiological (0.5 mg) dose of exogenous melatonin appears to shift the endogenous circadian pacemaker in humans according to a PRC (Lewy et al 1992) that is approximately 12 h out of phase with the PRCs to light (Fig. 2) (Czeisler et al 1989, Honma & Honma 1988, Minors et al 1991, Wever 1989). In other words, melatonin appears to act like a dark pulse. Regulated in part by the acute suppressant effect of light, endogenous melatonin stimulating the suprachiasmatic nuclei (SCN) provides a second (indirect) way in which the light–dark cycle can entrain the endogenous circadian pacemaker. Melatonin production at the twilight transitions may help entrainment of the endogenous circadian pacemaker, as well as shift its phase when the light–dark cycle is altered. Night-time melatonin production may also stabilize steady-state entrainment of the pacemaker. Because shifting sleep has little effect on the pacemaker if the light–dark cycle is held constant (Hoban et al 1991, Kronauer et al 1993), it appears that melatonin and the light–dark cycle are the only major zeitgebers in humans.

The phase-shift responses to exogenous melatonin also provide the basis for proper timing of its administration in the treatment of circadian phase disorders. Disorders that have previously responded to the phase-shifting effect of appropriately timed exposure to bright light should also respond to appropriately timed melatonin administration. A low dose should have minimal side effects and should be less soporific than higher doses. This is fortunate, because the PRC indicates that melatonin is most effective when administered during the day. Indeed, one important implication of the melatonin PRC is that it diminishes the decade-old association of melatonin with sleep, which occurs only in diurnal animals. Melatonin is associated with the *dark* phase, not the *sleep* phase. However, some individuals (perhaps those with low melatonin levels) might be especially sensitive to the soporific effects of melatonin. In these individuals, melatonin could possibly induce sleep at night, particularly at pharmacological doses.

Melatonin and the human circadian pacemaker

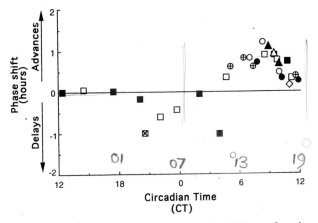

FIG. 2. Phase shifts of the dim light melatonin onset (DLMO) as a function of circadian time (CT) for a total of 30 trials in nine subjects, providing the first evidence for a human melatonin phase response curve (PRC). Each of the nine subjects has a separate symbol. Exogenous melatonin was administered at various times with respect to the time of endogenous melatonin production (CT 14 = baseline DLMO for each trial). The time of administration is shown as CT by convention and because of inter-individual variability in sleep-wake cycles, perceived light-dark cycles and internal circadian time. On average, CT 0 = 07.00 h clock time. Two subjects [J.H. (□) and S.E. (■)] each participated in seven trials: when internal CT is referenced to the baseline DLMO, plots of the data for these two subjects nearly superimpose. From Lewy et al 1992, with permission.

The melatonin PRC explains why previous attempts to shift human circadian rhythms with melatonin were not more robust or consistent (Arendt et al 1985, Mallo et al 1988, Wever 1989). Melatonin was not administered at the optimal, or perhaps even the correct, time. The variability of responses among these individuals may have resulted from administration of melatonin with respect to clock time rather than an individual's circadian time, which we have found to be preferable. As mentioned above, the DLMO is an excellent marker for circadian time. Indeed, exposure to bright light is also best scheduled according to circadian time.

The brighter the light, the greater is its phase-shifting effect. The optimal dose of melatonin for phase-shifting has not yet been determined, but it appears to be quite low. Because taking a melatonin capsule is more convenient than adjusting the (bright) light-dark cycle, clinical use of melatonin appears to have great potential, whether given alone or in combination with scheduling periods of darkness and/or bright light.

Phase resetting of transmeridional air travellers

Bright light and melatonin can be used to realign mismatches between circadian rhythms and the light-dark cycle and between forced sleep-wake periods

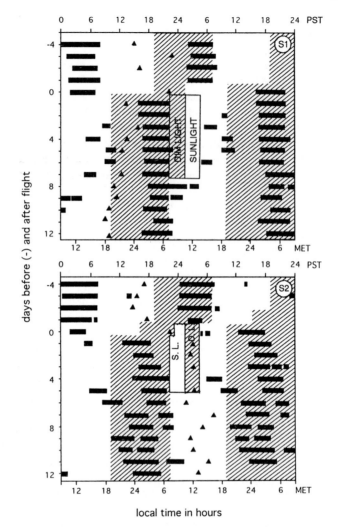

FIG. 3. Shifting of the endogenous circadian pacemaker by exposure to bright light. Subject 1 (*top*), exposed to sunlight during the phase advance portion of the PRC, entrained by the sixth day after west-to-east travel across nine time zones. Solid bar, sleep; ▲, oral temperature peak; PST, Pacific standard time; MET, middle European time. Subject 2 (*bottom*), exposed to sunlight during the phase delay portion of the PRC, had still not entrained by the 12th day after travel. From Daan & Lewy 1984, redrawn with permission.

and the other circadian rhythms more tightly coupled to the endogenous circadian pacemaker. Such mismatches occur among air travellers and shift workers. Air travellers have the benefit of being able to use bright natural daylight to aid readjustment; shift workers often do not, as will be discussed below.

People who have travelled rapidly across time zones can force themselves to try to sleep at the correct time. However, daytime alertness and the quality of their sleep are usually poor for several days, until the circadian rhythms that are more tightly coupled to the endogenous circadian pacemaker readjust to the new time zone. After the discovery that bright light could shift human circadian rhythms in a manner that was consistent with a hypothetical PRC (Lewy et al 1983), it became possible to devise schedules detailing when people should expose themselves to bright light (or avoid it in the case of travel across more than six time zones) (Daan & Lewy 1984): indeed, when a subject is exposed to sunlight at the 'wrong' time, the endogenous circadian pacemaker will shift in the direction opposite to that desired (Fig. 3). When a person travels across fewer than six time zones, sunlight is conveniently available to stimulate the light PRC on the advance portion after eastward travel or on the delay portion after westward travel if the person simply goes outdoors in the morning or in the late afternoon, respectively. When one travels across more than six time zones, morning light should be avoided for the first few days after going east and evening light avoided after going west; sunlight should be sought out during the middle of the day. After a few days, the light PRC adjusts so that either morning or late afternoon is the best time to expose oneself to sunlight, depending on the direction of travel. A nomogram has been devised to guide air travellers' bright light exposure times (Fig. 4).

As with light, melatonin administration is best scheduled according to its PRC. Melatonin administration times should ideally change in accordance with the daily shifts in the melatonin PRC until steady-state re-entrainment has been achieved. (Although there are both human and animal studies that support instantaneous phase resetting of the pacemaker's light PRC, it seems to us more likely that the light, and melatonin, PRCs readjust gradually in humans.) This is illustrated in Table 1, which designates the clock time at which melatonin should be taken so that the treatment is reasonably close to CT 7 when travelling east or to CT 0 when travelling west.

Several assumptions were made when devising this table. One was that the individual's baseline DLMO was at 21.00 h and normal waking-up time at 07.00 h. Another assumption was that the individual's rate of response to the melatonin treatment would be one to three hours per day—that is, that the endogenous circadian pacemaker and therefore the melatonin PRC would shift at about this rate. To minimize the possibility of crossing over from advance responses to delay responses (or vice versa), administration times have been scheduled to change at the rate of one hour per day.

In addition to indicating that the time of post-travel administration should vary on a daily basis, Table 1 also illustrates the point that many of these administration times are not during the hours of sleep. Fortunately, the 0.5 mg dose is minimally, if at all, soporific. Although some of the administration times do occur during sleep, having to wake up and take a pill is not unduly disruptive.

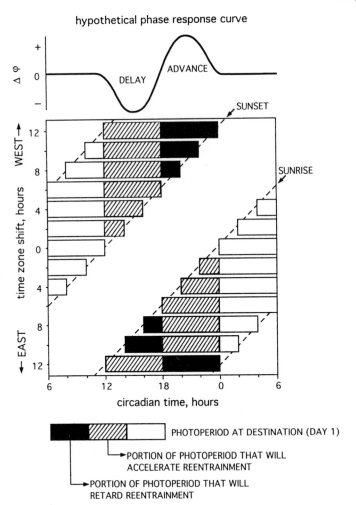

FIG. 4. Proposed times when bright light exposure should occur and when bright light exposure should be avoided in the first few days after transmeridional flight. For example, after a two-hour west-to-east trip, exposure to bright light should begin at dawn and should (optimally) be of two hours' duration. After a 10 h west-to-east trip, however, bright light should be avoided until four hours after sunrise when one should be exposed (optimally) for six hours. From Daan & Lewy 1984, redrawn with permission.

Nevertheless, it is hoped that delayed-release formulations will obviate this inconvenience, which is still minor in comparison with bright light treatment, particularly when scheduled during the sleep period.

For pre-travel adjustment, bright light must be obtained using artificial light fixtures. Unfortunately, almost all treatment times would be during the sleep

TABLE 1 Melatonin administration times upon arrival at destination according to the melatonin phase response curve

Time difference	Day 1	Day 2	Day 3	Day 4	Day 5	Day 6	Day 7
1 h east	3.00 p.m.						
2 h east	4.00 p.m.	3.00 p.m.					
3 h east	5.00 p.m.	4.00 p.m.					
4 h east	6.00 p.m.	5.00 p.m.	3.00 p.m.				
5 h east	7.00 p.m.	6.00 p.m.	4.00 p.m.	3.00 p.m.			
6 h east	8.00 p.m.	7.00 p.m.	5.00 p.m.	4.00 p.m.	3.00 p.m.		
7 h east	9.00 p.m.	8.00 p.m.	6.00 p.m.	5.00 p.m.	4.00 p.m.	3.00 p.m.	
8 h east	10.00 p.m.	9.00 p.m.	7.00 p.m.	6.00 p.m.	5.00 p.m.	4.00 p.m.	3.00 p.m.
9 h east	11.00 p.m.	10.00 p.m.	8.00 p.m.	7.00 p.m.	6.00 p.m.	5.00 p.m.	4.00 p.m.
10 h east	midnight	11.00 p.m.	9.00 p.m.	8.00 p.m.	7.00 p.m.	6.00 p.m.	5.00 p.m.
11 h east	1.00 a.m.	midnight	10.00 p.m.	9.00 p.m.	8.00 p.m.	7.00 p.m.	6.00 p.m.
			11.00 p.m.	10.00 p.m.	9.00 p.m.	8.00 p.m.	7.00 p.m.
12 h west	7.00 p.m.	8.00 p.m.	9.00 p.m.	10.00 p.m.	11.00 p.m.	midnight	1.00 a.m.
11 h west	8.00 p.m.	9.00 p.m.	10.00 p.m.	11.00 p.m.	midnight	1.00 a.m.	2.00 a.m.
10 h west	9.00 p.m.	10.00 p.m.	11.00 p.m.	midnight	1.00 a.m.	2.00 a.m.	3.00 a.m.
9 h west	10.00 p.m.	11.00 p.m.	midnight	1.00 a.m.	2.00 a.m.	3.00 a.m.	4.00 a.m.
8 h west	11.00 p.m.	midnight	1.00 a.m.	2.00 a.m.	3.00 a.m.	4.00 a.m.	5.00 a.m.
7 h west	midnight	1.00 a.m.	2.00 a.m.	3.00 a.m.	4.00 a.m.	5.00 a.m.	6.00 a.m.
6 h west	1.00 a.m.	2.00 a.m.	3.00 a.m.	4.00 a.m.	5.00 a.m.	6.00 a.m.	
5 h west	2.00 a.m.	3.00 a.m.	4.00 a.m.	5.00 a.m.	6.00 a.m.		
4 h west	3.00 a.m.	4.00 a.m.	5.00 a.m.	6.00 a.m.			
3 h west	4.00 a.m.	5.00 a.m.	6.00 a.m.				
2 h west	5.00 a.m.	6.00 a.m.					
1 h west	6.00 a.m.						

hours, making disruption of sleep unavoidable (unless it turns out that twilight simulators are effective). Pre-travel melatonin treatment times, which may also vary slightly on a daily basis, are conveniently during the day.

Phase resetting of shift workers

Air travellers shift their rhythms rather easily (although sometimes in the wrong direction). Night workers' rhythms generally do not shift. They are often 'permanently' set either for night-work or for their days off work.

There were nine subjects in our first study of melatonin rhythms in night workers (Sack et al 1992). These individuals worked five consecutive nights, then were off work for two days. After five nights of work, we found that with one exception the DLMO had shifted into the day. We did not obtain baseline off-work DLMOs, so we were not sure whether or not these individuals were successful in shifting their rhythms back and forth, or if they were 'permanently' shifted to the night-work schedule. It now appears that the latter was probably the case, as suggested by our second shift-work investigation, in which six subjects were studied (Sack et al 1994). They worked a '7–70' schedule, seven 10 h nights (21.00 to 07.00 h), followed by a week off work. Subjects were given two weeks of placebo followed by two weeks of melatonin (0.5 mg) or vice versa. For simplicity, subjects were instructed to take the capsules at bedtime.

One might expect seven days to be sufficient to reset these night-workers' pacemakers, but this apparently was not the case: without melatonin treatment, very few shifted significantly (Fig. 5). One subject adjusted completely to both melatonin and placebo, and another subject failed to shift at all on placebo and shifted only marginally on melatonin. The other four subjects shifted significantly more on melatonin than on placebo, an average of eight hours compared with 0.8 h: for the six subjects, the shifts were on average 7.5 and 2.3 h, respectively ($P \leqslant 0.02$).

There were marked carry-over effects, due to the robust phase-resetting effects of melatonin. Bias analysis, however, indicated that our results would have been even more robust had wash-out weeks between treatments been scheduled. More consistently robust results would also probably have been obtained had melatonin administration times been scheduled according to circadian time, ideally after assessment of pretreatment DLMOs.

Statistical analysis of the effects of melatonin treatment during the off-work week could be done in only a few subjects, because the DLMOs of subjects on placebo were generally permanently at the normal night-time phase. However, subjects who continued treatment with melatonin after adjusting to the work week on melatonin treatment readily readjusted to the off-work schedule. Clearly, melatonin treatment appears to work well for both on-work and off-work schedules.

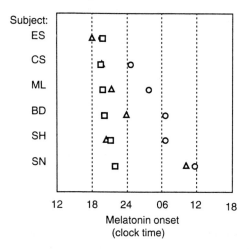

FIG. 5. Clock-time dim light melatonin onsets (DLMOs) of six night-shift workers obtained in three conditions: (1) Off work (□): receiving placebo, no treatment, or in the case of SH and ML, 0.5 mg melatonin at bedtime; (2) placebo (△): after seven consecutive night shifts, sleeping during the day and receiving placebo or no treatment; and (3) melatonin treatment (○): after seven consecutive night shifts, sleeping during the day and receiving melatonin, 0.5 mg at bedtime (between 08.00 and 10.00). Three profiles for each subject are included here. Subjects received two weeks of melatonin and two weeks of placebo in the following order: melatonin first, off-week first (ML and SH); placebo first, work-week first (ES, CS and BD); and placebo first, off-week first (SN). Three subjects (SN, BD and CS) had an intervening time of two weeks between treatments. Pre-baseline studies were done in two subjects (SH and ML). From Sack et al 1994, with permission.

Several questions arise from this study. Why did the work-week DLMOs of the night-workers in our earlier study occur during the day, when the DLMOs in this study remained at the usual evening phase in the placebo or untreated conditions? Apparently, five nights of work alternating with two days off work is a situation in which it is easier to remain entrained to the night-work schedule. In the 7-70 situation, it would appear that one might be as likely to have a daytime DLMO permanently as to have a night-time DLMO permanently; but for various reasons, the latter seems to occur more often.

Why do night-workers often fail to shift 8–10 h when air travellers typically accomplish this within several days? Night-workers are not generally exposed to bright light at work. Thus, night-time light is not bright enough to cause robust phase shifts and, furthermore, is not sufficiently bright to suppress endogenous melatonin production. Air travellers, in contrast, are exposed to sunlight that is sufficiently intense to suppress endogenous melatonin production as well as to induce phase shifts (albeit sometimes in the wrong direction).

Does melatonin act simply by stimulating the SCN according to the melatonin PRC, or does it have another circadian function? The concurrence of darkness (and sleep?) and melatonin may have a mutually reinforcing effect, strengthening the zeitgeber signal. Without exogenous melatonin treatment, the pacemaker apparently continues to interpret exposure to sunlight at 07.00 h as 'morning' light, even though the worker is departing for home in order to go to sleep. Without exogenous melatonin during the day, endogenous melatonin at night may win the competition as the overriding 'night zeitgeber'. Melatonin administration during daytime darkness (and sleep) possibly helps the pacemaker perceive this as the new 'night'. This second circadian function of melatonin is consistent with the fact that melatonin production is not directly induced by daytime darkness. Thus, melatonin may help the pacemaker discriminate the night zeitgeber between night-time and daytime darkness.

Of course, it is also possible that melatonin's only circadian effect is the one explained by the melatonin PRC. Perhaps exogenous melatonin delays the light PRC so that the cross-over point between advances and delays shifts past the 07.00 h light exposure, causing exposure to light at that time to stimulate the delay zone of the light PRC. However, merely 'tipping the balance' would not sufficiently explain our results if it were shown that some of these night-workers adjusted by shifting in the *advance* direction with the help of exogenous melatonin administration. Recent findings in at least one night-worker indicate adjustment to the work-week schedule by a robust phase advance. In any event, it makes sense to consider reducing endogenous melatonin production (photically or pharmacologically) to aid the resetting of the endogenous circadian pacemaker. It is also clear that melatonin administration can be an effective treatment without the accompaniment of scheduled exposure to bright light.

Conclusions

Clearly, administration of exogenous melatonin and exposure to bright light can be used separately or together to reset the endogenous circadian pacemaker. Scheduled darkness may also be important. We conclude with an intriguing speculation: if our night-time discriminator hypothesis is correct, darkness should be scheduled not only to avoid stimulation of the light PRC by light and to provide dark stimulation of the dark PRC (if one exists), but also to enhance the phase-shifting effects of melatonin administration. Melatonin administration should be timed according to the melatonin PRC; darkness should then be scheduled to occur at least during the time of melatonin administration. This can be accomplished by the use of dark or red goggles during periods of wakefulness. We have recently demonstrated that red goggles prevent suppression of night-time melatonin production by bright light. As an initial test of this technique, we gave goggles to a subject who had previously been in one of our melatonin PRC studies. In the previous study, when he was given

0.5 mg melatonin at about CT 1-2, he responded with a phase delay of 25 min. In a second study repeated almost identically, except that he wore goggles for six hours after he took melatonin, the delay was 61 min. If similar findings are obtained in a systematic study, the use of goggles plus melatonin may prove to be a potent method of resetting the endogenous circadian pacemaker in humans. If melatonin has a soporofic effect, one also wonders whether that could be demonstrated more easily when melatonin administration is combined with darkness.

Measurement of the circadian rhythm of melatonin production in humans has resulted in the discovery that bright light is a more effective zeitgeber than indoor light of ordinary intensity. Exposure to bright light in the morning causes phase advances; exposure in the evening causes phase delays. Exogenous administration of melatonin also can phase shift the endogenous circadian pacemaker. The PRCs to light and melatonin are about 12 h out of phase with each other. Melatonin taken in the morning causes a phase delay; taken in the afternoon it causes a phase advance. Thus, circadian phase disorders can be treated with exposure to bright light, exogenous melatonin or the combination. These disorders include advanced and delayed sleep phase syndrome, as well as maladaptation to air travel and shift work. Winter depression might also be treated with appropriately timed melatonin administration, if the phase-shift hypothesis for this disorder is correct. Melatonin administration may one day be about as effective as bright light in causing phase shifts and is much more convenient. Convenience and compliance are very important factors for successful implementation of phase-shifting interventions. Consequently, melatonin administration may become the treatment of choice for many individuals with circadian phase disorders.

Acknowledgements

We wish to thank the nursing staff of the Oregon Health Sciences University (OHSU) General Clinical Research Center (GCRC) and to acknowledge the assistance of Saeeduddin Ahmed, Jeannie M. Latham Jackson, Clifford M. Singer, Mary Cardoza, Neil Anderson, Lynette Currie, Richard Boney, Joanne Otto, Aaron Clemmons and Katherine Pratt. Supported by Public Health Service research grants MH 40161, MH00703, MH47089, MH01005, PO1 AG10794 and M01 RR00334 (OHSU GCRC), and grants from the Lighting Research Institute and the National Alliance for Research on Schizophrenia and Depression.

References

Arendt J, Bojkowski C, Folkard S et al 1985 Some effects of melatonin and the control of its secretion in humans. In: Photoperiodism, melatonin and the pineal. Pitman, London (Ciba Found Symp 117) p 266-283

Czeisler CA, Kronauer RE, Allan JS et al 1989 Bright light induction of strong (type 0) resetting of the human circadian pacemaker. Science 244:1328-1333

Czeisler CA, Johnson MP, Duffy JF, Brown EN, Ronda JM, Kronauer RE 1990 Exposure to bright light and darkness to treat physiologic maladaptation to night work. N Engl J Med 322:1253–1259

Daan S, Lewy AJ 1984 Scheduled exposure to daylight: a potential strategy to reduce 'jet lag' following transmeridian flight. Psychopharmacol Bull 20:566–568

Eastman CI 1987 Bright light in work–sleep schedules for shift workers: application of circadian rhythm principles. In: Rensing L, an der Heiden U, Mackey MC (eds) Temporal disorder in human oscillatory systems. Springer-Verlag, New York, p 176–185

Eastman CI 1990 Circadian rhythms and bright light: recommendations for shift work. Work & Stress 4:245–260

Eastman CI, Lahmeyer HW, Watell LG, Good GD, Young MA 1992 A placebo-controlled trial of light treatment for winter depression. J Affective Disord 26: 211–222

Hoban TM, Lewy AJ, Sack RL, Singer CM 1991 The effects of shifting sleep two hours within a fixed photoperiod. J Neural Transm Gen Sect 85:61–68

Honma K, Honma S 1988 A human phase response curve for bright light pulses. Jpn J Psychiatry & Neurol 42:167–168

Illnerová H, Vanecek J 1982 Two-oscillator structure of the pacemaker controlling the circadian rhythm of N-acetyltransferase in the rat pineal gland. J Comp Physiol A Sens Neural Behav Physiol 145:539–548

Kronauer RE, Duffy JF, Czeisler CA 1993 Inversion of the sleep/wake cycle in a dim light environment has no significant effect on the circadian pacemaker. Sleep Res 22:626

Lewy AJ, Markey SP 1978 Analysis of melatonin in human plasma by gas chromatography negative chemical ionization mass spectrometry. Science 201:741–743

Lewy AJ, Sack RL 1989 The dim light melatonin onset (DLMO) as a marker for circadian phase position. Chronobiol Int 6:93–102

Lewy AJ, Wehr TA, Goodwin FK, Newsome DA, Markey SP 1980 Light suppresses melatonin secretion in humans. Science 210:1267–1269

Lewy AJ, Kern HA, Rosenthal NE, Wehr TA 1982 Bright artificial light treatment of a manic-depressive patient with a seasonal mood cycle. Am J Psychiatry 139:1496–1498

Lewy A, Sack RL, Fredrickson RH, Reaves M, Denney D, Zielske DR 1983 The use of bright light in the treatment of chronobiologic sleep and mood disorders: the phase-response curve. Psychopharmacol Bull 19:523–525

Lewy AJ, Sack RL, Singer CL 1984 Assessment and treatment of chronobiologic disorders using plasma melatonin levels and bright light exposure: the clock–gate model and the phase response curve. Psychopharmacol Bull 20:561–565

Lewy AJ, Sack RL, Singer CM 1985a Immediate and delayed effects of bright light on human melatonin production: shifting 'dawn' and 'dusk' shifts the dim light melatonin onset (DLMO). Ann NY Acad Sci 453: 253–259

Lewy AJ, Sack RL, Singer CM 1985b Melatonin, light and chronobiological disorders. In: Photoperiodism, melatonin and the pineal. Pitman, London (Ciba Found Symp 117) p 231–252

Lewy AJ, Sack RL, Miller S, Hoban TM 1987 Antidepressant and circadian phase-shifting effects of light. Science 235:352–354

Lewy AJ, Sack RL, Singer CM, White DM, Hoban TM 1988 Winter depression and the phase shift hypothesis for bright light's therapeutic effects: history, theory and experimental evidence. J Biol Rhythms 3:121–134

Lewy AJ, Ahmed S, Jackson JML, Sack RL 1992 Melatonin shifts circadian rhythms according to a phase–response curve. Chronobiol Int 9:380–392

Mallo C, Zaidan R, Faure A, Brun J, Chazot G, Claustrat B 1988 Effects of a four-day nocturnal melatonin treatment on the 24 h plasma melatonin, cortisol and prolactin profiles in humans. Acta Endocrinol 119:474–480

Minors DS, Waterhouse JM, Wirz-Justice A 1991 A human phase–response curve to light. Neurosci Lett 133:36–40

Sack RL, Lewy AJ, White DM, Singer CM, Fireman MJ, Vandiver R 1990 Morning vs evening light treatment for winter depression. Arch Gen Psychiatry 47:343–351

Sack RL, Blood ML, Lewy AJ 1992 Melatonin rhythms in night shift workers. Sleep 15:434–441

Sack RL, Blood ML, Lewy AJ 1994 Melatonin administration promotes circadian adaptation to night-shift work. Sleep Res 23:509

Shanahan TL, Czeisler CA 1991 Light exposure induces equivalent phase shifts of the endogenous circadian rhythms of circulating plasma melatonin and core body temperature in men. J Clin Endocrinol & Metab 73:227–235

Wehr T, Moul D, Barbato G et al 1993 Conservation of photoperiod-responsive mechanisms in humans. Am J Physiol 265:R846–R857

Wever R 1989 Light effects on human circadian rhythms. A review of recent Andechs experiments. J Biol Rhythms 4:161–185

DISCUSSION

Arendt: I really appreciate the elegant work Dr Lewy has done refining the original phase advance that we saw, which was reported in the Ciba Foundation Symposium *Photoperiodism, melatonin and the pineal* held in 1985 (Arendt et al 1985). We gave, in our original experiments, 2 mg (a suprarphysiological dose) of exogenous melatonin at 17.00 h daily. This induced a one- to three-hour advance in the onset of endogenous production, visible when exogenous and endogenous melatonin were distinguishable, in comparison with placebo, but with less effect on the declining phase. We have always assumed that melatonin would cause phase delays if taken in the subjective morning for the following reasons. Melatonin, in many respects, acts like darkness. Giving it in the late afternoon or the evening should to some extent be equivalent to extending darkness into the late afternoon; giving it in the morning would be equivalent to extending darkness into the morning hours. Advancing or delaying darkness in this way similarly advances and delays the onset and termination of sleep.

Using this approach, we have treated people with melatonin for various problems, such as are associated with jet lag and shift work, for about seven years. Our practical approach doesn't involve the dim light melatonin onset because this is actually impossible to measure in the field. For westward travel, for example, we give no pre-flight early morning treatment, because in our experience taking it in the morning before a flight is more likely to make you sleepy than at other times of the day. We advise people to take melatonin after the flight for four days at local bedtime, which, over a sufficient number of time zones, should fall within a phase-delaying window. This is by no means ideal and a delayed-release pill would be desirable. Treatment for eastward flights is easier. We give pre-flight treatment in the late afternoon for one to two days, and in our original published study on jet lag we achieved the maximum phase-advancing position according to the recent phase response curve (PRC). After the flight, melatonin is taken at bedtime for four days, which is still in a phase-advancing position.

Combined data from all of our controlled and uncontrolled studies have recently been analysed (Arendt et al 1994). This consists of self-rated jet lag on a visual analogue scale (0 = insignificant, 100 = very bad) in groups of people, most of whom are travelling for their own reasons and not for experimental reasons. There was no control over exposure to natural light or other environmental factors, but they were given explicit instructions as to when they should take melatonin or the placebo. From 386 people on melatonin and 85 on placebo eastwards and westwards over all time zones, there was a 60% reduction in perceived jet lag eastwards and a 40% reduction westwards. The data from controlled studies, with journeys on placebo or melatonin in the same individual over the same number of time zones, give similar results. Without the refinement of using individual circadian phase to time treatment, which clearly would be desirable theoretically, this is a successful approach.

We have also used melatonin in Guildford policemen in a phase-delaying mode, and have substantially improved their daytime sleep when they are on the night shift and their alertness during the night. We continued to give them melatonin at bedtime (phase advance) on the rest days following the night shift to shift them back again, and again sleep was improved (Folkard et al 1994).

Czeisler: I would like to compliment Al Lewy and Bob Sack on the work they have done, particularly that on night-shift workers. I want to add one note of caution to the conclusion that Dr Lewy has drawn that melatonin alone can cause phase advances and phase delays equivalent to those caused by light. I agree that these results support the notion that there could be some synergistic action of light and melatonin which, owing to an uncharacterized interaction, results in a greater phase shift than anticipated. Alternatively, as Dr Lewy suggested, melatonin may first be shifting the light PRC, then the light exposure during travel home may be acting on the circadian pacemaker directly, because night-shift workers are generally exposed to bright light in the morning on their way home. A melatonin-induced change in the timing of the breakpoint of the light PRC could be very important, because the most light-sensitive circadian phase is ordinarily just before the night-shift workers normal departure time from work. Because the melatonin-induced PRC Dr Lewy has reported was measured in sighted people living in the outside world, rather than in controlled laboratory conditions, it's hard to tell what the effect of melatonin alone is. If its effects are as powerful as those of light, why is daily melatonin administration unable to entrain the circadian system of most blind people, whose period is on average only half an hour different from 24 h? If melatonin could induce the eight-hour phase shifts that Dr Lewy is concluding it can from his results, why can't melatonin entrain the circadian pacemaker of totally blind people?

Lewy: As I said, one possible reason why we are causing such robust phase shifts with melatonin is that we are delaying the light PRC such that the cross-over point between delays and advances has moved past the time that night workers depart for home, in which case morning sunlight would then be stimulating the delay zone and not the advance zone of the light PRC. Nevertheless, we are able to cause robust phase shifts with melatonin without specifically scheduling light exposure. The same light conditions without administration of melatonin were generally not associated with a large phase shift in circadian rhythm.

Czeisler: I like the idea that it tips the balance, but, from a theoretical perspective, I question the conclusion that melatonin alone can cause phase advances and phase delay shifts equivalent to those induced by light.

Lewy: The phase shifts in your light PRC (Czeisler et al 1989) are of greater magnitude than those in other light PRCs (Honma & Honma 1988, Minors et al 1991, Wever 1989). I think this is because you inverted the sleep–wake (and light–dark) cycles in your subjects. This is what happens to night workers and could be one reason why the large phase shifts we see in them are due primarily to melatonin administration alone. In other words, phase shifts due to either melatonin or light are greater when the sleep–wake cycle is inverted than when it is held constant (Lewy et al 1992) or when it shifts very slightly (Honma & Honma 1988, Minors et al 1991, Wever 1989).

I agree with your comment about blind people. Since we can induce robust phase shifts in blind people with melatonin, why are we not more successful in entraining them? Perhaps melatonin is more effective in individuals who can perceive a light–dark cycle that has been shifted 12 h.

Kronauer: You have shown phase shifts induced by application of melatonin for four days. What is the shift per day?

Lewy: When administered at the correct circadian time, melatonin induced phase shifts were 15 min/day in our four-day administration paradigm in sighted people whose sleep–wake and light–dark cycles were held constant.

Kronauer: That is what you might call the effect of melatonin alone.

Lewy: In our study to produce the PRC we thought it was important to hold the sleep–wake and light–dark cycles constant, so as to attribute any resulting phase shifts solely to melatonin administration. When melatonin has to compete against the light–dark cycle the phase shifts due to melatonin are relatively small (about 15 min/day). When we used (in our night workers) a research design similar to yours (inversion of the sleep–wake and light–dark cycles), we saw large phase shifts, comparable to what you have shown with bright light. Because melatonin can induce large phase shifts in both directions, I don't think they are due simply to a synergistic effect between melatonin and the ambient light–dark cycle, except in so far as concurrent darkness may enhance the phase-shifting effects of melatonin (and, possibly, vice versa).

Kronauer: What is the light level in the work area? You called it 'dim'.

Lewy: Light intensity at work varied greatly, but was fairly dim, generally less than 200–300 lux.

Czeisler: As we discussed earlier, light of that intensity could have a direct effect on the human circadian pacemaker, and the entraining effect of light should be considered when evaluating the response of those shift workers to their inverted schedules.

Menaker: I have been concerned for some time about the potential side effects of melatonin treatment. Why is this considered a safe thing to do? After all, melatonin is a hormone which operates at very low doses, and we have really no idea that its effects on the circadian system are its only effects. I would be concerned about taking the hormone for jet lag, which, although it can be annoying, is actually not a disease.

Arendt: Since we began giving melatonin in 1981 we have always asked people to report side effects. Those which have been reported more than once in jet lag studies are sleepiness (8.9%), headache (2.9%) and 'fuzziness/giddiness' (2.9%). There was also 0.9% nausea. Overall, there is no evidence for major problems. We are often asked whether melatonin affects the reproductive system. In 1981 we gave 2 mg doses in the late afternoon to our colleagues on a semi-chronic, daily basis for a month, and measured as many anterior pituitary hormones as we could and took self-rated measures of mood and alertness (Arendt et al 1985). There were no significant effects on luteinizing hormone, follicle-stimulating hormone, T_4, testosterone, growth hormone or cortisol, though the timing of the prolactin rhythm was advanced (Wright et al 1986). The administration time was chosen deliberately to mimic the kind of effects one sees in animals when they are put into a short photoperiod, so that we would maximize any effects of melatonin on the human reproductive system.

I have as yet seen little evidence that melatonin does anything but useful things. It appears to stimulate the immune system (Maestroni et al 1989), it may even have some anti-ageing effects (Trentini et al 1992) and there are at least two reports of its use as an adjunct to cancer chemotherapy in attempts to prolong survival time (Lissoni et al 1992, 1993).

Menaker: Let me remind you that Madame Curie carried a tube of radium around in the breast pocket of her lab coat for 20 years, so that she could show it to people in the dark, before she finally died of aplastic pernicious anaemia at the relatively young age of 66 (Reid 1974).

Arendt: I have been taking melatonin nearly every night for nearly 15 years, and I'm still OK.

Waterhouse: Nevertheless, it is fair to say that at this moment there are no epidemiological data on shift workers who have taken melatonin for an extended period.

Arendt: That's a fair comment. The longest proper study that I know of is one of ours on a blind man who has been taking 5 mg melatonin daily since 1988 (Arendt & Aldhous 1988).

When David Sugden first did a study of acute toxicity, he couldn't find an oral LD_{50} for melatonin in rats (Sugden 1983). Melatonin has really low toxicity according to published data.

References

Arendt J, Aldhous M 1988 Synchronisation of a disturbed sleep–wake cycle in a blind man by melatonin treatment. Lancet I:772–773

Arendt J, Bojkowski C, Folkard S et al 1985 Some effects of melatonin and the control of its secretion in humans. In: Photoperiodism, melatonin and the pineal. Pitman, London (Ciba Found Symp 117) p 266–283

Arendt J 1994 Clinical perspectives for melatonin and its agonists. Biol Psychiatry 35:1–2

Czeisler CA, Kronauer RE, Allan JS et al 1989 Bright light induction of strong (type 0) resetting of the human circadian pacemaker. Science 244:1328–1332

Folkard S, Arendt J, Clark M 1994 Can melatonin improve shift-workers tolerance of the night-shift? Some preliminary findings. Chronobiol Int, in press

Honma K, Honma S 1988 A human phase response curve for bright light pulses. Jpn J Psychiatry 42:167–168

Lewy AJ, Ahmed A, Jackson JML, Sack RL 1992 Melatonin shifts circadian rhythms according to a phase-response curve. Chronobiol Int 9:380–392

Lissoni P, Barni S, Ardizzoia A et al 1992 Randomized study with the pineal hormone melatonin versus supportive care alone in advanced nonsmall cell lung cancer resistant to a first line chemotherapy containing cisplatin. Oncology 49:336–339

Lissoni P, Barni S, Rovelli F et al 1993 Neuroimmunotherapy of advanced solid neoplasms with single evening subcutaneous injection of low-dose interleukin-2 and melatonin: preliminary results. Eur J Cancer 29A:185–189

Maestroni GJM, Conti A, Pierpaoli W 1989 Melatonin stress and the immune system. Pineal Res Rev 7:268

Minors DS, Waterhouse JM, Wirz-Justice A 1991 A human phase–response curve to light. Neurosci Lett 133:36–40

Reid R 1974 Marie Curie. Saturday Review Press, EP Dutton & Co, Chicago, IL

Sugden D 1983 Psychopharmacological effects of melatonin in mouse and rat. J Pharmacol Exp Ther 227:587–591

Trentini GP, Genazzini AR, Criscuolo M et al 1992 Melatonin treatment delays reproductive aging of female rat via the opiatergic system. Neuroendocrinology 56:364–370

Wever RA 1989 Light effects on human circadian rhythms. A review of recent Andechs experiments. J Biol Rhythms 4:161–184

Wright J, Aldhous M, Franey C, English J, Arendt J 1986 The effects of exogenous melatonin on endocrine functions in man. Clin Endocrinol 24:375–382

General discussion II

Kronauer: I would like to introduce something which few people seem to have noticed in Winfree's book (1980). Although this book is full of lovely material and interesting ideas, I find it difficult to read. On p 83, when he discusses what he calls 'the rules of the ring' (talking about phase-only, one-dimensional systems) he gives a brief proof related to the display of resetting data in the format of the phase transition curve in which final stimulus phase is plotted against initial stimulus phase. He shows that the phase transition curve may not have a negative slope if the system is a continuous, one-dimensional one.

Consider the case where a stimulus produces a phase advance. After the stimulus has done its work, the stimulus occurs at a later time in the rhythm. If the phase transition curve has a negative slope this says that if the stimulus is timed to occur earlier (earlier initial phase) it will end up at a later phase of the rhythm. I call this phenomenon leap-frogging. Winfree's proof says that the one-dimensional system cannot leap-frog. Translating this slope condition over to the conventional phase response curve (PRC) tells you that a one-dimensional system cannot have a slope more negative than -1. Thus, if an experiment gives a PRC with a slope of less than -1 the system cannot be one-dimensional (phase only).

The relevance of this to our earlier discussion is that the computer-derived single-episode PRC has two segments in which the average slope is about -1.5. This means that the curve in the upper panel of Fig. 1 (p 298) which Dr Daan has proposed as an acceptable single pulse curve represents a system which cannot be phase only—it violates the condition which Winfree showed a decade ago.

Daan: Slopes of -1 are widespread in PRCs.

Kronauer: Right, which says that there is a second variable. You do not have a one-dimensional system.

Ralph: I would like to make a brief comment about the issue of PRC distortion. I know several people can't believe that this is actually a controversial issue. The issue, as I see it, is whether or not the human PRC is type 0 or type 1 under different situations, or whether large, type 0-like phase shifts can be explained from the type 1 curve. There are several published examples in which a light-induced PRC has been distorted by a prior stimulus. Mrosovsky (1991) and Joy & Turek (1992) have published a PRC produced by light pulses that were preceded by locomotor activity. Pritchard & Lickey's data (1981) in *Aplysia* provide an example where a type 1, low amplitude PRC can be converted to type 0-like resetting by extended light pulses.

General discussion II

Those are three examples in which a PRC to light can be distorted by a single, prior phase-shifting stimulus. It would not be surprising if the human phase response behaved in the same way, especially after three applications of light. I do not believe that we can infer the changes in the shape of the underlying PRC simply by examining the cumulative response following three applications of the stimulus.

Czeisler: The pejorative word 'distorted' is unfortunate. Modulated might be better. To go back to the notion that we are dealing with a one-variable system is actually a backward step. There's plenty of evidence that another variable—amplitude—must be considered when evaluating these rhythms and their responses to light. Amplitude has been too-long neglected, as Winfree has been arguing for years (Winfree 1973, 1980, 1987). Dr Jeff Elliot's data indicate that increasing the ratio of activity:rest time in the hamster results in type 0 PRCs. The results Dr Turek has shown from ageing animals indicate that large phase shifts consistent with the type 0 model are possible. This is consistent with the idea that two variables are required to describe the resetting properties of the mammalian circadian pacemaker. The fact that a second light pulse gives a response different from that one would predict from two individual pulses simply means that the system is being modified in such a way as to reveal that it can be driven through the region of singularity. As Strogatz (1990) pointed out, repeated exposure to a stimulus which induced type 1 resetting can never yield a type 0 curve in a truly phase-only system (i.e., type 1 × type 1 × type 1 = type 1). There must be a second variable, amplitude, if you get type 0 resetting from repeated application of a weak stimulus, no matter how many times it is repeated.

Miller: There are some interesting relevant animal data from Gander & Lewis (1983), which I have replotted as a phase dose–response surface (Miller 1993). Gander uses long durations of light as stimuli, so perhaps I should refer to this as a phase duration–response surface but, at longer durations at least, the surface does seem to imply there is type 0 resetting in that particular rodent. What intrigues me is that we have a stimulus dimension, intensity dimension, duration dimension, whatever you want to call it, which is increasing linearly, with a qualitative change in topology. We have to address what this means at the level of the retina and at the level of the suprachiasmatic nucleus (SCN). Are we super-saturating receptors? Are we doing something qualitatively different when we make this transition from type 1 to type 0?

I'm also wondering about the generality of this intensity-produced qualitative shift. Is it possible that non-photic zeitgebers could also produce type 0 resetting under some circumstances?

Mrosovsky: We can get very large non-photic amplitudes in the *tau* mutant hamster, but I don't think anybody has seen type 0 resetting with non-photic stimuli.

Turek: We are hearing comparisons of a lot of apples and pears. Dr Lewy says that in response to a single treatment with melatonin he gets only small phase-shifts, and makes comparisons with the three five-hour light pulses that Dr Czeisler's group have given, and multiplies by a factor of 10 to try to compare the melatonin PRC with the three-light PRC. We are lashing ourselves too hard. I am not sure there is that much real discrepancy or disagreement. There are lots of things that we still don't understand about phase shifting. Let me give an example from our work on hamsters. If you shift the phase of the light–dark cycle by eight hours, animals can take 10–15 days to re-entrain. If you give them a single injection of triazolam, which can induce only a two-hour phase shift, you can get complete re-entrainment in one or two days. The shift in the system, 8–10 h, in response to the two stimuli is greater than the sum of the phase shifts brought about by the two stimuli individually. There are obviously lots of things going on that we don't understand. When we are using different paradigms to shift the clock, additivity is not enough; there is integration and synergy that we've paid little attention to.

Takahashi: Dr Czeisler mentioned that the sum of three light pulses to phase-only oscillators can give you only type 1 resetting. I agree with that analytical solution, but Winfree (1980), in the volume that Dr Kronauer mentioned, also points out that if you have a population of phase-only oscillators, their behaviour can approach type 0. We've been doing some interesting population simulations and have found, to our surprise, that if we define the input to each unit oscillator, by giving weak inputs to the oscillator so that individually each oscillator has only a type 1 response, the population of oscillators gives a type 0 system PRC. This depends strongly on the coupling strength and the standard deviation of the population. You get what Dr Kronauer calls leap-frogging in this system.

I would like to suggest an additional level of complexity that hasn't been taken into account here. The human system is unlikely to be composed of one phase-only oscillator. We already know it's two-dimensional.

Kronauer: What you are saying is that the ensemble can behave in a type 0 way. Winfree gave that as a possible explanation for type 0 resetting.

Czeisler: He even said that the ensemble's behaviour may be a reflection of an interaction between 10 000 SCN cells. When they get out of phase, the amplitude reduction would be the same as a change in the amplitude of the collective output, even if each individual cell can show only type 1 resetting.

We do have to consider multiple stimuli, because this is what systems are exposed to in nature. Trying to interpret the results of experiments in which animals are studied in a month of darkness punctuated by a single light stimulus in terms of what's happening in nature, when we know that the results are not additive, is a grave error.

Lewy: I would like to know Dr Kronauer's view on two studies that tested his amplitude hypothesis, one by Eastman (1992) and one by Wirz-Justice et al (1993). Eastman, in consultation with Dr Kronauer, tested suppression of circadian amplitude by centring bright light around the temperature minimum, which she calls 'squashing' (as opposed to 'nudging' in which amplitude is not crushed but light is timed to cause a phase shift). I think she found that the rate of re-entrainment was the same for the two techniques. Furthermore, she observed no 'phase jump' using the squashing technique which would be expected according to the Kronauer model. Instead, with both techniques there were transient shifts during adjustment to the new phase position. Now, it is true that Eastman did not use a constant routine; however, in her study several days of ambulatory temperature monitoring were corrected for masking by sleep and appear to provide a valid assessment of circadian phase. Wirz-Justice et al (1993) scheduled bright light exposure during the middle of the day but were not able to enhance amplitude.

Kronauer: You are correct that Eastman did not use a constant routine. If one acutely inverts the light–dark or sleep–wake cycle, and does nothing to the endogenous rhythm, because the evoked and endogenous temperature rhythms have approximately the same amplitude, the first result one observes is 'squashing' of the temperature rhythm although one has done nothing whatsoever to the endogenous rhythm component. We have had a lot of difficulty getting the timing just right in order to get strongly reduced amplitude. The system doesn't want to behave in that way; it wants to veer off to the side of the zero amplitude singular point. I would seriously doubt that Eastman could verify whether she did indeed squash the endogenous rhythm.

Lewy: I think she did find evidence of a squashed rhythm, but if you are correct, if it is difficult to crush amplitude, this would bring into serious doubt whether such a technique could ever be developed into a practical treatment.

Roenneberg: I agree with Fred Turek that it's too early to quarrel about these things, but when we are making a hypothesis, we should not take people's data, pick the raisins out and leave the rest on the plate if it doesn't fit the hypothesis. We have heard that ageing hamsters are a good example fitting the hypothesis that a decreased amplitude gives rise to increased phase shifting of the oscillator, but at the same time they do not respond to non-photic stimuli. However, amplitude reduction of the oscillator should change the response to all stimuli, photic and non-photic. Thus, the ageing hamster is a counter example of the low amplitude hypothesis.

Daan: I have never believed that a pacemaker with no amplitude can exist. My point is, can we explain phase-shifting data, as observed, on the basis of a simple phase-only system? I think we can with consecutive pulses in humans. However, certainly at some point there will be data showing that the system is more complex than one with only instantaneous phase resetting. This is clear, for example, in *Drosophila*, if you give them a second pulse within four hours of the first. The human pacemaker will also eventually turn out to be more complex than following phase-only resetting, but we have to prove that.

References

Eastman CI 1992 High-intensity light for circadian adaptation to a 12-h shift of the sleep schedule. Am J Physiol 263:428–436
Gander PH, Lewis RD 1983 Phase resetting action of light on the circadian activity rhythm of *Rattus exulans*. Am J Physiol 245:R10–R17
Joy JE, Turek FW 1992 Combined effects of agents with different phase response curves: phase shifting effects of triazolam and light. J Biol Rhythms 7:51–63
Miller JD 1993 On the nature of the circadian clock in mammals. Am J Physiol 264:R821–R832
Mrosovsky N 1991 Double pulse experiments with nonphotic and photic phase-shifting stimuli. J Biol Rhythms 6:167–179
Pritchard RG, Lickey ME 1981 *In vitro* resetting of the circadian clock in the *Aplysia* eye. I. Importance of efferent activity in optic nerve. J Neurosci 1:835–839
Strogatz SH 1990 Interpreting the human phase response curve to multiple bight-light exposures. J Biol Rhythms 5:169–174
Winfree AT 1973 Resetting the amplitude of *Drosophila*'s circadian chronometer. J Comp Physiol A Sens Neural Behav Physiol 85:105–140
Winfree AT 1980 The geometry of biological time. Springer-Verlag, New York
Winfree AT 1987 The timing of biological clocks. Scientific American, New York
Wirz-Justice A, Graw P, Kruchi K, Haug H, Leonhardt G, Brunner D 1993 Effect of light on unmasked circadian rhythms in winter depression. In: Wetterberg L, Beck Friis J (eds) Light and biological rhythms in man. Pergamon Press, Oxford, p 385–393

Index of contributors

Non-participating co-authors are indicated by asterisks. Entries in bold type indicate papers; other entries refer to discussion contributions.

Indexes compiled by Liza Weinkove

Arendt, J., 86, 169, 209, 252, 291, 317, 320
Armstrong, S. M., 104, 105, 111, 144, 170, 171, 194, 195, 207, 233, 248, 249, 293, 294, 295, 296
*Aronson, B. D., **3**

*Bauer, V. K., **303**
*Bell-Pedersen, D., **3**
Block, G. D., 20, 41, 44, 45, 47, **51**, 60, 61, 62, 63, 64, 65, 107, 109, 114, 115, 151, 172, 192
*Blood, M. L., **303**

Cassone, V. M., 83, 84, 101, 102, 104, 107, 108, 109, 110, 145, 190, 227, 250
*Crosthwaite, S., **3**
*Cutler, N. L., **303**
Czeisler, C. A., 61, 82, 86, 101, 104, 110, 115, 129, 130, 170, 195, 227, 232, 233, **254**, 290, 291, 292, 293, 294, 295, 296, 298, 302, 318, 320, 323, 324

Daan, S., 129, 149, 170, 232, 233, 298, 300, 322, 325
*Ding, J. M., **134**
Dunlap, J. C., **3**, 18, 19, 20, 21, 22, 23, 24, 42, 44, 47, 48, 49, 102, 112, 229, 250

*Ebling, F. J. P., **175**
Edmunds, L. N., 19, 22, 41, 42, 43, 46, 47, 48, 64, 82, 83, 108, 128, 131, 132, 150, 151, 230, 231

*Faiman, L. E., **134**

*Gallman, E. A., **134**
*Garceau, N. Y., **3**
*Gebauer, G., **26**
*Geusz, M., **51**
Gillette, M. U., 42, 43, 46, 48, 63, 103, 107, 108, 113, **134**, 144, 145, 146, 147, 148, 149, 150, 151, 192, 208, 228
*Grosse, J., **175**

Hastings, M. H., 23, 43, 45, 84, 108, 109, 145, 151, **175**, 190, 191, 192, 193, 194, 195, 196
*Herbert, J., **175**
*Hurd, M. W., **67**

*Johnson, K., **3**

*Kallies, A., **26**
*Khalsa, S., **51**
Kronauer, R. E., 23, 83, 84, 110, 114, 115, 128, 129, 131, 233, 294, 297, 298, 300, 301, 319, 320, 322, 324, 325

Lemmer, B., 42, 43, 44, 48, 49, 207, 208, 209, 229, 230, **235**, 247, 248, 249, 250, 251, 252
Lewy, A. J., 100, 107, 108, 150, 173, 210, 232, 233, 290, 291, **303**, 319, 320, 324, 325
*Lindgren, K., **3**
*Liu, C., **134**
Loros, J. J., **3**, 22, 47, 81, 113, 114, 116, 130

*Maywood, E. S., **175**
*McArthur, A. J., **134**

*Medanic, M., **134**
Meijer, J., 85, 148, 169
Menaker, M., 17, 23, 61, 84, 86, 87, 100, 103, 104, 105, 111, 130, 131, 132, 150, 168, 169, 191, 192, 194, 208, 209, 227, 228, 230, 231, 320
*Merrow, M., **3**
*Michel, S., **51**
Mikkelsen, J. D., 65, 102, 111, 167, 172, 173, **175**, 207, 226
Miller, J. D., 45, 46, 48, 62, 63, 65, 85, 100, 101, 102, 103, 109, 110, 111, 113, 128, 146, 147, 149, 167, 169, 173, 192, 193, 194, 209, 229, 230, 252, 323
*Mohsenzadeh, S., **26**
Moore, R. Y., 61, 86, **88**, 100, 101, 102, 103, 104, 107, 110, 111, 148, 167, 210, 226, 229, 231, 232, 234
Mrosovsky, N., 21, 60, 110, 111, 148, 149, **154**, 167, 168, 169, 170, 172, 173, 191, 230, 232, 249, 251, 323

*Penev, P., **212**

Ralph, M. R., 24, 63, **67**, 81, 82, 83, 84, 85, 86, 87, 103, 104, 105, 112, 192, 228, 322
Redfern, P. H., 44, 100, 102, 105, 193, 231, 249
Rensing, L., 23, **26**, 41, 42, 43, 44, 46, 47, 48, 49, 62, 114, 130
Reppert, S. M., 65, 105, 145, 194, **198**, 207, 208, 209, 210, 211

Roenneberg, T., 47, 62, 65, 100, **117**, 128, 129, 130, 131, 132, 170, 173, 227, 252, 325

*Sack, R. L., **303**
*Sumova, A., **175**

Takahashi, J. S., 19, 20, 21, 24, 44, 45, 47, 62, 110, 112, 115, 116, 130, 131, 190, 191, 192, 195, **212**, 227, 250, 251, 292, 297, 298, 324
*Tcheng, T. K., **134**
*Thomas, K. H., **303**
Turek, F. W., 23, 63, 64, 82, 107, 110, 146, 168, 169, 171, 172, 191, 194, 195, 208, **212**, 226, 227, 228, 229, 230, 231, 233, 234, 247, 248, 296, 300, 301, 323

*Van Reeth, O., **212**

Waterhouse, J. M., **1**, 22, 43, 44, 104, 107, 108, 112, 113, 128, 130, 131, 148, 172, 210, 231, 249, 292, 293, 295, 296, 300, 301, 302, 320
*Weber, E. T., **134**
*Whitmore, D., **51**

*Zee, P., **212**
*Zhang, Y., **212**

Subject index

activity
  age-related changes in responses, 217–220, 232–233
  blood pressure rhythms and, 249–250
  dark pulse PRCs and, 163–164, 173–174, 218
  light-induced phase shifts and, 196–197
  neuropeptide Y-induced phase shifts and, 160
  phase-shifting effects, 155–156, 157–158, 167–168, 172
  restriction, 170–171, 173, 217–218
  see also arousal; running
activity rhythms, 262
  age-related changes, 212–234
  photic effects in humans, 259
  see also sleep–wake cycles
ageing, 212–234, 325
  brain monoamines and, 221
  light responses and, 214–216, 217, 226–227
  non-photic responses and, 217–220
  rate of re-entrainment and, 216
  SCN transplantation studies, 73–74, 75, 87, 221–223, 227–228
aggregation rhythm, *Gonyaulax*, 118, 121, 125, 173
  oscillator controlling, 126, 128, 129, 130
air travellers, transmeridional
  phase resetting, 307–312, 313, 317–318
  see also jet lag
Alzheimer's disease, 213, 231–232
amplitude, circadian rhythm, 323
  effects of ageing, 213, 232–233, 325
  Gonyaulax, 128–129
  humans, 280–281, 324–325
anisomycin, 110, 220
anti-asthmatic drugs, 238, 239–241
anticancer drugs, 48–49, 238, 239, 248, 251
AP-1 binding, 167, 176–177, 190, 191, 192

Aplysia, 44–45, 47, 65, 109–110, 322
arousal, 175–176, 182, 184–187, 194–196
asthma, 236, 239–241

basal retinal neurons, *Bulla*, 52–53, 57–58, 112, 113
  entrainment, 53–54
  mechanisms of rhythm expression, 54–55
  mechanisms of rhythm generation, 55–57
benzodiazepines, 63, 217, 218, 233–234, 239
  see also triazolam
β_2-sympathomimetic drugs, 241
bioluminescence rhythm, *Gonyaulax*, 117, 120, 121, 125
  oscillator controlling, 126, 128, 129, 130
blind people, 260–261, 295–296, 303, 318–319
blind pregnant rats, 199, 201
blood pressure rhythms, 241–242, 243, 248–250
body temperature rhythm, 169, 172
  effects of ageing, 233
  light-induced shifts, 265–268, 269, 274–275, 324–325
  as marker of circadian phase, 264, 265, 290, 292
  SCN transplantation and, 104
  sleep–wake cycles and, 263, 296–297
brain slices, hypothalamic, 134–153
bright light
  effects on *Gonyaulax*, 132
  human circadian responses, 261, 265–281, 297–300, 305–306, 315, 324–325
  phototherapy, 3, 261, 281, 306
  transmeridional air travellers, 307–312
Bulla eye, 47, 51–66, 107, 109, 169
  circadian rhythm, 52
  entrainment of circadian rhythm, 53–54

Bulla eye (*cont.*)
   localization of retinal pacemaker, 52–53, 112
   temperature-induced changes, 41
   two-loop model, 57–58
   see also basal retinal neurons, *Bulla*

calcium (Ca^{2+})
   extracellular, *Bulla* basal retinal neurons, 56, 58
   fluxes, *Bulla* basal retinal neurons, 54, 58, 63
   -induced phase shifts, 47–48
   intracellular, 47–48, 145
      Bulla basal retinal neurons, 54, 58
      Neurospora, 29, 44
calcium channels
   Bulla basal retinal neurons, 54, 55–56
   Neurospora, 44
calmodulin, 28, 65
5-carboxamidotryptamine (5-CT), 193
cardiovascularly active drugs, 238, 239, 244, 249–250
catecholamines, 213–214, 229–230
ccg (clock-controlled genes), 5, 6
cel mutant, 37
chloride channels, 56
chlorophyll, 118, 126, 131
chronopharmacology, 235–253
   asthma, 239–241
   hypertension, 241–244, 248–250
   peptic ulcer, 237, 240
circadian system
   avian, organization, 69
   mammalian, 88–106
      entrainment pathways, 89–92
      output pathways, 95–97
      pacemaker *see* suprachiasmatic nucleus
clocks, circadian *see* pacemakers
cocaine, 202, 207, 209–210
conidiation, *Neurospora*, 10, 12–13, 20
constant routine technique, 264, 265–275
corticosteroids, 171, 195
cortisol, serum, 185
   effects of light, 259–261, 268, 274–275
creatine, 121–122, 123, 125, 129
CREB (cyclic AMP response-element-binding protein), 184, 227, 282–283
crisp mutant, 42
cyclic AMP (cAMP), 44–47, 147, 209
   circadian rhythms in levels, 41–42, 138

   -induced phase shifts, 46
   in vitro SCN, 136–138, 140–141, 145–146, 151
   Neurospora, 29–31
   non-photic entrainment and, 184
   temperature-induced changes, *Neurospora*, 28, 29, 30, 42–43
cyclic GMP (cGMP), 44–45
   circadian rhythms in levels, 42, 43, 138
   phase shifting effects *in vitro*, 137–138, 141
cycloheximide (CHX), 33, 36, 39, 62, 220

D_1 dopamine-receptor-related phosphoprotein (DARRP-32), 207
D_1 dopamine receptors, 102, 202, 203
dantrolene, 44
dark pulses
   Gonyaulax, 123–125, 173–174
   PRCs/phase shifts induced by, 107, 156, 217
      activity and, 163–164, 173–174, 218
      age-related changes, 218, 219, 220
darkness, melatonin therapy and, 314–315
dead zone, 57–58
2-deoxyglucose (2-DG) autoradiography, 199
depolarization, 107–108
   Bulla basal retinal neurons, 53–54, 57–58, 61, 63
   SCN neurons, 61–62
depression, 261, 306, 315
desynchronization
   fetal clocks within litter, 199
   internal *see* internal desynchronization
dichlorophenyldimethylurea (DCMU), 118–119, 122
diencephalon, 68
dim light melatonin onset (DLMO), 303, 307, 317
   as marker of circadian phase, 291–293, 304
   shift workers, 312–313
dimethoxybromoamphetamine (DOB), 193, 194
dopamine
   age-related changes, 213–214
   fetal clock entrainment, 203, 207–209
   Drosophila, 62, 150, 262, 279, 325
   see also Per; *per* gene

Subject index

drug treatment
  chronopharmacology, 235–253
  elderly, 233–234

egr-1/Egr-1, 176, 177–178, 182, 184
EGTA, 54, 58, 65–66
elderly people, 213, 231–234
  endogenous circadian phase and amplitude (ECPA) assessment, 264–265
  light-induced phase shifts, 266, 267
  see also ageing
enalaprit, 244
endogenous circadian phase and amplitude (ECPA) assessment, 264–265, 277
enkephalin, 90
entrainment
  anatomical pathways in mammals, 89–92, 97, 100
  Bulla basal retinal neurons, 53–54
  fetal clock, 198–211
  neural basis in mammals, 175–197
  non-photic, age-related changes, 220
  see also non-photic phase shifting; photic entrainment
Euglena, 42, 43, 46, 47–48, 128, 151
exercise
  neuropeptide Y release, 161–162
  see also activity
eye, 68, 78
  mammalian
    fetal clock entrainment, 201
    intrinsic oscillator, 94, 111
  Xenopus, 208–209
  see also Bulla eye

feedback loops, 61
  Bulla basal retinal neurons, 57–58
  frq regulation, 9–16, 19–21, 24
fetal suprachiasmatic nucleus
  circadian rhythm, 95, 96, 199
  entrainment, 198–211
  transplantation studies *see* suprachiasmatic nucleus (SCN), transplantation studies
food restriction, 94–95, 111
  fetal clock entrainment, 201, 203, 207, 208
formulation, drug, 244, 251
forskolin, 137, 146, 151
c-*fos*, 146, 150, 172–173, 282

  co-localization with neuropeptide Y, 161
  fetal SCN entrainment, 202–203, 207, 208
  neuropeptide Y regulation, 167
  non-photic phase shifts and, 158, 159, 182, 184, 190
  photic entrainment and, 176, 177–178, 179, 180, 189–192, 196–197
  tau mutants, 183, 196–197
  temperature-induced expression, 46
FRQ, 5, 6, 9, 24
  as clock component, 9–16, 19–21
  synthesis and degradation rates, 38
  transcriptional activation, 9
frq gene, 5–16, 18–24, 115
  elevated expression, 12–13, 19–20
  patterns of expression, 10, 11
  phylogenetic conservation, 7–9
  protein phosphorylation sites, 31, 42
  regulation, 9–16, 19–21, 24
  stepped decreases in expression, 13–16
  structure, 6
  transcripts, 6–7, 9, 24
frq mutants, 4, 5, 7
  period lengths, 4, 18–19, 22–23
  PRCs, 27
  temperature compensation, 37, 38–39, 44
fura-2, 44, 48, 54

GABA (γ-aminobutyric acid), 61–62
  Bulla eye, 63
  intergeniculate leaflet, 90, 138–139
  mammalian SCN, 93, 100
ganglion cells, retinal, 89–90, 97, 100–101
gastrin-releasing peptide (GRP), 177
gastro-intestinal tract rhythms, 237, 244, 245, 247–248
Gelasinospora cerealis, 8
geniculohypothalamic tract (GHT), 61–62, 90–92, 97, 110
glutamate, 61–62, 139, 172–173, 282
  phase shifting effects, 102, 141, 148–149
  photic entrainment and, 179–183
goggles, dark/red, 314–315
golden (Syrian) hamster, 262, 324
  ageing, 73–74, 75, 87, 212–234, 325
  fetal clock entrainment, 202
  non-photic phase shifting, 154–174, 182–183, 184–187

golden (Syrian) hamster (*cont.*)
  origin, 171–172
  photic entrainment, 176–182, 190
  PRCs, 60
  SCN transplantation studies, 70–76, 81–87
  strain differences, 194–195
  tau mutants see *tau* mutant hamsters
Gonyaulax (polyedra), 38, 117–133, 262
  different receptor for each oscillator, 125–126
  effects of creatine, 121–122, 123, 125, 129
  effects of dark pulses, 123–125, 173–174
  two light receptors, 118–121, 131

H_2 histamine receptor blockers, 237, 240
hamster see golden hamster; Siberian hamster
heart rate rhythm, 242, 250
heat-shock proteins (Hsps), 43–44
  chick pineal, 45
  Neurospora, 32, 33–34, 35, 36, 38
hepatic blood flow, 245
homeotherms, 44, 45
hormones, entrainment of fetal clock, 200–201
5-HT see serotonin
5-HT$_{1A}$ receptors, 140, 148, 186–187, 193–194
5-HT$_{1C}$ receptors (now 5-HT$_{2C}$), 186, 187, 193
5-HT$_2$ receptors (now 5-HT$_{2A}$), 186
5-HT$_7$ receptors, 140, 193–194
humans
  circadian rhythms, 236
  clinical chronopharmacology, 235–253
  effects of ageing, 213, 231–234
  effects of light, 254–302, 305–306, 324–325
  melatonin therapy see melatonin, therapy
  SCN development, 209, 210
hydroxymercuribenzoate (HMB), 33, 36
5-hydroxytryptamine see serotonin
hyperpolarization, 107–108
  Bulla basal retinal neurons, 58, 61, 64–65
  SCN neurons, 62
hypertension, 236, 241–244, 248–249
hypothalamus
  age-related changes, 213–214

brain slices, 134–153
see also suprachiasmatic nucleus

immediate-early genes (IEG), 175–197
  expression in SCN, 176–177
  neuroanatomy of cells expressing, 177–179
  non-photic entrainment and, 182–183, 184–187
  photic entrainment and, 176–182, 189–192
  tau mutants, 183–184, 196–197
  see also c-*fos*
immunocytochemistry, 161, 177–179
inositol trisphosphate (InsP$_3$), 42–43, 47, 48
  cAMP-induced changes, 30, 31
  melatonin-induced changes, 145
  temperature-induced changes, 28, 29, 30
intergeniculate leaflet (IGL), 76, 90–91, 97
  afferent pathways, 89–90, 95, 96, 167–168
  c-*fos* expression, 158, 159
  efferent pathways, 95, 97
  lesions, 158, 167–168
  neuropeptide Y/c-*fos* co-localization, 161
  SCN projections, 90, 96, 100, 101, 138–139
intergeniculate leaflet–geniculohypothalamic tract, 61–62, 90–92
internal desynchronization, 128, 262–264
  endogenous circadian phase and amplitude (ECPA) assessment, 264–265
  forced, 263, 295, 297
  Gonyaulax, 125, 128
  spontaneous, 263
internal synchronization, 262
ionic conductances, 55–56, 151–152
isosorbide-5-mononitrate (IS-5-MN), 244

jet lag, 170, 257, 306, 315
  melatonin therapy and, 309–312, 317–318
  phototherapy, 281
  see also air travellers, transmeridional
jun-B, 176, 184

ketanserin, 185, 186, 193–194
Kronauer's mathematical model, 270–273, 296

Subject index

lateral geniculate nucleus (LGN), 158, 164
lateral septal area, 100
Leucophaea maderae, 279
light
  entrainment *see* photic entrainment
  intensity
    Gonyaulax period length and, 118, 119, 129, 131
    Gonyaulax phase shifts and, 120, 121, 129–132
    human circadian rhythms and, 261, 270–275, 293–295
    pharmacological blockade and, 190
    work, 313, 320
    see also bright light
  non-photic interactions, 168, 172–173
  ordinary indoor, 257, 268, 270–274, 294–295, 315
  pulses, period effects in *Gonyaulax*, 123–125
  receptors *see* photoreceptors
light-induced phase shifts/PRCs, 108–109, 175–176
  ageing hamsters, 214–216, 217, 227
  Gonyaulax, 118–121, 125–126, 128–131
    effects of creatine, 121–122, 123, 125, 129
    light intensity and, 120, 121, 129–132
  humans, 254–302, 305–306
    experimental strategies, 261–275
    type 0 resetting to a 3-cycle stimulus, 275–281, 297–300
  rats, 148–149
  tau mutant hamsters, 183–184, 196–197, 227
  see also photic entrainment
luteinizing hormone (LH), pre-ovulatory surge, 172, 213

meals, timing of, 247–248
mecamylamine, 182, 190, 191
melatonin, 303–321
  assays, 291–292, 304
  breast milk, 209, 210
  chick pineal, 45
  dim light onset *see* dim light melatonin onset
  entrainment of fetal clock, 200, 201, 203, 208, 210–211
  PRCs, 107, 108–109, 144, 157

  humans, 306–307, 318, 319–320, 323–324
  in vitro SCN, 141–142
  receptors in SCN, 61, 139, 145–146, 201, 209
  rhythmic secretion
    effects of ageing, 233
    effects of light, 261, 268, 269, 274, 290–291, 293, 305
    as marker of circadian phase, 291–293
  pineal transplants, 69, 105
  signal transduction pathway, 145–146
  therapy, 282, 306–315, 317–321
    air travellers, 309–312, 313, 317–318
    hamsters with SCN transplants, 110
    shift workers, 312–314, 318–320
    side-effects, 320–321
Mesocricetus auratus see golden hamster
metamphetamine, 94–95, 111
MK-801, 168, 180, 190, 191
molluscan eye, 46, 51–66
  see also Bulla eye
monoamines, brain, 221, 222, 229–230
mouse, 170–171, 212–213

neuropeptide Y (NPY), 101, 156, 186
  antibodies, 161, 162
  exercise-associated release, 161–162
  in vitro phase shift effects, 140, 141–142, 144, 149
  in vivo phase shift effects, 91, 148, 149, 160
  intergeniculate leaflet (IGL), 90, 138–139, 161
  non-photic phase shifting and, 158–163, 164, 167, 172–173
Neurospora crassa, 62, 112, 113
  circadian system components, 5–6
  genetic basis of circadian clock, 3–25, 115
  mutants, 4, 18–19, 22–23
    see also frq mutants
  PRCs, 23–24, 27, 64
  temperature-induced changes, 26–50
    compensation after longer exposures, 34, 37–39
    intracellular pathways, 28–37, 41–48
    phase shifts, 27, 41
  wc (blind) mutants, 29, 30, 47
Neurospora discreta, 8

Neurospora intermedia, 8
nickel chloride, 54
nifedipine, 44, 244
night-shift workers *see* shift workers
NMDA (*N*-methyl-D-aspartate) receptors, 179–182
nocturnal mammals, 255, 259
non-NMDA receptors, 179–180
non-photic phase shifting, 154–174, 175–176, 323
   activity-responce curves, 157–158
   c-*fos* and, 158, 159, 182, 184, 190
   immediate-early genes and, 182–183, 184–187
   neural basis, 184–187, 193–194
   neuropeptide Y and, 158–163, 164, 167, 172–173
   photic interactions, 168, 172–173
non-steroidal anti-inflammatory drugs, 238, 239
noradrenaline, 104, 105
novelty-induced wheel running, 154–164, 168, 169–170

8-OH-DPAT, 63, 147, 193, 194
oscillators, 112–113, 114–115
   Gonyaulax, 125–126, 128, 129
   see also pacemakers
oxygen evolution, 122, 123

pacemakers
   anatomical substrates, 67–68
   Bulla eye *see* basal retinal neurons, *Bulla*
   components within a cell, 5–6, 112, 113
   endogenous human
      effects of light, 254–302, 305–306
      markers of output, 262–264, 290–293, 304
   entrainment of fetal, 198–211
   evolution, 17–18
   mammalian non-SCN, 94–95, 111–112
   properties of component proteins, 113–116
   transplantation studies, 68–76, 81–87, 102–105
   versus oscillators, 112–113
   vertebrate, 67–87
   see also eye; pineal gland; suprachiasmatic nucleus
paraventricular nucleus of hypothalamus (PVN), 101, 135–136, 178, 179

paraventricular thalamic nucleus (PVT), 95, 96, 178
peptic ulcer, 237, 240
peptide histidine-isoleucine (PHI), 177, 179
Per, 27, 31, 42
per gene, 10, 19–20, 24, 115
period length ([gt])
   after-effects, 123–125
   age-related changes, 213, 230
   Gonyaulax
      dark versus light pulses, 123–125
      effects of creatine, 121, 122
      light intensity and, 118, 119, 129, 131
   humans, 262–263, 294, 295–296
   in vitro SCN slices, 151
   Neurospora mutants, 4, 18–19, 22–23
   quantal elements, 18–19
   tau mutant hamsters, 70, 71, 183
   transfer by SCN transplantation, 70–76, 81–86
pharmacokinetics, 236, 237, 239, 249, 251–252
phase, circadian
   disorders in humans, 305–306, 315
   markers in humans, 262–264, 291–293, 304
   transfer by pineal transplantation, 68–69
   transfer by SCN transplantation, 70
phase response curves (PRCs)
   break-points
      Gonyaulax, 119–120, 128–129
      humans, 277
   Bulla eye, 56, 58, 60–63, 65–66, 107, 109
   dead zone, 57–58
   Neurospora, 23–24, 27, 64
   singularity point, 280
   type 0 (high amplitude), 108, 255, 323
      humans, 275–281, 297–300, 322, 324
   type 1 (low amplitude), 108, 255, 279
   type 2, 47
   types, 58, 60–63, 64–65, 107–110, 156, 322–324
   see also specific phase-shifting stimuli
Phodopus sungorus, 177, 178
phosphodiesterases, 43
phospholipase C, 145
phosphorylation, protein, 43
   CREB, 184, 227
   Neurospora, 29–31, 42

Subject index

photic entrainment, 109
  age-related changes, 213, 214, 215, 216
  Bulla basal retinal neurons, 53–54
  immediate-early genes and, 176–182, 189–192
  neural substrates, 141, 177–182
  pathways in mammals, 89–92, 97, 100
  rat pups of SCN-lesioned dams, 199–200
  see also light-induced phase shifts/PRCs
photoreceptors
  Gonyaulax, 118–121, 125–126, 131
  mediating entrainment in mammals, 90, 97, 100–101, 282
  Neurospora, 47
photosynthesis, 118–119, 122, 126
phototherapy, 3, 261, 281, 306
pineal gland, 68, 76–78
  afferent pathways, 100
  chick, 44–45, 46
  entrainment of fetal clock, 200
  melatonin secretion *see* melatonin, rhythmic secretion
  transplantation studies, 68–69, 104–105
plants, 21–22, 231, 262
potassium (K^+) channels, 151–152
  Bulla basal retinal neurons, 55–56, 63
PRC *see* phase response curves
preterm babies, 210
prokaryotes, 17, 18
propranolol, 233, 244
protein(s)
  degradation
    inhibitors, 33, 36
    temperature compensation in *Neurospora*, 34, 38–39
    temperature effects in *Neurospora*, 28, 32, 33–36
  properties of pacemaker, 113–116
  synthesis *see* translation
  translocation, *Neurospora*, 36
protein kinase A, 31, 43
protein kinase C, 145, 146
psychotropic drugs, 238, 239

qa-2 gene/promoter system, 11–12, 21
quantal periods, 18–19

ranitidine, 237, 240
raphe nuclei, 92, 138, 139, 186
rat, 177–178, 180
  ageing, 212–213, 226, 228
  entrainment of fetal clock, 198–211
  hypertension models, 242, 243, 249
  hypothalamic brain slices, 134–153
  SCN, 92–93
  stress-induced phase shifts, 170, 171, 172, 194
reserpine, 221, 222, 229–230
rest–activity cycles *see* activity rhythms; sleep–wake cycles
retinal ganglion cells, 89–90, 97, 100–101
retinal neurons, basal *see* basal retinal neurons, *Bulla*
retinohypothalamic tract (RHT), 88, 89–90, 91, 97, 139, 259
  lesions, 89, 101
  substance P, 101–102
retrochiasmatic area (RCA), 89, 90–91, 95, 96, 167–168
ritanserin, 193–194
running, 170–171, 172, 217
  cold-induced, 169, 170
  enforced, 169, 170, 182
  novelty-induced, 154–164, 168, 169–170

saline injections, 182, 184–185, 194, 195
SCN *see* suprachiasmatic nucleus
second messengers
  Neurospora, 28, 29–31
  see also calcium; cyclic AMP; cyclic GMP; inositol trisphosphate
serotonin (5-HT), 63, 92, 138, 193–194
  Aplysia eye, 109–110
  non-photic entrainment pathways, 185–187
  phase-shifting effects
    in vitro SCN, 92, 139–141, 144, 147–148
    in vivo, 149
shift workers, 257, 263, 306, 315
  melatonin therapy, 312–314, 318–320
  phototherapy, 281
Siberian hamster, 177, 178
singularity point, 280
SKF-38393, 202, 208
sleep disorders, circadian rhythm, 281–282, 306, 315
sleep–wake cycles
  amplitude, age-related changes, 213
  blind people, 260–261
  elderly people, 231, 232, 233–234
  versus physiological rhythms, 262–263, 268, 294, 296–297

social cues, 255–258, 261
sodium channels, SCN, 93
Sordaria fimicola, 8–9
sparrow, house, 68–69, 102, 104–105
spontaneously hypertensive rat (SHR), 242, 243
stress-induced phase shifts, 170–171, 194–196
subparaventricular zone (SPVZ), 95, 96
substance P, 101–102, 145
suprachiasmatic nucleus (SCN), 64, 68, 69, 88, 92–95
　afferent pathways, 89–92, 97, 138–139
　age-related changes, 213, 226–227, 228–229, 232
　avian, 69
　D_1 dopamine receptors, 102
　depolarization/hyperpolarization, 61–62, 108
　development, 93, 199, 209, 210
　evidence for pacemaker function, 92
　fetal *see* fetal suprachiasmatic nucleus
　functional organization, 93, 94, 102–103, 146
　human, markers of output, 263–264
　immediate-early gene expression, 176–177
　in vitro studies, 63, 134–153, 199, 228–229
　　cAMP and cGMP rhythms, 42, 43, 138
　　sensitivities to neuromodulators, 139–142, 144–146, 147–148
　　spontaneous circadian rhythm, 135, 136, 137
　　temperature compensation, 45–46
　　versus *in vivo*, 148–149
　lesions, 70, 163
　　pregnant rats, 199–200, 201
　melatonin receptors, 61, 139, 145–146, 201, 209
　morphological organization, 92–93
　neuropeptide Y injections into, 91, 148, 149, 160
　non-photic inputs, 154–174
　output pathways, 95–97, 100, 101, 102–104, 178–179
　photic inputs, 89–92, 100, 101, 177, 259, 282–283
　transplantation studies, 70–76, 81–87, 102–105, 110, 227–228

cell mixtures, 74–76, 77, 78, 81–86
communication issues, 73–74, 75
transfer of period, 70, 72
transfer of phase, 70
triazolam-induced phase shifts, 163, 221–223, 228
versus *Bulla* eye, 63
swimming, forced, 172
Syrian hamster *see* golden hamster

τ *see* period length
tau mutant hamsters, 60, 191, 196–197, 227, 323
　immediate-early gene expression, 183–184, 196–197
　period length, 70, 71, 183
　SCN transplantation studies, 70–76, 81–87
temperature compensation, 21–22, 27, 44–46
　Neurospora, 34, 37–39
　Neurospora mutants, 4, 22–23, 37, 38–39, 44
temperature-induced changes
　Bulla eye, 41
　chick pineal, 45
　Drosophila, 150
　hamster, 169
　Neurospora see Neurospora crassa, temperature-induced changes
　suprachiasmatic nucleus, 45–46
temperature rhythm *see* body temperature rhythm
temporal domains, 135, 142, 147
terbutaline, 241
tetraethylammonium (TEA), 55, 63
tetrodotoxin, 93, 103
Thalassomyxa australis, 22
theophylline, 239–241
thyroid-stimulating hormone, 274–275
transcription, 62
　Bulla pacemaker cells, 56–57, 65
transgenic TGR(mRen-2)27 hypertensive rat, 242, 243, 249
translation (protein synthesis), 62, 65–66
　Bulla pacemaker, 56–57, 65–66
　inhibitors, 33, 36, 107, 220
　temperature compensation in *Neurospora*, 34, 38–39
　temperature effects in *Neurospora*, 28, 32, 33–36

Subject index

transplantation studies, pacemakers, 68–76, 81–87, 102–105
triazolam, 172, 182, 324
  activity mediating effects, 155–156, 157–158, 168
  age-related changes in responses, 218–220, 233–234
  anatomical pathways mediating responses, 110, 158
  chronopharmacology, 251–252
  effects on period ([gt]), 295
  SCN transplant studies, 163, 221–223, 228

tyrosine, 203, 207, 208

vasoactive intestinal polypeptide (VIP), 45, 93, 104, 173, 177
vasopressin (arginine vasopressin), 45, 93, 177

white collar (*ws-1/wc-2*) *Neurospora* mutants, 29, 30, 47
whole cell patch recordings, 136

Xenopus retina, 208–209